Tensors

Anadijiban Das

Tensors

The Mathematics of Relativity Theory and Continuum Mechanics

Springer

Anadijiban Das
Department of Mathematics and
Pacific Institute for the Mathematical Sciences
Simon Fraser University
8888 University Avenue
Burnaby, V5A 1S6
BC, Canada

ISBN 978-1-4419-2410-0 e-ISBN 978-0-387-69469-6

Printed on acid-free paper.

9 8 7 6 5 4 3 2 1

springer.com

Dedicated To Sri Sarada Devi

Preface

Tensor algebra and tensor analysis were developed by Riemann, Christoffel, Ricci, Levi-Civita and others in the nineteenth century. The special theory of relativity, as propounded by Einstein in 1905, was elegantly expressed by Minkowski in terms of tensor fields in a flat space-time.

In 1915, Einstein formulated the general theory of relativity, in which the space-time manifold is curved. The theory is aesthetically and intellectually satisfying. The general theory of relativity involves tensor analysis in a pseudo-Riemannian manifold from the outset. Later, it was realized that even the pre-relativistic particle mechanics and continuum mechanics can be elegantly formulated in terms of tensor analysis in the three-dimensional Euclidean space. In recent decades, relativistic quantum field theories, gauge field theories, and various unified field theories have all used tensor algebra analysis exhaustively.

This book develops from abstract tensor algebra to tensor analysis in various differentiable manifolds in a mathematically rigorous and logically coherent manner. The material is intended mainly for students at the fourth-year and fifth-year university levels and is appropriate for students majoring in either mathematical physics or applied mathematics.

The first chapter deals with tensor algebra, or algebra of multilinear mappings in a general field $\mathcal{F}$. (The background vector space need not possess an inner product or dot product.). The second chapter restricts the algebraic field to the set of real numbers $\mathbb{R}$. Moreover, it is assumed that the underlying real vector space is endowed with an inner product (or dot product). Chapter 3 defines and investigates a differentiable manifold without imposing any other structure. Chapter 4 discusses tensor analysis in a general differentiable manifold. Differential forms are introduced and investigated. Next, a connection form indicating parallel transport is brought forward. As a logical consequence, the fourth-order curvature tensor is generated. Chapter 5 deals with Riemannian and pseudo-Riemannian manifolds. Tensor analysis, in terms of coordinate components as well as orthonormal components, is exhaustively investigated. In Chapter 6, *special* Riemannian and pseudo-Riemannian manifolds are studied. Flat manifolds, spaces of constant curvature, Einstein spaces, and conformally flat spaces are explored. Hypersurfaces and submanifolds embedded in higher-dimensional manifolds are discussed in chapter 7. Extrinsic curvature tensors are defined in all cases. Moreover, Gauss and Codazzi-Mainardi equations are derived.

We would like to elaborate on the notation used in this book. The letters i, j, k, l, m, n, etc., are used for the subscripts and superscripts of a tensor field in the *coordinate* basis. However, we use the letters a, b, c, d, e, f, etc., for subscripts and superscripts of the same tensor field relative to an *orthonormal* basis. The numerical enumeration of coordinate components v^i of a vector field is given by $v^1, v^2, \ldots, v^N$. However, numerical elaboration of orthonormal components of the same vector field is furnished by $v^{(1)}, v^{(2)}, \ldots, v^{(N)}$ (to avoid confusion). Similar distinctions are made for tensor field components. The *flat* metric components are denoted either by d_{ij} or d_{ab}. (The usual symbol $\boldsymbol{\eta}_{..}$ is reserved only for the totally antisymmetric pseudotensor of Levi-Civita.) The generalized Laplacian in the N-dimension is denoted by Δ.

I would like to acknowledge my gratitude to several people for various reasons. During my stay at the Dublin Institute for Advanced Studies from 1958 to 1961, I learned a lot of classical tensor analysis from the late Professor J. L. Synge, F. R. S.. Professor W. Noll, a colleague of mine at Carnegie-Mellon University from 1963 to 1966, introduced me to the abstract tensor algebra, or the algebra of multilinear mappings. My research projects and teachings on general relativity for many years have consolidated the understanding of tensors. Dr. Andrew DeBenedictis has kindly read the proof, edited and helped with computer work. Mrs. Judy Borwein typed from chapter 1 to chapter 5 and edited the text diligently and flawlessly. Mrs. Sabine Lebhart typed the difficult chapter 7 and appendices. She also helped in the final editing. Mr. Robert Birtch drew thirty-four figures of the book. Last but not least, my wife, Mrs. Purabi Das, was a constant source of encouragement.

Contents

List of Figures

Chapter 1

Finite-Dimensional Vector Spaces and Linear Mappings

1.1 Fields

We shall discuss briefly objects called numbers or scalars. The set of certain numbers is called a **field**. (This is *distinct* from a physical field in the universe.) A field $\mathcal{F}$ is endowed with two operations, namely **addition** and **multiplication**. The axioms for these operations are listed below.

A1. $\alpha + \beta$ is in $\mathcal{F}$ for all α, β in $\mathcal{F}$.

A2. $\beta + \alpha = \alpha + \beta$ for all α, β in $\mathcal{F}$.

A3. $(\alpha + \beta) + \gamma = \alpha + (\beta + \gamma)$ for all α, β, γ in $\mathcal{F}$.

A4. There exists a unique number 0 (called zero) such that $\alpha + 0 = \alpha$ for all α in $\mathcal{F}$.

A5. To every α in $\mathcal{F}$ there corresponds a number $-\alpha$ such that

$$\alpha - \alpha := \alpha + (-\alpha) = 0. \tag{1.1}$$

M1. $\alpha\beta$ is in $\mathcal{F}$ for all α, β in $\mathcal{F}$.

M2. $\beta\alpha = \alpha\beta$ for all α, β in $\mathcal{F}$.

M3. $(\alpha\beta)\gamma = \alpha(\beta\gamma)$ for all α, β, γ in $\mathcal{F}$.

M4. There exists a unique number 1 such that $\alpha 1 = \alpha$ for all α in $\mathcal{F}$.

M5. To every non-zero number α there corresponds a number α^{-1} such that $\alpha\alpha^{-1} = 1$.

D1. $\alpha(\beta + \gamma) = (\alpha\beta) + (\alpha\gamma)$ for all α, β, γ in $\mathcal{F}$.

Three common examples of a field are (i) $\mathbb{R}$, the set of all real numbers; (ii) $\mathbb{C}$, the set of all complex numbers; and (iii) $\mathbb{Q}$, the set of all (real) rational numbers.

The set of all real numbers $\mathbb{R}$ is endowed with ordering as well as completeness. The set of all complex numbers $\mathbb{C}$ does not have ordering. However, it is complete. It is also algebraically closed. (That is, every nth-degree polynomial equation has n solutions, counting multiplicities.) The field of all (real) rational numbers has ordering, but it is incomplete. The minimum number of elements a field can have is two. The fields $\mathbb{R}$ and $\mathbb{C}$ are mostly used in mathematical physics.

Example 1.1.1 Suppose that $\alpha\beta = 0$ for two numbers α and β in a field $\mathcal{F}$. Prove that $\alpha = 0$, $\beta = 0$, or both.

Proof. Let us prove the statement above by a contradiction. Assume that $\alpha \neq 0$ and $\beta \neq 0$. Then, by axioms M5, M3, A5, and D1, we have

$$\begin{aligned}\beta &= 1\beta = (\alpha^{-1}\alpha)\beta = \alpha^{-1}(\alpha\beta) = \alpha^{-1}(0) = \alpha^{-1}[\alpha + (-\alpha)] \\ &= (\alpha^{-1}\alpha) + [\alpha^{-1}(-\alpha)] = 1 - 1 = 0.\end{aligned}$$

Thus we have reached a contradiction. Therefore, we have proved that $\alpha\beta = 0$ implies $\alpha = 0$, $\beta = 0$, or both. ■

Exercises 1.1

1. Let $\mathbb{R}^2$ be the set of all ordered pairs of real numbers (α, β) with addition and multiplication rules:

$$\begin{aligned}(\alpha, \beta) + (\gamma, \delta) &:= (\alpha + \gamma, \beta + \delta), \\ (\alpha, \beta)(\gamma, \delta) &:= (\alpha\gamma - \beta\delta, \alpha\delta + \beta\gamma).\end{aligned}$$

Prove that $\mathbb{R}^2$ with the rules above constitutes a field.

(*Remark:* The field above is isomorphic to the complex field $\mathbb{C}$.)

2. Determine whether or not the following sets constitute a field.

(i) The set of all real numbers of the form $m + n\sqrt{2}$, where m and n are integers.

(ii) The set of all real numbers of the form $a + b\sqrt[3]{5} + c\sqrt[3]{25}$, where a, b, c are real rational numbers.

(iii) The set of all complex numbers of the form $a + bi$, where a, b are (real) rational numbers.

1.2 Finite-Dimensional Vector Spaces

We are now in a position to define a vector space. A vector space presupposes a particular field $\mathcal{F}$. The numbers or scalars used in the definition are the elements of $\mathcal{F}$. A vector space $\mathcal{V}$ over the field $\mathcal{F}$ is a set of elements called vectors. The set $\mathcal{V}$ is endowed with two compositions, namely addition and scalar multiplication. The axioms for vector addition and scalar multiplication of vectors are furnished below.

A1. $\vec{\mathbf{a}} + \vec{\mathbf{b}}$ belongs to $\mathcal{V}$ for all $\vec{\mathbf{a}}$, $\vec{\mathbf{b}}$ in $\mathcal{V}$.

A2. $\vec{\mathbf{b}} + \vec{\mathbf{a}} = \vec{\mathbf{a}} + \vec{\mathbf{b}}$ for all $\vec{\mathbf{a}}, \vec{\mathbf{b}}$ in $\mathcal{V}$.

A3. $(\vec{\mathbf{a}} + \vec{\mathbf{b}}) + \vec{\mathbf{c}} = \vec{\mathbf{a}} + (\vec{\mathbf{b}} + \vec{\mathbf{c}})$ for all $\vec{\mathbf{a}}, \vec{\mathbf{b}}, \vec{\mathbf{c}}$ in $\mathcal{V}$.

A4. There exists a unique vector $\vec{\mathbf{0}}$ (the zero vector) such that $\vec{\mathbf{a}} + \vec{\mathbf{0}} = \vec{\mathbf{a}}$ for all $\vec{\mathbf{a}}$ in $\mathcal{V}$.

A5. To every vector $\vec{\mathbf{a}}$, there corresponds a unique (negative) vector $-\vec{\mathbf{a}}$ such that $\vec{\mathbf{a}} - \vec{\mathbf{a}} := \vec{\mathbf{a}} + (-\vec{\mathbf{a}}) = \vec{\mathbf{0}}$. (1.2)

M1. $\alpha\vec{\mathbf{a}}$ belongs to $\mathcal{V}$ for all α in $\mathcal{F}$ and all $\vec{\mathbf{a}}$ in $\mathcal{V}$.

M2. $\alpha(\beta\vec{\mathbf{a}}) = (\alpha\beta)\vec{\mathbf{a}}$ for all α, β in $\mathcal{F}$ and all $\vec{\mathbf{a}}$ in $\mathcal{V}$.

M3. $1\vec{\mathbf{a}} = \vec{\mathbf{a}}$ for all $\vec{\mathbf{a}}$ in $\mathcal{V}$.

M4. $\alpha(\vec{\mathbf{a}} + \vec{\mathbf{b}}) = (\alpha\vec{\mathbf{a}}) + (\alpha\vec{\mathbf{b}})$ for all α in $\mathcal{F}$ and all $\vec{\mathbf{a}}, \vec{\mathbf{b}}$ in $\mathcal{V}$.

M5. $(\alpha + \beta)\vec{\mathbf{a}} = (\alpha\vec{\mathbf{a}}) + (\beta\vec{\mathbf{a}})$ for all α, β in $\mathcal{F}$ and all $\vec{\mathbf{a}}$ in $\mathcal{V}$.

(See the book by Halmos [17].)

If $\mathcal{F} = \mathbb{R}$, we call $\mathcal{V}$ a real vector space. If $\mathcal{F} = \mathbb{C}$, the vector space $\mathcal{V}$ is said to be complex. In mathematical physics, we usually encounter both real and complex vector spaces.

Example 1.2.1 Let us consider the set of complex numbers $\mathbb{C}$. We interpret complex numbers as vectors in the field $\mathcal{F} = \mathbb{C}$ by the following rules:

$$\begin{aligned} \vec{\mathbf{a}} &:= \alpha, \qquad \alpha \in \mathbb{C}, \\ \vec{\mathbf{a}} + \vec{\mathbf{b}} &:= \alpha + \beta, \\ \vec{\mathbf{0}} &:= 0, \\ -\vec{\mathbf{a}} &:= -\alpha, \\ \lambda\vec{\mathbf{a}} &:= \lambda\alpha. \end{aligned}$$

By these rules, $\mathbb{C}$ is a complex vector space. □

Example 1.2.2 Consider $\mathbb{R}^N$, the set of all ordered N-tuples of real numbers. ($N \in \mathbb{Z}^+$, the set of all positive integers.) We define the vector addition and scalar multiplication in the field $\mathcal{F} := \mathbb{R}$ by the following rules:

$$\begin{aligned} \vec{\mathbf{a}} &:= (\alpha_1, \alpha_2, \ldots, \alpha_N), \\ \vec{\mathbf{a}} + \vec{\mathbf{b}} &:= (\alpha_1 + \beta_1, \alpha_2 + \beta_2, \ldots, \alpha_N + \beta_N), \\ \vec{\mathbf{0}} &:= (0, 0, \ldots, 0), \\ -\vec{\mathbf{a}} &:= (-\alpha_1, -\alpha_2, \ldots, -\alpha_N), \\ \lambda\vec{\mathbf{a}} &:= (\lambda\alpha_1, \lambda\alpha_2, \ldots, \lambda\alpha_N). \end{aligned} \tag{1.3}$$

All the ten rules (1.2) of vector space can be verified. Thus, $\mathbb{R}^N$ with (1.3) is a real vector space. □

Example 1.2.3 Consider $C^0[(a,b) \subset \mathbb{R}; \mathbb{R}]$, the set of all continuous, real-valued functions over the interval $(a,b) \subset \mathbb{R}$. We can define the vector addition and scalar multiplication in the real field by the following rules:

$$\begin{aligned} \vec{\mathbf{f}} &:= f(x), \qquad x \in (a,b), \\ \vec{\mathbf{f}} + \vec{\mathbf{g}} &:= f(x) + g(x), \\ \lambda\vec{\mathbf{f}} &:= \lambda f(x), \\ \vec{\mathbf{0}} &:= 0, \\ -\vec{\mathbf{f}} &:= -f(x). \end{aligned}$$

By the rules above, $C^0[(a,b) \subset \mathbb{R}; \mathbb{R}]$ is a real vector space. (This is an example of an *infinite-dimensional* vector space.) □

Consider, in Newtonian physics, the motion of a massive particle in space. The instantaneous velocity, acceleration, and momentum of the particle are three examples of vectors in physics.

Now we shall define a **vector subspace** $\mathcal{W}$ of $\mathcal{V}$. The subset $\mathcal{W} \subset \mathcal{V}$ is a vector subspace provided $\lambda\vec{\mathbf{a}} + \mu\vec{\mathbf{b}}$ belongs to $\mathcal{W}$ for all λ, μ in $\mathcal{F}$ and all $\vec{\mathbf{a}}, \vec{\mathbf{b}}$ in $\mathcal{W}$.

Example 1.2.4 Consider $\mathcal{V} = \mathbb{R}^N$. Let $\mathcal{W}$ be a proper subset of $\mathbb{R}^N$ such that it consists of the vectors of the form $\vec{\mathbf{a}} := (0, \alpha_2, \alpha_3, \ldots, \alpha_N)$. By (1.3), for two vectors $\vec{\mathbf{a}}, \vec{\mathbf{b}}$ in $\mathcal{W}$, We have the linear combination

$$\lambda\vec{\mathbf{a}} + \mu\vec{\mathbf{b}} = (0, \lambda\alpha_2 + \mu\beta_2, \lambda\alpha_3 + \mu\beta_3, \ldots, \lambda\alpha_N + \mu\beta_N).$$

The vector above is obviously in $\mathcal{W}$. Thus, $\mathcal{W}$ is a vector subspace. □

Example 1.2.5 Consider again $\mathcal{V} = \mathbb{R}^N$. Let $\mathcal{W}^{\#}$ be a proper subset of $\mathbb{R}^N$ such that it consists of the vectors of the form

$$\vec{\mathbf{a}} := (\alpha_1, \alpha_2, \ldots, \alpha_N) \text{ with } \sum_{k=1}^{N} (\alpha_k)^2 = 1.$$

The zero vector $\vec{\mathbf{0}} = (0, 0, \ldots, 0)$ does not belong to $\mathcal{W}^{\#}$. Thus, $\mathcal{W}^{\#}$ is *not* a vector subspace. □

Now, we define a linear combination of vectors. The vector $\vec{\mathbf{a}} := \sum_{k=1}^{N} \alpha^k \vec{\mathbf{a}}_k$, where $\alpha^1, \ldots, \alpha^k$ are in $\mathcal{F}$, is a **linear combination** of vectors $\vec{\mathbf{a}}, \ldots, \vec{\mathbf{a}}_N$. (Here, in α^k, k is a superscript, *not* a power or exponent.)

A subset $\mathcal{W}$ of $\mathcal{V}$ such that it consists of vectors of the form $\vec{\mathbf{a}} := \sum_{j=1}^{k} \alpha^j \vec{\mathbf{a}}_j$ is said to be *spanned* (or *generated*) by vectors $\vec{\mathbf{a}}_1, \ldots, \vec{\mathbf{a}}_k$. Such a subset can be proved to be a vector subspace.

Consider now k non-zero vectors $\vec{\mathbf{a}}_1, \ldots, \vec{\mathbf{a}}_k$ and the vector equation

$$\sum_{j=1}^{k} \mu^j \vec{\mathbf{a}}_j = \vec{\mathbf{0}}. \tag{1.4}$$

If the equation above implies that $\mu^1 = \mu^2 = \cdots = \mu^k = 0$, the vectors $\vec{\mathbf{a}}_1, \vec{\mathbf{a}}_2, \ldots, \vec{\mathbf{a}}_k$ are called **linearly independent**. Otherwise, they are **linearly dependent**. (The concept of linear independence is the abstraction of vectors pointing to different directions, ignoring reflections.)

Example 1.2.6 Consider N vectors in $\mathcal{V} = \mathbb{R}^N$ given by

$$\vec{\mathbf{e}}_1 := (1, 0, \ldots, 0), \; \vec{\mathbf{e}}_2 := (0, 1, \ldots, 0), \ldots, \vec{\mathbf{e}}_N := (0, 0, \ldots, 1). \tag{1.5}$$

The vector equation $\sum_{j=1}^{N} \mu^j \vec{\mathbf{e}}_j = \vec{\mathbf{0}}$ yields $(\mu^1, \mu^2, \ldots, \mu^N) = (0, 0, \ldots, 0)$. Therefore, $\mu^1 = \mu^2 = \cdots = \mu^N = 0$. Thus, these vectors are linearly independent. □

Example 1.2.7 The vectors of N-tuples

$$\vec{\mathbf{a}}_1 := (1, 0, 0, \ldots, 0),\ \vec{\mathbf{a}}_2 := (1, 1, 0, \ldots, 0), \ldots, \vec{\mathbf{a}}_N := (1, 1, 1, \ldots, 1)$$

are linearly independent. □

Example 1.2.8 Two vectors of N-tuples

$$\vec{\mathbf{a}}_1 := (1, -1, 0, \ldots, 0) \text{ and } \vec{\mathbf{a}}_2 := (-\sqrt{2}, \sqrt{2}, 0, \ldots, 0)$$

are linearly dependent (and scalar multiples of each other). □

A **basis set** of vectors $\{\vec{\mathbf{e}}_1, \vec{\mathbf{e}}_2, \ldots, \vec{\mathbf{e}}_N\}$ for $\mathcal{V}$ is a spanning as well as a linearly independent set.

Example 1.2.9 Let N vectors in $\mathbb{R}^N$ be defined by $\vec{\mathbf{e}}_1 := (1, 0, \ldots, 0), \vec{\mathbf{e}}_2 := (0, 1, \ldots, 0), \ldots, \vec{\mathbf{e}}_N := (0, 0, \ldots, 1)$. The set $\{\vec{\mathbf{e}}_1, \vec{\mathbf{e}}_2, \ldots, \vec{\mathbf{e}}_N\}$ is a basis set for $\mathbb{R}^N$. Furthermore, this special set of vectors is called the **standard basis**. □

Example 1.2.10 Let $\widehat{\vec{\mathbf{e}}}_1 := (1, 0, 0, \ldots, 0)$, $\widehat{\vec{\mathbf{e}}}_2 := (1, \frac{1}{2}, 0, \ldots, 0), \ldots, \widehat{\vec{\mathbf{e}}}_N := (1, \frac{1}{2}, \frac{1}{3}, \ldots, \frac{1}{N})$. The set $\{\widehat{\vec{\mathbf{e}}}_1, \widehat{\vec{\mathbf{e}}}_2, \ldots, \widehat{\vec{\mathbf{e}}}_N\}$ is a basis set for $\mathbb{R}^N$. □

In the usual vector calculus, the basis $\{\vec{\mathbf{i}}, \vec{\mathbf{j}}, \vec{\mathbf{k}}\}$ is the standard basis.The number of vectors in a basis set is called the **dimension** of $\mathcal{V}$ and is denoted by $\dim(\mathcal{V})$.

Example 1.2.11 $\dim(\mathbb{R}^N) = N$, $N \in \mathbb{Z}^+$. □

Example 1.2.12 The smallest number of vectors in a vector space is *one*. Such a vector space is the singleton set $\{\vec{\mathbf{0}}\}$. The dimension of this vector space is defined to be $\dim\{\vec{\mathbf{0}}\} := 0$. (In the sequel, we *avoid* such a vector space.) □

Every vector $\vec{\mathbf{a}}$ in $\mathcal{V}$ can be expressed as a linear combination of basis vectors (since a basis set must be a spanning set). Therefore, relative to a basis set $\{\vec{\mathbf{e}}_1, \vec{\mathbf{e}}_2, \ldots, \vec{\mathbf{e}}_N\}$, a vector $\vec{\mathbf{a}}$ admits the linear combination

$$\vec{\mathbf{a}} = \alpha^1 \vec{\mathbf{e}}_1 + \alpha^2 \vec{\mathbf{e}}_2 + \cdots + \alpha^N \vec{\mathbf{e}}_N. \tag{1.6}$$

The scalars $\alpha^1, \alpha^2, \ldots, \alpha^N$ are called the **components** of the vector $\vec{\mathbf{a}}$ *relative* to the basis set $\{\vec{\mathbf{e}}_1, \vec{\mathbf{e}}_2, \ldots, \vec{\mathbf{e}}_N\}$.

Theorem 1.2.13 *Let the components be* $\alpha^1, \ldots, \alpha^N$ *of a vector* $\vec{\mathbf{a}}$ *relative to the basis set* $\{\vec{\mathbf{e}}_1, \ldots, \vec{\mathbf{e}}_N\}$. *Then the* N*-tuple* $(\alpha^1, \ldots, \alpha^N)$ *is unique.*

Proof. Let us assume to the contrary that there exist other scalars $\beta^1, \ldots, \beta^N$ such that

$$\vec{\mathbf{a}} = \sum_{k=1}^{N} \alpha^k \vec{\mathbf{e}}_k = \sum_{k=1}^{N} \beta^k \vec{\mathbf{e}}_k, \qquad \alpha^k - \beta^k \not\equiv 0.$$

From the vector equation above, we obtain that

$$\sum_{k=1}^{N} (\alpha^k - \beta^k)\vec{\mathbf{e}}_k = \vec{\mathbf{0}}.$$

By the linear independence of the basis vector, we must have

$$\alpha^k - \beta^k \equiv 0.$$

Thus a contradiction is reached and the theorem is proved. ∎

Now we shall explain the **Einstein summation convention**. In a mathematical expression, wherever two repeated Roman (or Greek, or other) indices are present, the sum over the repeated index is *implied*. For example, we write

$$u^k v_k := \sum_{k=1}^{N} u^k v_k = \sum_{j=1}^{N} u^j v_j =: u^j v_j.$$

The summation indices are called **dummy indices** since they can be replaced by other indices over the same range. Dummy indices that repeat more than twice are *not* allowed in the summation convention. This restriction is necessary to avoid wrong answers; for example, in case we write

$$u^k v_k u^k v_k = \sum_{k=1}^{N} u^k v_k u^k v_k \neq \sum_{k=1}^{N} \sum_{j=1}^{N} u^k v_k u^j v_j = u^k v_k u^j v_j = (u^k v_k)^2.$$

Thus we can have inconsistencies. We shall use the summation convention in the sequel.

Now we shall define the **Kronecker delta**. It is defined by the scalars

$$\delta^i_j := \begin{cases} 1 & \text{for} \quad i = j, \\ 0 & \text{for} \quad i \neq j. \end{cases} \tag{1.7}$$

The $N \times N$ matrix (in the field $\mathcal{F}$) with entries δ^i_j is the **unit matrix**. In other words, $[\delta^i_j] = [I]$.

Example 1.2.14 Using the summation convention,

$$
\begin{aligned}
\delta^1_j\alpha^j &= \delta^1_1\alpha^1 + \delta^1_2\alpha^2 + \cdots + \delta^1_N\alpha^N = \alpha^1,\\
\delta^i_j\alpha^j &= \alpha^i,\\
\delta^i_j\delta^j_k &= \delta^i_k,\\
\delta^i_j\delta^j_k\delta^k_i &= N = \dim(\mathcal{V}).
\end{aligned}
$$

□

Let us consider a change of basis sets without altering the vectors. This is a **passive transformation**. Let $\{\vec{\mathbf{e}}_1, \ldots, \vec{\mathbf{e}}_N\}$ and $\{\widehat{\mathbf{e}}_1, \ldots, \widehat{\mathbf{e}}_N\}$ be two basis sets for $\mathcal{V}$. By the spanning properties, there must exist scalars λ^k_i and μ^j_k such that

$$\widehat{\mathbf{e}}_i = \lambda^k_i\, \vec{\mathbf{e}}_k, \quad \vec{\mathbf{e}}_k = \mu^j_k\, \widehat{\mathbf{e}}_j. \tag{1.8}$$

Theorem 1.2.15 *Let* $\{\vec{\mathbf{e}}_1, \ldots, \vec{\mathbf{e}}_N\}$ *and* $\{\widehat{\mathbf{e}}_1, \ldots, \widehat{\mathbf{e}}_N\}$ *be two basis sets and* $\vec{\mathbf{a}} = \alpha^i\vec{\mathbf{e}}_i = \widehat{\alpha}^j\widehat{\mathbf{e}}_j$ *be an arbitrary vector in* $\mathcal{V}$. *Then*

(i)
$$\lambda^k_i\, \mu^j_k = \mu^k_i\lambda^j_k = \delta^j_i; \tag{1.9}$$

(ii)
$$\widehat{\alpha}^i = \mu^i_k\, \alpha^k, \quad \alpha^i = \lambda^i_k\, \widehat{\alpha}^k. \tag{1.10}$$

Proof. **(i)** From (1.8), it follows that

$$\widehat{\mathbf{e}}_i = \lambda^k_i\vec{\mathbf{e}}_k = \lambda^k_i\left(\mu^j_k\widehat{\mathbf{e}}_j\right),$$

or

$$\left(\delta^j_i - \lambda^k_i\mu^j_k\right)\widehat{\mathbf{e}}_j = \vec{\mathbf{0}}.$$

By the linear independence of basis vector $\widehat{\mathbf{e}}_j$'s, we have the coefficients

$$\delta^j_i - \lambda^k_i\mu^j_k \equiv 0.$$

Thus, one of equations (1.9) is proved. The other equation in (1.10) can be proved similarly.

(ii) By the equation

$$\widehat{\alpha}^i\widehat{\mathbf{e}}_i = \alpha^k\vec{\mathbf{e}}_k = \alpha^k\left(\mu^i_k\widehat{\mathbf{e}}_i\right)$$

and the *uniqueness* of the components of theorem 1.2.13, the first of the equations (1.10) follows. The second equation in (1.10) can be proved similarly. ■

Example 1.2.16 Consider two basis sets in $\mathbb{R}^4$ given by

$$\vec{\mathbf{e}}_1 := (1,0,0,0),\ \vec{\mathbf{e}}_2 := (0,1,0,0),\ \vec{\mathbf{e}}_3 := (0,0,1,0),\ \vec{\mathbf{e}}_4 := (0,0,0,1);$$

$$\widehat{\vec{\mathbf{e}}}_1 := (1,0,0,0),\ \widehat{\vec{\mathbf{e}}}_2 := (1,1,0,0),\ \widehat{\vec{\mathbf{e}}}_3 := (1,1,1,0),\ \widehat{\vec{\mathbf{e}}}_4 := (1,1,1,1).$$

The 4×4 transformation matrices are

$$\left[\lambda^k_i\right] = \begin{bmatrix} 1 & 1 & 1 & 1 \\ 0 & 1 & 1 & 1 \\ 0 & 0 & 1 & 1 \\ 0 & 0 & 0 & 1 \end{bmatrix}, \quad \left[\mu^j_k\right] = \begin{bmatrix} 1 & -1 & 0 & 0 \\ 0 & 1 & -1 & 0 \\ 0 & 0 & 1 & -1 \\ 0 & 0 & 0 & 1 \end{bmatrix},$$

$$\det\left[\lambda^k_i\right] = \det\left[\mu^j_k\right] = 1.$$

(Such basis sets are called **tetrads** in relativistic physics.) □

Example 1.2.17 Let us consider the two-dimensional complex vector space $\mathbb{C}^2$ and the transformation of basis vectors furnished by

$$\begin{aligned}
&\widehat{\vec{\mathbf{e}}}_1 = (\lambda \cosh\phi)\vec{\mathbf{e}}_1 + (\mu \sinh\phi)\vec{\mathbf{e}}_2, \\
&\widehat{\vec{\mathbf{e}}}_2 = (\mu^{-1} \sinh\phi)\vec{\mathbf{e}}_1 + (\lambda^{-1} \cosh\phi)\vec{\mathbf{e}}_2, \\
&\lambda\mu \neq 0, \\
&\det\left[\lambda^k_i\right] = \det\left[\mu^j_k\right] = 1.
\end{aligned}$$

(Such basis sets are called **spinor dyads** in relativistic physics.) □

Exercises 1.2

1. Prove that complex vectors (α^1, α^2) and (β^1, β^2) are linearly dependent if and only if $\alpha^1\beta^2 - \alpha^2\beta^1 = 0$.

2. In $\mathbb{R}^4$, a subspace $\mathcal{U}$ is spanned by $(-1, 1, -2, 3)$ and $(-1, -1, -2, 0)$. Another subspace $\mathcal{W}$ is spanned by $(0, 2, 0, 3)$, $(1, 0, 1, 0)$, and $(1, -\frac{1}{3}, 2, -2)$. Obtain the dimension of the vector subspace $\mathcal{U} \cap \mathcal{W}$.

1.3 Linear Mappings of a Vector Space

Let $\mathcal{V}$ and $\widehat{\mathcal{V}}$ be two vector spaces in the same field $\mathcal{F}$. A **linear mapping** (or **transformation**) $\mathbf{L}$ from $\mathcal{V}$ into $\widehat{\mathcal{V}}$ is defined to be such that

$$\mathbf{L}(\lambda\vec{\mathbf{a}} + \mu\vec{\mathbf{b}}) = [\lambda\mathbf{L}(\vec{\mathbf{a}})] + [\mu\mathbf{L}(\vec{\mathbf{b}})] \tag{1.11}$$

for all λ, μ in $\mathcal{F}$ and all $\vec{\mathbf{a}}, \vec{\mathbf{b}}$ in $\mathcal{V}$. If $\widehat{\mathcal{V}} = \mathcal{V}$, the mapping $\mathbf{L}$ is called a **linear operator**.

Example 1.3.1 (i) Consider a linear mapping $\mathbf{L}$ from $\mathcal{V}$ into $\widehat{\mathcal{V}}$ given by

$$\mathbf{L}\,(\vec{\mathbf{a}}) = \vec{\mathbf{0}}$$

for all $\vec{\mathbf{a}}$ in $\mathcal{V}$. This mapping is called the **zero** or **null mapping** and is denoted by $\mathbf{L} = \mathbf{O}$. □

Example 1.3.2 The linear operator $\mathbf{I}$ with the property

$$\mathbf{I}\,(\vec{\mathbf{a}}) = \vec{\mathbf{a}}$$

for all $\vec{\mathbf{a}}$ in $\mathcal{V}$ is called the **identity operator**. □

Remark: The transition from classical mechanics to quantum mechanics is achieved by replacing scalar dynamical variables by the corresponding linear operators on the Hilbert (vector) space.

Consider a linear operator $\mathbf{L}$ acting on $\mathcal{V}$. In the case where it is not the identity operator, $\mathbf{L}$ transforms most of the vectors into different vectors. (That is why it is called an **active transformation**.) Especially, a set of basis vectors $\{\vec{\mathbf{e}}_1, \ldots, \vec{\mathbf{e}}_N\}$ undergoes the following transformation:

$$\mathbf{L}\,(\vec{\mathbf{e}}_j) = \widehat{\vec{\mathbf{e}}}_j = \lambda_j^k\, \vec{\mathbf{e}}_k. \tag{1.12}$$

The $N \times N$ matrix $[L] := \left[\lambda_j^k\right]$ is called the **representation matrix** of the operator $\mathbf{L}$.

Two vector spaces $\mathcal{V}$ and $\widehat{\mathcal{V}}$ in $\mathcal{F}$ are **isomorphic** provided there exists a one-to-one and onto mapping $\mathcal{I}$ such that

$$\mathcal{I}(\alpha\vec{\mathbf{a}} + \beta\vec{\mathbf{b}}) = [\alpha\mathcal{I}(\vec{\mathbf{a}})] + [\beta\mathcal{I}(\vec{\mathbf{b}})] \tag{1.13}$$

for all α, β in $\mathcal{F}$ and all $\vec{\mathbf{a}}, \vec{\mathbf{b}}$ in $\mathcal{V}$. ($\mathcal{I}$ is of course a linear mapping.)

Exercises 1.3

1. Let $\mathbf{A}$ and $\mathbf{B}$ be two linear operators acting on $\mathcal{V}$. The addition, scalar multiplication, etc., are defined by

$$\begin{aligned}
&(\mathbf{A} + \mathbf{B})\,(\vec{\mathbf{a}}) := \mathbf{A}\,(\vec{\mathbf{a}}) + \mathbf{B}\,(\vec{\mathbf{a}}) && \text{for all } \vec{\mathbf{a}} \text{ in } \mathcal{V},\\
&\mathbf{O}\,(\vec{\mathbf{a}}) := \vec{\mathbf{o}} && \text{for all } \vec{\mathbf{a}} \text{ in } \mathcal{V},\\
&(-\mathbf{A})\,(\vec{\mathbf{a}}) := -\left[(\mathbf{A})\,(\vec{\mathbf{a}})\right] && \text{for all } \vec{\mathbf{a}} \text{ in } \mathcal{V},\\
&[\lambda\mathbf{A}]\,(\vec{\mathbf{a}}) := \lambda\,[\mathbf{A}\,(\vec{\mathbf{a}})] && \text{for all } \vec{\mathbf{a}} \text{ in } \mathcal{V}, \text{ and all } \lambda \text{ in } \mathcal{F}.
\end{aligned}$$

Prove that, under these rules, the set of all linear operators on $\mathcal{V}$ constitutes a vector space.

2. Let **A** and **B** be two linear operators on $\mathcal{V}$ with $N \times N$ matrix representations $\left[\alpha^i_j\right]$ and $\left[\beta^j_k\right]$. Prove that the composite linear mapping $\mathbf{A} \circ \mathbf{B}$ has the matrix representation $\left[\alpha^i_j\, \beta^j_k\right] = \left[\alpha^i_j\right] \left[\beta^j_k\right]$.

(Here, the summation convention is used.)

3. Prove that every N-dimensional vector space over the field $\mathcal{F}$ is isomorphic to $\mathcal{F}^N$.

1.4 Dual or Covariant Vector Spaces

A function $\tilde{\mathbf{u}}$ from $\mathcal{V}$ into $\mathcal{F}$ such that

$$\tilde{\mathbf{u}}(\alpha\vec{\mathbf{a}} + \beta\vec{\mathbf{b}}) = [\alpha\tilde{\mathbf{u}}(\vec{\mathbf{a}})] + [\beta\tilde{\mathbf{u}}(\vec{\mathbf{b}})] \tag{1.14}$$

for all α, β in $\mathcal{F}$ and all $\vec{\mathbf{a}}, \vec{\mathbf{b}}$ in $\mathcal{V}$ is called a **covariant vector**. (Other names for a covariant vector are **dual** vector, **covector**, **linear form**, or **linear functional**.)

Example 1.4.1 The zero covariant vector $\tilde{\mathbf{o}}$ is defined by the mapping

$$\tilde{\mathbf{o}}\,(\vec{\mathbf{a}}) := 0 \qquad \text{for all } \vec{\mathbf{a}} \text{ in } \mathcal{V}.$$

It can be proved that $\tilde{\mathbf{o}}$ is a unique mapping. □

Example 1.4.2 Let $\mathcal{V} = \mathbb{R}^N$. Moreover, let a mapping $\tilde{\mathbf{u}}$ be defined by

$$\begin{aligned}
&\tilde{\mathbf{u}}(\vec{\mathbf{a}}) \equiv \tilde{\mathbf{u}}(\alpha^1, \alpha^2, \ldots, \alpha^N) := \alpha^1, \\
&\tilde{\mathbf{u}}(\mu\vec{\mathbf{a}} + \nu\vec{\mathbf{b}}) = (\mu\alpha^1) + (\nu\beta^1) = [\mu\tilde{\mathbf{u}}(\vec{\mathbf{a}})] + [\nu\tilde{\mathbf{u}}(\vec{\mathbf{b}})].
\end{aligned}$$

Therefore, $\tilde{\mathbf{u}}$ is a covariant vector. □

Remark: In Newtonian physics, the gradient of the gravitational potential at a spatial point is a covariant vector.

The addition, scalar multiplication, etc., for covariant vectors are defined as

$$\begin{aligned}
&[\tilde{\mathbf{u}} + \tilde{\mathbf{v}}](\vec{\mathbf{a}}) := \tilde{\mathbf{u}}(\vec{\mathbf{a}}) + \tilde{\mathbf{v}}(\vec{\mathbf{a}}), \\
&\tilde{\mathbf{o}}(\vec{\mathbf{a}}) := 0, \\
&[-\tilde{\mathbf{u}}]\,(\vec{\mathbf{a}}) := -[\tilde{\mathbf{u}}\,(\vec{\mathbf{a}})], \\
&[\lambda\tilde{\mathbf{u}}]\,(\vec{\mathbf{a}}) := \lambda[\tilde{\mathbf{u}}(\vec{\mathbf{a}})],
\end{aligned} \tag{1.15}$$

for all $\vec{\mathbf{a}}$ in $\mathcal{V}$. The set $\tilde{\mathcal{V}}$ of all covariant vectors under the rules (1.15) constitutes a vector space. The set $\tilde{\mathcal{V}}$ is called the **covariant** or **dual vector space**.

Lemma 1.4.3 *Let* $\{\vec{\mathbf{e}}_1, \cdots, \vec{\mathbf{e}}_N\}$ *be a basis set for* $\mathcal{V}$ *and* $(\alpha_1, \ldots, \alpha_N)$ *be a prescribed* N*-tuple. Then there exists a unique covariant vector* $\tilde{\mathbf{a}}$ *such that* $\tilde{\mathbf{a}}(\vec{\mathbf{e}}_i) = \alpha_i$ *for* $i \in \{1, \cdots, N\}$.

Proof. Let $\vec{\mathbf{b}} = \beta^j \vec{\mathbf{e}}_j$ be an arbitrarily chosen vector. (We are using the summation convention!) Let us define a covariant vector $\tilde{\mathbf{a}}$ by the equation

$$\tilde{\mathbf{a}}(\vec{\mathbf{b}}) \equiv \tilde{\mathbf{a}}(\beta^j \vec{\mathbf{e}}_j) := \alpha_j \beta^j.$$

Then

$$\tilde{\mathbf{a}}(\vec{\mathbf{e}}_i) = \tilde{\mathbf{a}}(\delta_i^j \vec{\mathbf{e}}_j) = \alpha_j \delta_i^j = \alpha_i.$$

The equation above shows the existence of $\tilde{\mathbf{a}}$.

To prove the uniqueness of our choice, we assume to the contrary the existence of another covariant vector $\tilde{\mathbf{a}}'$ such that

$$\tilde{\mathbf{a}}'(\vec{\mathbf{b}}) = \alpha_j \beta^j$$

and $\tilde{\mathbf{a}}' \neq \tilde{\mathbf{a}}$. Therefore,

$$\begin{aligned} [\tilde{\mathbf{a}}' - \tilde{\mathbf{a}}](\vec{\mathbf{b}}) &= \tilde{\mathbf{a}}'(\vec{\mathbf{b}}) - \tilde{\mathbf{a}}(\vec{\mathbf{b}}) \\ &= \alpha_j \beta^j - \alpha_j \beta^j \equiv 0 \end{aligned}$$

for all $\vec{\mathbf{b}}$ in $\mathcal{V}$. By the uniqueness of the zero covariant vector $\tilde{\mathbf{o}}$, we must have $\tilde{\mathbf{a}}' - \tilde{\mathbf{a}} = \tilde{\mathbf{o}}$. This is the contradiction. Thus the lemma is proved. ■

Theorem 1.4.4 *Let* $\{\vec{\mathbf{e}}_1, \ldots, \vec{\mathbf{e}}_N\}$ *be a basis set for* $\mathcal{V}$. *Then there exists a unique covariant basis set* $\{\tilde{\mathbf{e}}^1, \ldots, \tilde{\mathbf{e}}^N\}$ *for* $\tilde{\mathcal{V}}$ *such that*

$$\tilde{\mathbf{e}}^j(\vec{\mathbf{e}}_i) = \delta_i^j \tag{1.16}$$

for all i, j *in* $\{1, \ldots, N\}$.

Proof. By the preceding lemma, there exists a unique set of covariant vectors such that

$$\tilde{\mathbf{e}}^j(\vec{\mathbf{e}}_i) = \delta_i^j.$$

To prove the linear independence, consider the covariant vector equation

$$\mu_j \tilde{\mathbf{e}}^j = \tilde{\mathbf{o}}.$$

By (1.14), we have

$$\left[\mu_j \tilde{\mathbf{e}}^j\right](\vec{\mathbf{e}}_i) = \mu_j \left[\tilde{\mathbf{e}}^j(\vec{\mathbf{e}}_i)\right] = \mu_j \delta^j_i = \mu_i = 0$$

for every i in $\{1, \cdots, N\}$. Therefore, the $\tilde{\mathbf{e}}_j$'s are linearly independent by (1.4).

To prove the spanning property, choose an arbitrary covariant vector $\tilde{\mathbf{a}}$. Let $\tilde{\mathbf{a}}(\vec{\mathbf{e}}_i) = \alpha_i$. Therefore, by (1.14),

$$\tilde{\mathbf{a}}\left(\vec{\mathbf{b}}\right) \equiv \tilde{\mathbf{a}}\left(\beta^i \vec{\mathbf{e}}_i\right) = \beta^i \alpha_i \tag{1.17}$$

for an arbitrary vector $\vec{\mathbf{b}}$. By (1.14), (1.15), and (1.16), we have

$$\left[\alpha_i \tilde{\mathbf{e}}^i\right]\left(\vec{\mathbf{b}}\right) = \alpha_i \left[\tilde{\mathbf{e}}^i\left(\beta^j \vec{\mathbf{e}}_j\right)\right] = \alpha_i \beta^j \delta^i_j = \alpha_i \beta^i. \tag{1.18}$$

Comparing (1.17) and (1.18), and recalling the uniqueness of $\tilde{\mathbf{o}}$, we conclude that

$$\tilde{\mathbf{a}} = \alpha_i \, \tilde{\mathbf{e}}^i.$$

The equation above proves the spanning property. Thus $\{\tilde{\mathbf{e}}^1, \ldots, \tilde{\mathbf{e}}^N\}$ is the basis set. ■

Corollary 1.4.5

$$\dim\left(\tilde{\mathcal{V}}\right) = \dim(\mathcal{V}). \tag{1.19}$$

The proof is obvious.

Now we shall deal with the transformation of covariant components under a change of basis vectors (*passive transformation* in (1.8)).

Theorem 1.4.6 *Let $\{\vec{\mathbf{e}}_2, \ldots \vec{\mathbf{e}}_N\}$ and $\{\widehat{\vec{\mathbf{e}}}_1, \ldots \widehat{\vec{\mathbf{e}}}_N\}$ be the two basis sets for $\mathcal{V}$ such that*

$$\widehat{\vec{\mathbf{e}}}_i = \lambda^k_i \, \vec{\mathbf{e}}_k, \quad \vec{\mathbf{e}}_k = \mu^j_k \, \widehat{\vec{\mathbf{e}}}_j.$$

Then the corresponding covariant basis sets $\{\tilde{\mathbf{e}}^1, \ldots, \tilde{\mathbf{e}}^N\}$ and $\{\widehat{\tilde{\mathbf{e}}}^1, \ldots, \widehat{\tilde{\mathbf{e}}}^N\}$ transform as

$$\widehat{\tilde{\mathbf{e}}}^a = \mu^a_b \tilde{\mathbf{e}}^b, \quad \tilde{\mathbf{e}}^a = \lambda^a_b \, \widehat{\tilde{\mathbf{e}}}^b. \tag{1.20}$$

Proof. By spanning properties of the covariant basis vectors, there exist scalars α^i_j and β^i_j such that

$$\widehat{\tilde{\mathbf{e}}}^i = \alpha^i_j \tilde{\mathbf{e}}^j, \quad \tilde{\mathbf{e}}^i = \beta^i_j \widehat{\tilde{\mathbf{e}}}^j.$$

By (1.16), (1.15), and (1.14), we have

$$\delta^i_k = \tilde{\mathbf{e}}^i(\vec{\mathbf{e}}_k) = \left[\beta^i_j \widehat{\tilde{\mathbf{e}}}^j\right]\left(\mu^a_k \widehat{\vec{\mathbf{e}}}_a\right) = \beta^i_j \mu^a_k \left[\widehat{\tilde{\mathbf{e}}}^j\left(\widehat{\vec{\mathbf{e}}}_a\right)\right] = \beta^i_j \mu^a_k \delta^j_a = \beta^i_j \mu^j_k.$$

But from (1.10) we know that $\lambda^i_j \mu^j_k = \delta^i_k$. By the uniqueness of an inverse matrix, we must have $\beta^i_j = \lambda^i_j$. Similarly, we can prove that $\alpha^i_j = \mu^i_j$. ■

Corollary 1.4.7 *Under the change of basis sets given by* (1.8) *and* (1.20), *the components of a covariant vector* $\tilde{\mathbf{w}} = w_i \tilde{\mathbf{e}}^i = \widehat{w}_j \, \widehat{\tilde{\mathbf{e}}}^j$ *transform as the following:*

$$\widehat{w}_j = \lambda^i_j \, w_i, \qquad w_i = \mu^j_i \, \widehat{w}_j. \tag{1.21}$$

Proof. By (1.20) and (1.15), we have

$$w_i \tilde{\mathbf{e}}^i = \widehat{w}^j \, \widehat{\tilde{\mathbf{e}}}^j = \widehat{w}_j [\mu^j_i \, \tilde{\mathbf{e}}^i] = (\mu^j_i \widehat{w}_j) \tilde{\mathbf{e}}^i.$$

By the uniqueness of components, we obtain $w_i = \mu^j_i \widehat{w}_j$. Similarly, we can prove that $\widehat{w}_j = \lambda^i_j w_i$. ∎

(Compare and contrast the transformation rules in (1.10) and (1.21).)

Example 1.4.8 Consider the (**spinor**) vector space $\mathbb{C}^2$ and two basis sets (**dyads**):

$$\vec{\mathbf{e}}_1 = (1, 0), \quad \vec{\mathbf{e}}_2 = (0, 1);$$
$$\widehat{\vec{\mathbf{e}}}_1 = (i, 0), \quad \widehat{\vec{\mathbf{e}}}_2 = (i, i).$$

The transformation matrices are given by

$$\left[\lambda^j_i\right] = \begin{bmatrix} i & i \\ 0 & i \end{bmatrix}, \quad \left[\mu^j_i\right] = \begin{bmatrix} -i & i \\ 0 & -i \end{bmatrix}.$$

The covariant basis set $\{\tilde{\mathbf{e}}^1, \tilde{\mathbf{e}}^2\}$ is furnished by the rules

$$\tilde{\mathbf{e}}^i \left(\vec{\mathbf{e}}_j\right) = \delta^i_j, \quad \tilde{\mathbf{e}}^i \left(\alpha^1, \alpha^2\right) = \alpha^1 \delta^i_1 + \alpha^2 \delta^i_2,$$
$$\tilde{\mathbf{e}}^1 \left(\alpha^1, \alpha^2\right) = \alpha^1, \quad \tilde{\mathbf{e}}^2 \left(\alpha^1, \alpha^2\right) = \alpha^2,$$

for all $\left(\alpha^1, \alpha^2\right)$ in $\mathbb{C}^2$. In the transformed covariant basis $\left\{\widehat{\tilde{\mathbf{e}}}^1, \widehat{\tilde{\mathbf{e}}}^2\right\}$,

$$\begin{aligned} \widehat{\tilde{\mathbf{e}}}^1 \left(\alpha^1, \alpha^2\right) &= \left[\mu^1_1 \tilde{\mathbf{e}}^1 + \mu^1_2 \tilde{\mathbf{e}}^2\right] \left(\alpha^1, \alpha^2\right) \\ &= \mu^1_1 \alpha^1 + \mu^1_2 \alpha^2 = i \left(-\alpha^1 + \alpha^2\right), \end{aligned}$$

$$\begin{aligned} \widehat{\tilde{\mathbf{e}}}^2 \left(\alpha^1, \alpha^2\right) &= \left[\mu^2_1 \tilde{\mathbf{e}}^1 + \mu^2_2 \tilde{\mathbf{e}}^2\right] \left(\alpha^1, \alpha^2\right) \\ &= \mu^2_1 \alpha^1 + \mu^2_2 \alpha^2 = -i \alpha^2. \end{aligned}$$

Let a particular covariant vector $\tilde{\mathbf{w}}$ be given by

$$\tilde{\mathbf{w}} \left(\alpha^1, \alpha^2\right) := i \left(\alpha^1 + \alpha^2\right)$$

for all $\left(\alpha^1, \alpha^2\right)$ in $\mathbb{C}^2$. This covariant vector can be expressed as

$$\tilde{\mathbf{w}} = i \tilde{\mathbf{e}}^1 + i \tilde{\mathbf{e}}^2 = -\widehat{\tilde{\mathbf{e}}}^1 - 2 \widehat{\tilde{\mathbf{e}}}^2,$$
$$w_1 = w_2 = i, \quad \widehat{w}_1 = -1, \quad \widehat{w}_2 = -2.$$

Thus, the equations (1.21) are validated. □

Exercises 1.4

1. Consider the following two basis sets (or **triads**) in $\mathbb{R}^3$:

$$\{\vec{\mathbf{e}}_1, \vec{\mathbf{e}}_2, \vec{\mathbf{e}}_3\} := \{(1,0,0),(0,1,0),(0,0,1)\},$$
$$\{\widehat{\vec{\mathbf{e}}}_1, \widehat{\vec{\mathbf{e}}}_2, \widehat{\vec{\mathbf{e}}}_3\} := \{(1,0,0),(1,1,0),(1,1,1)\}.$$

Let a covariant vector $\tilde{\mathbf{u}}$ be defined by $\tilde{\mathbf{u}}(\alpha^i \vec{\mathbf{e}}_i) := \alpha^3 - \alpha^2$. Obtain explicitly the components of $\tilde{\mathbf{u}}$ relative to the corresponding bases $\{\tilde{\mathbf{e}}^1, \tilde{\mathbf{e}}^2, \tilde{\mathbf{e}}^3\}$ and $\{\widehat{\tilde{\mathbf{e}}}^1, \widehat{\tilde{\mathbf{e}}}^2, \widehat{\tilde{\mathbf{e}}}^3\}$.

2. Consider the standard basis set (or **tetrad**) in $\mathbb{R}^4$. Another tetrad is given by

$$\widehat{\vec{\mathbf{e}}}_1 = \vec{\mathbf{e}}_1, \;\; \widehat{\vec{\mathbf{e}}}_2 = \vec{\mathbf{e}}_2, \;\; \widehat{\vec{\mathbf{e}}}_3 = (\cosh\alpha)\vec{\mathbf{e}}_3 - (\sinh\alpha)\vec{\mathbf{e}}_4,$$
$$\widehat{\vec{\mathbf{e}}}_4 = -(\sinh\alpha)\vec{\mathbf{e}}_3 + (\cosh\alpha)\vec{\mathbf{e}}_4; \;\; \alpha \in \mathbb{R}.$$

Let a particular covariant vector be characterized by $\tilde{\mathbf{w}} := \widehat{w}_1 \widehat{\tilde{\mathbf{e}}}^1 + \widehat{w}_4 \widehat{\tilde{\mathbf{e}}}^4$. Obtain the four components w_i explicitly.

(*Remark:* This exercise is relevant in the special theory of relativity.)

3. (i) Prove that if a function $\tilde{\mathbf{u}}\left(\zeta^1, \zeta^2, \zeta^3\right) := \zeta^1 + \zeta^2 + \zeta^3$ for all complex vectors $\left(\zeta^1, \zeta^2, \zeta^3\right)$ in $\mathbb{C}^3$, then $\tilde{\mathbf{u}}$ is a covariant vector.

(ii) Using the covariant vector $\tilde{\mathbf{u}}$ above, find a basis set for the null space given by

$$\mathcal{N} := \left\{\left(\zeta^1, \zeta^2, \zeta^3\right) \in \mathbb{C}^3 \;:\; \tilde{\mathbf{u}}\left(\zeta^1, \zeta^2, \zeta^3\right) = 0\right\}.$$

Chapter 2

Tensor Algebra

2.1 Second-Order Tensors

Suppose that $\mathcal{V}$ is a vector space in the field $\mathcal{F}$. An ordered pair of vectors $\left(\vec{\mathbf{a}}, \vec{\mathbf{b}}\right)$ belong to the set $\mathcal{V} \times \mathcal{V}$ (the Cartesian product of $\mathcal{V}$ with itself).

A **second-order covariant tensor** is a function $\mathbf{T}_{\cdot\cdot}$ from $\mathcal{V} \times \mathcal{V}$ into $\mathcal{F}$ such that

$$\text{(i)} \qquad \mathbf{T}_{\cdot\cdot}\left(\lambda\vec{\mathbf{a}} + \mu\vec{\mathbf{b}}, \vec{\mathbf{c}}\right) = \left[\lambda\mathbf{T}_{\cdot\cdot}\left(\vec{\mathbf{a}}, \vec{\mathbf{c}}\right)\right] + \left[\mu\mathbf{T}_{\cdot\cdot}\left(\vec{\mathbf{b}}, \vec{\mathbf{c}}\right)\right], \tag{2.1}$$

$$\text{(ii)} \qquad \mathbf{T}_{\cdot\cdot}\left(\vec{\mathbf{a}}, \lambda\vec{\mathbf{b}} + \mu\vec{\mathbf{c}}\right) = \left[\lambda\mathbf{T}_{\cdot\cdot}\left(\vec{\mathbf{a}}, \vec{\mathbf{b}}\right)\right] + \left[\mu\mathbf{T}_{\cdot\cdot}\left(\vec{\mathbf{a}}, \vec{\mathbf{c}}\right)\right], \tag{2.2}$$

for all λ, μ in $\mathcal{F}$ and all $\vec{\mathbf{a}}, \vec{\mathbf{b}}, \vec{\mathbf{c}}$ in $\mathcal{V}$. (Note that $\mathbf{T}_{\cdot\cdot}$ is a linear function in *both slots*. That is why it is also called a **bilinear form**.)

Example 2.1.1 Let us define $\tilde{\mathbf{O}}_{\cdot\cdot}\left(\vec{\mathbf{a}}, \vec{\mathbf{b}}\right) := 0$ for all $\left(\vec{\mathbf{a}}, \vec{\mathbf{b}}\right)$ in $\mathcal{V} \times \mathcal{V}$. This is the (unique) second-order covariant **zero tensor**. □

Example 2.1.2 $\mathbf{T}_{\cdot\cdot}\left(\vec{\mathbf{a}}, \vec{\mathbf{b}}\right) = \mathbf{T}_{\cdot\cdot}\left(\alpha^i\, \vec{\mathbf{e}}_i,\, \beta^j\, \vec{\mathbf{e}}_j\right) := \alpha^1\beta^1$. □

Remark: The moment of inertia in the dynamics of a rigid body is a second-order covariant tensor.

We can define the addition and scalar multiplication of the second-order tensors by the following equations:

$$\left[\mathbf{T}_{\cdot\cdot} + \mathbf{G}_{\cdot\cdot}\right]\left(\vec{\mathbf{a}}, \vec{\mathbf{b}}\right) := \left[\mathbf{T}_{\cdot\cdot}\left(\vec{\mathbf{a}}, \vec{\mathbf{b}}\right)\right] + \left[\mathbf{G}_{\cdot\cdot}\left(\vec{\mathbf{a}}, \vec{\mathbf{b}}\right)\right],$$

$$[\lambda \mathbf{T}..]\left(\vec{\mathbf{a}}, \vec{\mathbf{b}}\right) := \lambda \left[\mathbf{T}.. \left(\vec{\mathbf{a}}, \vec{\mathbf{b}}\right)\right], \tag{2.3}$$

$$[-\mathbf{T}..]\left(\vec{\mathbf{a}}, \vec{\mathbf{b}}\right) := -\left[\mathbf{T}.. \left(\vec{\mathbf{a}}, \vec{\mathbf{b}}\right)\right],$$

for all $\vec{\mathbf{a}}, \vec{\mathbf{b}}$ in $\mathcal{V}$ and all λ in $\mathcal{F}$.

Theorem 2.1.3 *Under the rules* (2.3) *of addition and scalar multiplication and the definition of the second-order covariant zero tensor, the set* $\tilde{\mathcal{V}} \otimes \tilde{\mathcal{V}}$ *of all second-order covariant tensors constitutes a vector space.*

The proof is left to the reader.

Now, the **tensor product** (or **outer product**) between two covariant vectors $\tilde{\mathbf{u}}$ and $\tilde{\mathbf{v}}$ will be defined. It is the function $\tilde{\mathbf{u}} \otimes \tilde{\mathbf{v}}$ from $\mathcal{V} \times \mathcal{V}$ into $\mathcal{F}$ such that

$$[\tilde{\mathbf{u}} \otimes \tilde{\mathbf{v}}]\left(\vec{\mathbf{a}}, \vec{\mathbf{b}}\right) := [\tilde{\mathbf{u}}(\vec{\mathbf{a}})]\left[\tilde{\mathbf{v}}\left(\vec{\mathbf{b}}\right)\right] \tag{2.4}$$

for all $\vec{\mathbf{a}}, \vec{\mathbf{b}}$ in $\mathcal{V}$. Note that $\tilde{\mathbf{u}} \otimes \tilde{\mathbf{v}}$ is an example of a second-order covariant tensor.

Theorem 2.1.4 *The tensor product of covariant vectors satisfies the following equations:*

(*i*)
$$\tilde{\mathbf{a}} \otimes \left(\tilde{\mathbf{b}} + \tilde{\mathbf{c}}\right) = \left[\tilde{\mathbf{a}} \otimes \tilde{\mathbf{b}}\right] + [\tilde{\mathbf{a}} \otimes \tilde{\mathbf{c}}], \tag{2.5}$$

(*ii*)
$$\left(\tilde{\mathbf{a}} + \tilde{\mathbf{b}}\right) \otimes \tilde{\mathbf{c}} = [\tilde{\mathbf{a}} \otimes \tilde{\mathbf{c}}] + \left[\tilde{\mathbf{b}} \otimes \tilde{\mathbf{c}}\right], \tag{2.6}$$

(*iii*)
$$\lambda\left(\tilde{\mathbf{a}} \otimes \tilde{\mathbf{b}}\right) = (\lambda\tilde{\mathbf{a}}) \otimes \tilde{\mathbf{b}} = \tilde{\mathbf{a}} \otimes \left(\lambda\tilde{\mathbf{b}}\right), \tag{2.7}$$

(*iv*)
$$\tilde{\mathbf{a}} \otimes \tilde{\mathbf{o}} = \tilde{\mathbf{o}} \otimes \tilde{\mathbf{a}} = \tilde{\mathbf{O}}.., \tag{2.8}$$

for all λ *in* $\mathcal{F}$ *and all* $\tilde{\mathbf{a}}, \tilde{\mathbf{b}}, \tilde{\mathbf{c}}$ *in* $\tilde{\mathcal{V}}$.

Proof of part (i). The domains of the bilinear functions on both sides of (2.5) are identical, namely $\mathcal{V} \times \mathcal{V}$. Moreover, for two arbitrary vectors $\vec{\mathbf{u}}, \vec{\mathbf{v}}$ in $\mathcal{V}$, we have (by (2.4) and (1.15))

$$\left[\tilde{\mathbf{a}} \otimes \left(\tilde{\mathbf{b}} + \tilde{\mathbf{c}}\right)\right](\vec{\mathbf{u}}, \vec{\mathbf{v}}) = [\tilde{\mathbf{a}}(\vec{\mathbf{u}})]\left[\left(\tilde{\mathbf{b}} + \tilde{\mathbf{c}}\right)(\vec{\mathbf{v}})\right]$$
$$= [\tilde{\mathbf{a}}(\vec{\mathbf{u}})]\left[\tilde{\mathbf{b}}(\vec{\mathbf{v}}) + \tilde{\mathbf{c}}(\vec{\mathbf{v}})\right] = [\tilde{\mathbf{a}}(\vec{\mathbf{u}})]\left[\tilde{\mathbf{b}}(\vec{\mathbf{v}})\right] + [\tilde{\mathbf{a}}(\vec{\mathbf{u}})][\tilde{\mathbf{c}}(\vec{\mathbf{v}})]$$
$$= \left[\left(\tilde{\mathbf{a}} \otimes \tilde{\mathbf{b}}\right)(\vec{\mathbf{u}}, \vec{\mathbf{v}})\right] + [\tilde{\mathbf{a}} \otimes \tilde{\mathbf{c}}](\vec{\mathbf{u}}, \vec{\mathbf{v}})$$
$$= \left[\left(\tilde{\mathbf{a}} \otimes \tilde{\mathbf{b}}\right) + (\tilde{\mathbf{a}} \otimes \tilde{\mathbf{c}})\right](\vec{\mathbf{u}}, \vec{\mathbf{v}}).$$

Thus, part (i) is proved. ■

(Similarly, other parts can be proved.)

Example 2.1.5 Let two covariant vectors $\tilde{\mathbf{u}}$ and $\tilde{\mathbf{v}}$ be defined by

$$\tilde{\mathbf{u}}(\vec{\mathbf{a}}) = \tilde{\mathbf{u}}\left(\alpha^i \vec{\mathbf{e}}_i\right) := \alpha^1, \quad \tilde{\mathbf{v}}\left(\vec{\mathbf{b}}\right) = \tilde{\mathbf{v}}\left(\beta^j \vec{\mathbf{e}}_j\right) := \beta^N.$$

By (2.4),

$$[\tilde{\mathbf{u}} \otimes \tilde{\mathbf{v}}]\left(\vec{\mathbf{a}}, \vec{\mathbf{b}}\right) = [\tilde{\mathbf{u}}(\vec{\mathbf{a}})]\left[\tilde{\mathbf{v}}\left(\vec{\mathbf{b}}\right)\right] = \alpha^1 \beta^N,$$

$$[\tilde{\mathbf{u}} \otimes \tilde{\mathbf{v}}]\left(\vec{\mathbf{b}}, \vec{\mathbf{b}}\right) = \beta^1 \beta^N.$$ □

Let us consider the tensor products of basis covariant vectors $\tilde{\mathbf{e}}^a \otimes \tilde{\mathbf{e}}^b$, where a, b take values from 1 to N.

Theorem 2.1.6 *The set of tensor products of basis covariant vectors* $\{\tilde{\mathbf{e}}^a \otimes \tilde{\mathbf{e}}^b\}$, *where* $a, b \in \{1.\ldots, N\}$, *constitutes a basis set for* $\tilde{\mathcal{V}} \otimes \tilde{\mathcal{V}}$.

Proof. To prove the linear independence, consider the tensor equation

$$\mu_{ab}\left[\tilde{\mathbf{e}}^a \otimes \tilde{\mathbf{e}}^b\right] = \tilde{\mathbf{O}}.. \ .$$

The equation above, by (2.3), (2.4), and (1.16), implies that

$$\left[\mu_{ab}\left(\tilde{\mathbf{e}}^a \otimes \tilde{\mathbf{e}}^b\right)\right](\vec{\mathbf{e}}_c, \vec{\mathbf{e}}_d) = \mu_{ab}\left[\tilde{\mathbf{e}}^a(\vec{\mathbf{e}}_c)\right]\left[\tilde{\mathbf{e}}^b(\vec{\mathbf{e}}_d)\right] = \mu_{ab}\delta^a_c\delta^b_d = \mu_{cd} = 0$$

for all c, d. Thus, the linear independence is proved.

To prove the spanning property, consider an arbitrary tensor $\mathbf{T}..$ yielding scalars

$$\tau_{cd} := \mathbf{T}..(\vec{\mathbf{e}}_c, \vec{\mathbf{e}}_d). \tag{2.9}$$

By (2.1), (2.2), and (2.9),

$$\begin{aligned} \mathbf{T}..\left(\vec{\mathbf{a}}, \vec{\mathbf{b}}\right) &= \mathbf{T}..\left(\alpha^c \vec{\mathbf{e}}_c, \beta^d \vec{\mathbf{e}}_d\right) = \alpha^c\beta^d\mathbf{T}..(\vec{\mathbf{e}}_c, \vec{\mathbf{e}}_d) \\ &= \tau_{cd}\alpha^c\beta^d. \end{aligned} \tag{2.10}$$

Consider now the second-order covariant tensor $\tau_{cd}(\tilde{\mathbf{e}}^c \otimes \tilde{\mathbf{e}}^d)$. By (2.3) and (2.4),

$$\begin{aligned} \left[\tau_{cd}\left(\vec{\mathbf{e}}_c \otimes \vec{\mathbf{e}}_d\right)\right]\left(\vec{\mathbf{a}}, \vec{\mathbf{b}}\right) &= \tau_{cd}\left[\tilde{\mathbf{e}}^c(\vec{\mathbf{a}})\right]\left[\tilde{\mathbf{e}}^d\left(\vec{\mathbf{b}}\right)\right] \\ &= \tau_{cd}\alpha^c\beta^d. \end{aligned} \tag{2.11}$$

Subtracting (2.11) from (2.10), we obtain

$$\left[\mathbf{T}.. - \tau_{cd}\left(\tilde{\mathbf{e}}^c \otimes \tilde{\mathbf{e}}^d\right)\right]\left(\vec{\mathbf{a}}, \vec{\mathbf{b}}\right) = 0 = \tilde{\mathbf{O}}..\left(\vec{\mathbf{a}}, \vec{\mathbf{b}}\right)$$

for every ordered pair $(\vec{\mathbf{a}}, \vec{\mathbf{b}})$ in $\mathcal{V} \times \mathcal{V}$. By the uniqueness of the zero tensor $\tilde{\mathbf{O}}_{..}$, we must have

$$\mathbf{T}_{..} = \tau_{cd} \left(\tilde{\mathbf{e}}^c \otimes \tilde{\mathbf{e}}^d \right).$$

Thus, the spanning property is proved. Consequently, $\{\tilde{\mathbf{e}}^a \otimes \tilde{\mathbf{e}}^b\}$ is a basis set. ■

Corollary 2.1.7

$$\dim(\tilde{\mathcal{V}} \otimes \tilde{\mathcal{V}}) = N^2. \tag{2.12}$$

The proof is omitted.

The (unique) scalars τ_{cd} appearing in (2.9) are defined to be the **components** of $\mathbf{T}_{..}$ *relative* to the second-order covariant tensor basis $\{\tilde{\mathbf{e}}^c \otimes \tilde{\mathbf{e}}^d\}$.

Now, we shall discuss the transformation rules for the components of a second-order covariant tensor under a change of basis in $\mathcal{V}$. (See (1.8).)

Theorem 2.1.8 *Let $\{\vec{\mathbf{e}}_1, \ldots, \vec{\mathbf{e}}_N\}$ and $\{\widehat{\vec{\mathbf{e}}}_1, \ldots, \widehat{\vec{\mathbf{e}}}_N\}$ be two basis sets with transformation equations $\widehat{\vec{\mathbf{e}}}_a = \lambda^b_a \vec{\mathbf{e}}_b$, $\vec{\mathbf{e}}_a = \mu^b_a \widehat{\vec{\mathbf{e}}}_b$. Then, the components τ_{cd} of a second-order covariant tensor $\mathbf{T}_{..}$ undergo the following transformation equations:*

$$\begin{aligned} \widehat{\tau}_{ab} &= \lambda^c_a \lambda^d_b \tau_{cd}, \\ \tau_{ab} &= \mu^c_a \mu^d_b \widehat{\tau}_{cd}. \end{aligned} \tag{2.13}$$

Proof. Recall (1.20), yielding

$$\widehat{\tilde{\mathbf{e}}}^a = \mu^a_b \tilde{\mathbf{e}}^b, \quad \tilde{\mathbf{e}}^a = \lambda^a_b \widehat{\tilde{\mathbf{e}}}^b.$$

The second-order tensor $\mathbf{T}_{..}$ has the representations

$$\mathbf{T}_{..} = \tau_{cd} \tilde{\mathbf{e}}^c \otimes \tilde{\mathbf{e}}^d = \widehat{\tau}_{ab} \widehat{\tilde{\mathbf{e}}}^a \otimes \widehat{\tilde{\mathbf{e}}}^b.$$

By (1.20) and (2.3), we obtain

$$\begin{aligned} \tau_{cd} \tilde{\mathbf{e}}^c \otimes \tilde{\mathbf{e}}^d &= \widehat{\tau}_{ab} \left[(\mu^a_c \tilde{\mathbf{e}}^c) \otimes \left(\mu^b_d \tilde{\mathbf{e}}^d \right) \right] \\ &= \left[\mu^a_c \mu^b_d \widehat{\tau}_{ab} \right] \left[\tilde{\mathbf{e}}^c \otimes \tilde{\mathbf{e}}^d \right]. \end{aligned}$$

By the uniqueness of the components of $\mathbf{T}_{..}$, we get

$$\tau_{cd} = \mu^a_c \mu^b_d \widehat{\tau}_{ab}.$$

Similarly, we can prove the other equation in (2.13). ■

Example 2.1.9 Let $\mathcal{V}$ be a two-dimensional vector space in the field $\mathcal{F}$. A transformation between two basis sets is characterized by

$$\widehat{\vec{\mathbf{e}}}_1 = \vec{\mathbf{e}}_2, \quad \widehat{\vec{\mathbf{e}}}_2 = -\vec{\mathbf{e}}_1 .$$

Then the components of $\mathbf{T}..$ will transform as

$$\begin{aligned} &\widehat{\tau}_{11} = \lambda_1^c \lambda_1^d \tau_{cd} = \lambda_1^2 \lambda_1^2 \tau_{22} = \tau_{22}, \\ &\widehat{\tau}_{12} = -\tau_{21}, \;\; \widehat{\tau}_{21} = -\tau_{12}, \;\; \widehat{\tau}_{22} = \tau_{11}, \\ &\det\left[\widehat{\tau}_{ab}\right] = \det\left[\tau_{ab}\right]. \end{aligned}$$

□

The **transposition** $\mathbf{B}..^T$ of a second-order covariant tensor $\mathbf{B}..$ is defined by the equation

$$\mathbf{B}..^T\left(\vec{\mathbf{a}}, \vec{\mathbf{b}}\right) := \mathbf{B}..\left(\vec{\mathbf{b}}, \vec{\mathbf{a}}\right) \tag{2.14}$$

for every ordered pair of vectors $(\vec{\mathbf{a}}, \vec{\mathbf{b}})$ in $\mathcal{V} \times \mathcal{V}$.

A **symmetric** second-order covariant tensor $\mathbf{S}..$ is defined by the condition

$$\mathbf{S}..^T = \mathbf{S}.. \, . \tag{2.15}$$

The components of such a tensor satisfy

$$\sigma_{ji} := \mathbf{S}..\left(\vec{\mathbf{e}}_j, \vec{\mathbf{e}}_i\right) = \mathbf{S}..^T\left(\vec{\mathbf{e}}_i, \vec{\mathbf{e}}_j\right) = \mathbf{S}..\left(\vec{\mathbf{e}}_i, \vec{\mathbf{e}}_j\right) = \sigma_{ij}. \tag{2.16}$$

The $N \times N$ matrix $[\sigma_{ij}]$ is a symmetric matrix. It has $N + (N-1)\left(\frac{N}{2}\right) = \frac{N(N+1)}{2}$ linearly independent entries or components.

Example 2.1.10 The second-order covariant tensor

$$\mathbf{S}.. := \frac{1}{2}[(\tilde{\mathbf{u}} \otimes \tilde{\mathbf{v}}) + (\tilde{\mathbf{v}} \otimes \tilde{\mathbf{u}})] \tag{2.17}$$

is a symmetric tensor. □

Remark: The stress tensor at a point of a three-dimensional deformable body is a symmetric second-order tensor with six independent components.

An **antisymmetric** second-order covariant tensor $\mathbf{A}..$ is defined by the condition

$$\mathbf{A}..^T = -\mathbf{A}.. \, . \tag{2.18}$$

The components of such a tensor satisfy

$$\alpha_{ji} = -\alpha_{ij} \, . \tag{2.19}$$

The number of linearly independent components of a second-order antisymmetric tensor is $\frac{N(N-1)}{2}$.

Example 2.1.11 The second-order covariant tensor

$$\mathbf{A}_{..} := (\tilde{\mathbf{u}} \otimes \tilde{\mathbf{v}}) - (\tilde{\mathbf{v}} \otimes \tilde{\mathbf{u}}) \tag{2.20}$$

is antisymmetric. □

Remark: The vorticity tensor at a point of a fluid flow in space is an antisymmetric tensor with three independent components.

A second-order **contravariant** tensor $\mathbf{T}^{..}$ is a function from $\tilde{\mathcal{V}} \times \tilde{\mathcal{V}}$ into $\mathcal{F}$ such that

$$\text{(i)} \qquad \mathbf{T}^{..}(\lambda\tilde{\mathbf{u}} + \mu\tilde{\mathbf{v}}, \tilde{\mathbf{w}}) = [\lambda\mathbf{T}^{..}(\tilde{\mathbf{u}}, \tilde{\mathbf{w}})] + [\mu\mathbf{T}^{..}(\tilde{\mathbf{v}}, \tilde{\mathbf{w}})], \tag{2.21}$$

$$\text{(ii)} \qquad \mathbf{T}^{..}(\tilde{\mathbf{u}}, \lambda\tilde{\mathbf{v}} + \mu\tilde{\mathbf{w}}) = [\lambda\mathbf{T}^{..}(\tilde{\mathbf{u}}, \tilde{\mathbf{v}})] + [\mu\mathbf{T}^{..}(\tilde{\mathbf{u}}, \tilde{\mathbf{w}})], \tag{2.22}$$

for all λ, μ in $\mathcal{F}$ and all $\tilde{\mathbf{u}}, \tilde{\mathbf{v}}, \tilde{\mathbf{w}}$ in $\tilde{\mathcal{V}}$. Note that the function $\mathbf{T}^{..}$ is *bilinear*.

Example 2.1.12 The (unique) second-order contravariant zero tensor is defined by

$$\mathbf{O}^{..}(\tilde{\mathbf{u}}, \tilde{\mathbf{v}}) := 0 \quad \text{for all} \quad \tilde{\mathbf{u}}, \tilde{\mathbf{v}} \quad \text{in} \quad \tilde{\mathcal{V}}.$$ □

Example 2.1.13

$$\mathbf{T}^{..}(\tilde{\mathbf{a}}, \tilde{\mathbf{b}}) = \mathbf{T}^{..}(\alpha_i\tilde{\mathbf{e}}^i, \beta_j\tilde{\mathbf{e}}^j) := \alpha_N\beta_1.$$ □

Logically, we should denote the set of all second-order contravariant tensors as $\tilde{\tilde{\mathcal{V}}} \otimes \tilde{\tilde{\mathcal{V}}}$. But for a *finite-dimensional* vector space $\mathcal{V}$, the doubly dual space $\tilde{\tilde{\mathcal{V}}}$ is isomorphic (see (1.13)) to $\mathcal{V}$. Hence, we denote the set $\tilde{\tilde{\mathcal{V}}} \otimes \tilde{\tilde{\mathcal{V}}}$ by $\mathcal{V} \otimes \mathcal{V}$ in the sequel.

The addition, scalar multiplication, etc., of the second-order contravariant tensors are exactly analogous to (2.3). Under such rules, the set $\mathcal{V} \otimes \mathcal{V}$ becomes a vector space of dimension N^2.

The tensor product between two vectors (or contravariant vectors) is defined as the function $\vec{\mathbf{a}} \otimes \vec{\mathbf{b}}$ such that

$$\left[\vec{\mathbf{a}} \otimes \vec{\mathbf{b}}\right](\tilde{\mathbf{u}}, \tilde{\mathbf{v}}) := [\tilde{\mathbf{u}}(\vec{\mathbf{a}})]\left[\tilde{\mathbf{v}}\left(\vec{\mathbf{b}}\right)\right] \tag{2.23}$$

for all $\tilde{\mathbf{a}}, \tilde{\mathbf{b}}$ in $\tilde{\mathcal{V}}$. The tensor product $\vec{\mathbf{a}} \otimes \vec{\mathbf{b}}$ is a second-order contravariant tensor. The set of N^2 tensor products $\{\vec{\mathbf{e}}_a \otimes \vec{\mathbf{e}}_b\}$ forms a basis set for $\mathcal{V} \otimes \mathcal{V}$. The N^2 scalars τ^{ab} that appear in the linear combination

$$\mathbf{T}^{..} = \tau^{ab}(\vec{\mathbf{e}}_a \otimes \vec{\mathbf{e}}_b) \tag{2.24}$$

are (unique) components of $\mathbf{T}^{\cdot\cdot}$ relative to the basis set $\{\vec{\mathbf{e}}_a \otimes \vec{\mathbf{e}}_b\}$.

If the basis vectors transform according to

$$\widehat{\vec{\mathbf{e}}}_a = \lambda_a^b\, \vec{\mathbf{e}}_b, \quad \vec{\mathbf{e}}_a = \mu_a^b\, \widehat{\vec{\mathbf{e}}}_b,$$

the components of a second-order contravariant tensor $\mathbf{T}^{\cdot\cdot}$ transform as

$$\widehat{\tau}^{ab} = \mu_c^a\, \mu_d^b\, \tau^{cd}, \quad \tau^{ab} = \lambda_c^a\, \lambda_d^b\, \widehat{\tau}^{cd}. \tag{2.25}$$

(Compare and contrast the equation above with (2.13).)

The symmetric and antisymmetric second-order contravariant tensors can be defined analogously to (2.15) and (2.18), respectively. The number of independent components are exactly the same.

Remark: The angular momentum of a point particle (relative to a fixed point in space) is an antisymmetric second-order contravariant tensor in Newtonian physics.

Next we will define a **mixed second-order tensor**. Each of the bilinear functions

$$\mathbf{T}^{\cdot}{}_{\cdot} : \tilde{\mathcal{V}} \times \mathcal{V} \longrightarrow \mathcal{F},$$

$$\mathbf{T}_{\cdot}{}^{\cdot} : \mathcal{V} \times \tilde{\mathcal{V}} \longrightarrow \mathcal{F},$$

is called a second-order mixed tensor.

Example 2.1.14

$$\mathbf{O}^{\cdot}{}_{\cdot}(\tilde{\mathbf{u}}, \vec{\mathbf{b}}) := 0 \tag{2.26}$$

for all $\tilde{\mathbf{u}}$ and all $\vec{\mathbf{b}}$. Moreover,

$$\mathbf{O}_{\cdot}{}^{\cdot}(\vec{\mathbf{a}}, \tilde{\mathbf{v}}) := 0 \tag{2.27}$$

for all $\vec{\mathbf{a}}$ and all $\tilde{\mathbf{v}}$. These are called **zero mixed tensors**. □

Example 2.1.15 The **identity mixed tensors** $\mathbf{I}^{\cdot}{}_{\cdot}$ and $\mathbf{I}_{\cdot}{}^{\cdot}$ are defined by

$$\mathbf{I}^{\cdot}{}_{\cdot}(\tilde{\mathbf{e}}^a, \vec{\mathbf{e}}_b) := \delta_b^a, \tag{2.28}$$

$$\mathbf{I}_{\cdot}{}^{\cdot}(\vec{\mathbf{e}}_a, \tilde{\mathbf{e}}^b) := \delta_a^b. \tag{2.29}$$

Thus, we have

$$\mathbf{I}^{\cdot}{}_{\cdot}(\alpha_a\, \tilde{\mathbf{e}}^a, \beta^b\, \vec{\mathbf{e}}_b) = \alpha_a\, \beta^a = \mathbf{I}_{\cdot}{}^{\cdot}(\beta^a \vec{\mathbf{e}}_a, \alpha_b\, \tilde{\mathbf{e}}^b).$$ □

We define the tensor products by

$$[\vec{\mathbf{a}} \otimes \tilde{\mathbf{v}}](\tilde{\mathbf{u}}, \vec{\mathbf{b}}) := [\tilde{\mathbf{u}}(\vec{\mathbf{a}})]\,[\tilde{\mathbf{v}}(\vec{\mathbf{b}})], \tag{2.30}$$

$$[\tilde{\mathbf{u}} \otimes \vec{\mathbf{b}}](\vec{\mathbf{a}}, \tilde{\mathbf{v}}) := [\tilde{\mathbf{u}}(\vec{\mathbf{a}})]\,[\tilde{\mathbf{v}}(\vec{\mathbf{b}})], \tag{2.31}$$

for all $\vec{\mathbf{a}}, \vec{\mathbf{b}}$ and all $\tilde{\mathbf{u}}, \tilde{\mathbf{v}}$.

The sets $\{\vec{\mathbf{e}}_a \otimes \tilde{\mathbf{e}}^b\}$ and $\{\tilde{\mathbf{e}}^a \otimes \vec{\mathbf{e}}_b\}$ are basis sets for $\mathcal{V} \otimes \tilde{\mathcal{V}}$ and $\tilde{\mathcal{V}} \otimes \mathcal{V}$, respectively. The N^2-dimensional vector spaces $\mathcal{V}\otimes\tilde{\mathcal{V}}$ and $\tilde{\mathcal{V}}\otimes\mathcal{V}$ are isomorphic.

Suppose that transformation between two basis sets are furnished by the familiar equations

$$\widehat{\vec{\mathbf{e}}}_a = \lambda^b_{\ a}\, \vec{\mathbf{e}}_b, \quad \vec{\mathbf{e}}_a = \mu^b_{\ a} \widehat{\vec{\mathbf{e}}}_b.$$

Then, the components of the mixed tensors $\mathbf{T}^{\cdot}_{\ \cdot}$ and $\mathbf{T}_{\cdot}^{\ \cdot}$ transform as

$$\widehat{\tau}^a_{\ b} = \mu^a_{\ c}\, \lambda^d_{\ b}\, \tau^c_{\ d}, \quad \tau^a_{\ b} = \lambda^a_{\ c}\, \mu^d_{\ b}\, \widehat{\tau}^c_{\ d}, \tag{2.32}$$

$$\widehat{\tau}_a^{\ b} = \lambda^c_{\ a}\, \mu^b_{\ d}\, \tau_c^{\ d}, \quad \tau_a^{\ b} = \lambda^c_{\ a}\, \mu^b_{\ d}\, \widehat{\tau}_c^{\ d}. \tag{2.33}$$

Example 2.1.16 Let $\mathcal{V} := \mathbb{R}^2$, $\tilde{\mathbf{u}}(\alpha^1, \alpha^2) := \alpha^2$, and $\tilde{\mathbf{v}}(\beta^1, \beta^2) := \beta^1$. Consider the class of mixed second-order tensors $(\alpha^1, 2) \otimes \tilde{\mathbf{v}}$ in $\mathcal{V} \otimes \tilde{\mathcal{V}}$. We want to determine the (real) values of the function $(\alpha^1, 2) \otimes \tilde{\mathbf{v}}$ for the elements $(\tilde{\mathbf{u}}, (-2^{-1}, \beta^2))$ of $\tilde{\mathcal{V}} \times \mathcal{V}$. By (2.30), we have

$$[(\alpha^1, 2) \otimes \tilde{\mathbf{v}}]\,(\tilde{\mathbf{u}}, (-2^{-1}, \beta^2)) = [\tilde{\mathbf{u}}(\alpha^1, 2)]\,[\tilde{\mathbf{v}}(-2^{-1}, \beta^2)] \equiv (2)(-2^{-1}) = -1.$$

□

Exercises 2.1

1. Let $\mathcal{V}$ be a four-dimensional real vector space and let the matrix

$$[\lambda^a_{\ b}] := \begin{bmatrix} \cosh\beta & 0 & 0 & -\sinh\beta \\ 0 & 1 & 0 & 0 \\ 0 & 0 & 1 & 0 \\ -\sinh\beta & 0 & 0 & \cosh\beta \end{bmatrix}$$

represents the transformation of basis sets. If $\mathbf{F}_{\cdot\cdot} = \phi_{ab} \tilde{\mathbf{e}}^a \otimes \tilde{\mathbf{e}}^b$ is antisymmetric, obtain the transformed components $\widehat{\phi}_{ab}$ explicitly.

(*Remark:* This problem provides the transformation of the electromagnetic field components at an event under a Lorentz transformation.)

2. Prove that every second-order covariant tensor $\mathbf{T}_{..}$ can be decomposed *uniquely* as $\mathbf{T}_{..} = \mathbf{S}_{..} + \mathbf{A}_{..}$, where $\mathbf{S}_{..}$ is symmetric and $\mathbf{A}_{..}$ is antisymmetric.

3. Let a second-order mixed tensor $\mathbf{L}^{\cdot}{}_{\cdot}$ belong to $\mathcal{V} \otimes \tilde{\mathcal{V}}$. Prove that the function $\mathbf{L}$ defined by $\tilde{\mathbf{u}}[\mathbf{L}(\cdot)] := \mathbf{L}^{\cdot}{}_{\cdot}(\tilde{\mathbf{u}}, \cdot)$, where $\tilde{\mathbf{u}}$ is any covariant vector, is a linear operator on $\mathcal{V}$.

2.2 Higher-Order Tensors

Let r and s be two non-negative integers. An $(r+s)$th-order (or rank) mixed tensor ${}^r{}_s\mathbf{T}$ is a function from the set $\underbrace{\tilde{\mathcal{V}} \times \cdots \times \tilde{\mathcal{V}}}_{r} \times \underbrace{\mathcal{V} \times \cdots \times \mathcal{V}}_{s}$ into the field $\mathcal{F}$. Moreover, it has to satisfy the **multilinearity** conditions

$$\begin{aligned}
\text{(i)} \quad & {}^r{}_s\mathbf{T}(\tilde{\mathbf{u}}^1, \ldots, \lambda\tilde{\mathbf{u}}^k + \mu\tilde{\mathbf{v}}^k, \ldots, \tilde{\mathbf{u}}^r; \vec{\mathbf{a}}_1, \ldots, \vec{\mathbf{a}}_s) \\
&= \lambda\left[{}^r{}_s\mathbf{T}(\tilde{\mathbf{u}}^1, \ldots, \tilde{\mathbf{u}}^k, \ldots, \tilde{\mathbf{u}}^r; \vec{\mathbf{a}}_1, \ldots, \vec{\mathbf{a}}_s)\right] \\
&+ \mu\left[{}^r{}_s\mathbf{T}(\tilde{\mathbf{u}}^1, \ldots, \tilde{\mathbf{v}}^k, \ldots, \tilde{\mathbf{u}}^r; \vec{\mathbf{a}}_1, \ldots, \vec{\mathbf{a}}_s)\right],
\end{aligned} \tag{2.34}$$

$$\begin{aligned}
\text{(ii)} \quad & {}^r{}_s\mathbf{T}(\tilde{\mathbf{u}}^1, \ldots, \tilde{\mathbf{u}}^r; \vec{\mathbf{a}}_1, \ldots, \lambda\vec{\mathbf{a}}_j + \mu\vec{\mathbf{b}}_j, \ldots, \vec{\mathbf{a}}_s) \\
&= \lambda[{}^r{}_s\mathbf{T}(\tilde{\mathbf{u}}^1, \ldots, \tilde{\mathbf{u}}^r; \vec{\mathbf{a}}_1, \ldots, \vec{\mathbf{a}}_j, \ldots, \vec{\mathbf{a}}_s)] \\
&+ \mu[{}^r{}_s\mathbf{T}(\tilde{\mathbf{u}}^1, \ldots, \tilde{\mathbf{u}}^r; \vec{\mathbf{a}}_1, \ldots, \vec{\mathbf{b}}_j, \ldots, \vec{\mathbf{a}}_s)],
\end{aligned} \tag{2.35}$$

for all λ, μ in $\mathcal{F}$, all $\tilde{\mathbf{u}}^1, \ldots, \tilde{\mathbf{u}}^r$ in $\tilde{\mathcal{V}}$, all $\vec{\mathbf{a}}_1, \ldots, \vec{\mathbf{a}}_s$ in $\mathcal{V}$, all k in $\{1, \ldots, r\}$, and all j in $\{1, \ldots, s\}$. Here, the integer r stands for the **contravariant order** and the integer s stands for the **covariant order**. Moreover, the function ${}^r{}_s\mathbf{T}$ is *linear in each of the* $r+s$ *slots.*

Example 2.2.1 The (unique) $(r+s)$th-order zero tensor is defined by

$${}^r{}_s\mathbf{O}(\tilde{\mathbf{u}}^1, \ldots, \tilde{\mathbf{u}}^r; \vec{\mathbf{a}}_1, \ldots, \vec{\mathbf{a}}_s) := 0 \tag{2.36}$$

for all $\tilde{\mathbf{u}}^1, \ldots, \tilde{\mathbf{u}}^r$ in $\tilde{\mathcal{V}}$ and all $\vec{\mathbf{a}}_1, \ldots, \vec{\mathbf{a}}_s$ in $\mathcal{V}$. □

Example 2.2.2 Let $\tilde{\mathbf{b}}^k := \beta^k_1\,\tilde{\mathbf{e}}^1 + \cdots + \beta^k_N\,\tilde{\mathbf{e}}^N =: \beta^k_{i_k}\,\tilde{\mathbf{e}}^{i_k}$ and $\vec{\mathbf{a}}_j = \alpha^1_j\,\vec{\mathbf{e}}_1 + \cdots + \alpha^N_j\,\vec{\mathbf{e}}_N =: \alpha^{\ell_j}_j\,\vec{\mathbf{e}}_{\ell_j}$. Let us define a function by

$${}^r{}_s\mathbf{T}(\tilde{\mathbf{b}}^1, \ldots, \tilde{\mathbf{b}}^r; \vec{\mathbf{a}}_1, \ldots, \vec{\mathbf{a}}_s) := \left(\prod_{k=1}^{r} \beta^k_k\right)\left(\prod_{j=1}^{s} \alpha^j_j\right).$$

(*The summation convention is suspended on* k *and* j.) This function is an $(r+s)$th-order tensor. □

Remark: In the generalized Hooke's law connecting stress and strain at a point of an elastic body, an elastic tensor of order $(0+4)$ is involved.

There exist different but isomorphic $(r+s)$th-order mixed tensors for which the domain set $r+s$ copies of $\mathcal{V}$ and $\tilde{\mathcal{V}}$ are juxtaposed differently. We shall usually use the order of $r+s$ copies of $\mathcal{V}$ and $\tilde{\mathcal{V}}$ as in the definition in (2.34) and (2.35). *Lower-order tensors are denoted with dots* (as we have done for the second-order tensors in the preceding section). For example, we write

$$\mathbf{T}_{\cdot\cdot} \equiv {}^0{}_2\mathbf{T},\ \mathbf{T}^{\cdot\cdot} \equiv {}^2{}_0\mathbf{T},\ \mathbf{T}^{\cdot}_{\cdot} \equiv {}^1{}_1\mathbf{T},$$

$$\mathbf{R}^{\cdot}_{\cdot\cdot\cdot} \equiv {}^1{}_3\mathbf{R}.$$

There exists *a more general definition* of a higher-order tensor. The domain set for such a tensor is $\tilde{\mathcal{V}}_1 \times \cdots \times \tilde{\mathcal{V}}_r \times \mathcal{V}_{r+1} \times \cdots \times \mathcal{V}_{r+s}$, where vector spaces $\mathcal{V}_1, \ldots, \mathcal{V}_{r+s}$ may have *different dimensions* (but the *same* field $\mathcal{F}$). We shall mostly avoid such **hybrid tensors** except in section 7.3.

We shall now define the addition and scalar multiplication of $(r+s)$th-order tensors in (2.34) and (2.35). These are furnished by

$$\begin{aligned}
&[{}^r{}_s\mathbf{T} + {}^r{}_s\mathbf{W}](\tilde{\mathbf{u}}^1, \ldots, \tilde{\mathbf{u}}^r; \vec{\mathbf{a}}_1, \ldots, \vec{\mathbf{a}}_s) \\
&:= \left[{}^r{}_s\mathbf{T}(\tilde{\mathbf{u}}^1, \ldots, \tilde{\mathbf{u}}^r; \vec{\mathbf{a}}_1, \ldots, \vec{\mathbf{a}}_s)\right] + \left[{}^r{}_s\mathbf{W}(\tilde{\mathbf{u}}^1, \ldots, \tilde{\mathbf{u}}^r; \vec{\mathbf{a}}_1, \ldots, \vec{\mathbf{a}}_s)\right], \\
&{}^r{}_s\mathbf{O}(\tilde{\mathbf{u}}^1, \ldots, \tilde{\mathbf{u}}^r; \vec{\mathbf{a}}_1, \ldots, \vec{\mathbf{a}}_s) := 0, \\
&[-\,{}^r{}_s\mathbf{T}](\tilde{\mathbf{u}}^1, \ldots, \tilde{\mathbf{u}}^r; \vec{\mathbf{a}}_1, \ldots, \vec{\mathbf{a}}_s) := -\left[{}^r{}_s\mathbf{T}(\tilde{\mathbf{u}}^1, \ldots, \tilde{\mathbf{u}}^r; \vec{\mathbf{a}}_1, \ldots, \vec{\mathbf{a}}_s)\right], \\
&[\lambda\,{}^r{}_s\mathbf{T}](\tilde{\mathbf{u}}^1, \ldots, \tilde{\mathbf{u}}^r; \vec{\mathbf{a}}_1, \ldots, \vec{\mathbf{a}}_s) := \lambda\left[{}^r{}_s\mathbf{T}(\tilde{\mathbf{u}}^1, \ldots, \tilde{\mathbf{u}}^r; \vec{\mathbf{a}}_1, \ldots, \vec{\mathbf{a}}_s)\right],
\end{aligned} \tag{2.37}$$

for all λ in $\mathcal{F}$, all $\vec{\mathbf{a}}_1, \ldots, \vec{\mathbf{a}}_s$ in $\mathcal{V}$, and all $\tilde{\mathbf{u}}^1, \ldots, \tilde{\mathbf{u}}^r$ in $\tilde{\mathcal{V}}$. (Note that addition between two tensors of different orders is *not* permitted.)

The set of all $(r+s)$th-order tensors is denoted by $\underbrace{\mathcal{V} \otimes \cdots \otimes \mathcal{V}}_{r} \otimes \underbrace{\tilde{\mathcal{V}} \otimes \cdots \otimes \tilde{\mathcal{V}}}_{s}$. Since this is long and inconvenient notation, we shall follow instead the simpler expression

$${}^r{}_s\mathcal{T}(\mathcal{V}) := \underbrace{\mathcal{V} \otimes \cdots \otimes \mathcal{V}}_{r} \otimes \underbrace{\tilde{\mathcal{V}} \otimes \cdots \otimes \tilde{\mathcal{V}}}_{s}. \tag{2.38}$$

Example 2.2.3 According to (2.38) and consistency requirements, we have

$$
\begin{aligned}
&{}^{0}{}_{0}\mathcal{T}(\mathcal{V}) := \mathcal{F},\ {}^{1}{}_{0}\mathcal{T}(\mathcal{V}) \equiv \mathcal{V},\\
&{}^{0}{}_{1}\mathcal{T}(\mathcal{V}) = \tilde{\mathcal{V}},\ {}^{0}{}_{2}\mathcal{T}(\mathcal{V}) = \tilde{\mathcal{V}} \otimes \tilde{\mathcal{V}},\\
&{}^{1}{}_{1}\mathcal{T}(\mathcal{V}) = \mathcal{V} \otimes \tilde{\mathcal{V}}.
\end{aligned}
$$

□

Theorem 2.2.4 *Under the rules of addition and scalar multiplication in* (2.37), *the set* ${}^{r}{}_{s}\mathcal{T}(\mathcal{V})$ *of all* $(r+s)$ *th-order tensors forms a vector space.*

The proof is left to the reader.

The **tensor product** (or **outer product**) between two tensors ${}^{r}{}_{s}\mathbf{T}$ and ${}^{p}{}_{q}\mathbf{W}$ is defined by the mapping ${}^{r}{}_{s}\mathbf{T} \otimes {}^{p}{}_{q}\mathbf{W}$ from $\underbrace{\tilde{\mathcal{V}} \times \cdots \times \tilde{\mathcal{V}}}_{r+p} \times \underbrace{\mathcal{V} \times \cdots \times \mathcal{V}}_{s+q}$ into $\mathcal{F}$ such that

$$
\begin{aligned}
&[{}^{r}{}_{s}\mathbf{T} \otimes {}^{p}{}_{q}\mathbf{W}]\,(\tilde{\mathbf{u}}^1, \ldots, \tilde{\mathbf{u}}^r;\, \tilde{\mathbf{v}}^1, \ldots, \tilde{\mathbf{v}}^p;\, \vec{\mathbf{a}}_1, \ldots, \vec{\mathbf{a}}_s;\, \vec{\mathbf{b}}_1, \ldots, \vec{\mathbf{b}}_q)\\
&:= \left[{}^{r}{}_{s}\mathbf{T}(\tilde{\mathbf{u}}^1, \ldots, \tilde{\mathbf{u}}^r;\, \vec{\mathbf{a}}_1, \ldots, \vec{\mathbf{a}}_s)\right] \left[{}^{p}{}_{q}\mathbf{W}(\tilde{\mathbf{v}}^1, \ldots, \tilde{\mathbf{v}}^p;\, \vec{\mathbf{b}}_1, \ldots, \vec{\mathbf{b}}_q)\right]
\end{aligned}
\tag{2.39}
$$

for all $\vec{\mathbf{a}}_1, \ldots, \vec{\mathbf{a}}_s;\ \vec{\mathbf{b}}_1, \ldots, \vec{\mathbf{b}}_q$ in $\mathcal{V}$ and all $\tilde{\mathbf{u}}^1, \ldots, \tilde{\mathbf{u}}^r;\ \tilde{\mathbf{v}}^1, \ldots, \tilde{\mathbf{v}}^p$ in $\tilde{\mathcal{V}}$. The important properties of the tensor products are listed below.

Theorem 2.2.5 *The tensor product* $\otimes$ *satisfies the following equations:*

(*i*) ${}^{r}{}_{s}\mathbf{T} \otimes {}^{p}{}_{q}\mathbf{W}$ *belongs to* ${}^{r+p}{}_{s+q}\mathcal{T}(\mathcal{V})$;

(*ii*) $$({}^{r}{}_{s}\mathbf{T} + {}^{r}{}_{s}\mathbf{U}) \otimes {}^{p}{}_{q}\mathbf{W} = ({}^{r}{}_{s}\mathbf{T} \otimes {}^{p}{}_{q}\mathbf{W}) + ({}^{r}{}_{s}\mathbf{U} \otimes {}^{p}{}_{q}\mathbf{W})\,; \tag{2.40}$$

(*iii*) $${}^{r}{}_{s}\mathbf{T} \otimes ({}^{p}{}_{q}\mathbf{U} + {}^{p}{}_{q}\mathbf{W}) = ({}^{r}{}_{s}\mathbf{T} \otimes {}^{p}{}_{q}\mathbf{U}) + ({}^{r}{}_{s}\mathbf{T} \otimes {}^{p}{}_{q}\mathbf{W})\,; \tag{2.41}$$

(*iv*) $$(\lambda\, {}^{r}{}_{s}\mathbf{T}) \otimes ({}^{p}{}_{q}\mathbf{W}) = {}^{r}{}_{s}\mathbf{T} \otimes (\lambda\, {}^{p}{}_{q}\mathbf{W}) = \lambda\, ({}^{r}{}_{s}\mathbf{T} \otimes {}^{p}{}_{q}\mathbf{W})\,; \tag{2.42}$$

(*v*) $${}^{r}{}_{s}\mathbf{O} \otimes {}^{p}{}_{q}\mathbf{W} = {}^{r}{}_{s}\mathbf{T} \otimes {}^{p}{}_{q}\mathbf{O} = {}^{r+p}{}_{s+q}\mathbf{O}; \tag{2.43}$$

(*vi*) $$({}^{r}{}_{s}\mathbf{A} \otimes {}^{p}{}_{q}\mathbf{B}) \otimes {}^{j}{}_{k}\mathbf{C} = {}^{r}{}_{s}\mathbf{A} \otimes ({}^{p}{}_{q}\mathbf{B} \otimes {}^{j}{}_{k}\mathbf{C})\,; \tag{2.44}$$

(*vii*) $${}^{r}{}_{s}\mathbf{T} \otimes {}^{p}{}_{q}\mathbf{W} \not\equiv {}^{p}{}_{q}\mathbf{W} \otimes {}^{r}{}_{s}\mathbf{T}. \tag{2.45}$$

Proof of part (vi). Let vectors $\vec{\mathbf{a}}_1, \ldots, \vec{\mathbf{a}}_s\,;\ \vec{\mathbf{b}}_1, \ldots, \vec{\mathbf{b}}_q\,;\ \vec{\mathbf{c}}_1, \ldots, \vec{\mathbf{c}}_k\,;\ \tilde{\mathbf{u}}^1, \ldots, \tilde{\mathbf{u}}^r\,;\ \tilde{\mathbf{v}}^1, \ldots, \tilde{\mathbf{v}}^p\,;\ \tilde{\mathbf{w}}^1, \ldots, \tilde{\mathbf{w}}^j$ be chosen arbitrarily from $\mathcal{V}$ and $\tilde{\mathcal{V}}$. Using (2.39) several times, we obtain

$$\begin{aligned}
&\left[({}^r{}_s\mathbf{A} \otimes {}^p{}_q\mathbf{B}) \otimes {}^j{}_k\mathbf{C}\right] \\
&\qquad (\tilde{\mathbf{u}}^1, \ldots, \tilde{\mathbf{u}}^r; \tilde{\mathbf{v}}^1, \ldots, \tilde{\mathbf{v}}^p; \tilde{\mathbf{w}}^1, \ldots, \tilde{\mathbf{w}}^j; \vec{\mathbf{a}}_1, \ldots, \vec{\mathbf{a}}_s; \vec{\mathbf{b}}_1, \ldots, \vec{\mathbf{b}}_q; \vec{\mathbf{c}}_1, \ldots, \vec{\mathbf{c}}_k) \\
&= \left\{\left[{}^r{}_s\mathbf{A}\left(\tilde{\mathbf{u}}^1, \ldots, \tilde{\mathbf{u}}^r; \vec{\mathbf{a}}_1, \ldots, \vec{\mathbf{a}}_s\right)\right]\left[{}^p{}_q\mathbf{B}(\tilde{\mathbf{v}}^1, \ldots, \tilde{\mathbf{v}}^p; \vec{\mathbf{b}}_1, \ldots, \vec{\mathbf{b}}_q)\right]\right\} \\
&\qquad \left[{}^j{}_k\mathbf{C}(\tilde{\mathbf{w}}^1, \ldots, \tilde{\mathbf{w}}^j; \vec{\mathbf{c}}_1, \ldots, \vec{\mathbf{c}}_k)\right] \\
&= \left[{}^r{}_s\mathbf{A}(\tilde{\mathbf{u}}^1, \ldots, \tilde{\mathbf{u}}^r; \vec{\mathbf{a}}_1, \ldots, \vec{\mathbf{a}}_s)\right] \\
&\qquad \left\{\left[{}^p{}_q\mathbf{B}(\tilde{\mathbf{v}}^1, \ldots, \tilde{\mathbf{v}}^p; \vec{\mathbf{b}}_1, \ldots, \vec{\mathbf{b}}_q)\right]\left[{}^j{}_k\mathbf{C}(\tilde{\mathbf{w}}^1, \ldots, \tilde{\mathbf{w}}^j; \vec{\mathbf{c}}_1, \ldots, \vec{\mathbf{c}}_k)\right]\right\} \\
&= \left[{}^r{}_s\mathbf{A} \otimes ({}^p{}_q\mathbf{B} \otimes {}^j{}_k\mathbf{C})\right] \\
&\qquad (\tilde{\mathbf{u}}^1, \ldots, \tilde{\mathbf{u}}^r; \tilde{\mathbf{v}}^1, \ldots, \tilde{\mathbf{v}}^p; \tilde{\mathbf{w}}^1, \ldots, \tilde{\mathbf{w}}^j; \vec{\mathbf{a}}_1, \ldots, \vec{\mathbf{a}}_s; \vec{\mathbf{b}}_1, \ldots, \vec{\mathbf{b}}_q; \vec{\mathbf{c}}_1, \ldots, \vec{\mathbf{c}}_k).
\end{aligned}$$

Since the domains of the functions $({}^r{}_s\mathbf{A}\otimes {}^p{}_q\mathbf{B})\otimes {}^j{}_k\mathbf{C}$ and ${}^r{}_s\mathbf{A}\otimes({}^p{}_q\mathbf{B}\otimes {}^j{}_k\mathbf{C})$ are identical, the equation (2.44) (which is the associativity of the tensor product) is proved. ■

Because of the associativity, the expression $\vec{\mathbf{e}}_{i_1} \otimes \cdots \otimes \vec{\mathbf{e}}_{i_r} \otimes \vec{\mathbf{e}}^{j_1} \otimes \ldots \vec{\mathbf{e}}^{j_s}$ is meaningful. It is in fact a mixed tensor of $(r+s)$th order.

Theorem 2.2.6 *The set of N^{r+s} mixed tensors $\{\vec{\mathbf{e}}_{i_1} \otimes \cdots \otimes \vec{\mathbf{e}}_{i_r} \otimes \tilde{\mathbf{e}}^{j_1} \otimes \tilde{\mathbf{e}}^{j_s}\}_1^N$ is a basis set for ${}^r{}_s\mathcal{T}(\mathcal{V})$.*

The proof is left to the reader.

Corollary 2.2.7 $\dim\left[{}^r{}_s\mathcal{T}(\mathcal{V})\right] = N^{r+s}$.

The proof is skipped.

If we define the N^{r+s} components of ${}^r{}_s\mathbf{T}$ by

$$\tau^{i_1\ldots i_r}_{j_1\ldots j_s} := {}^r{}_s\mathbf{T}(\tilde{\mathbf{e}}^{i_1}, \ldots, \tilde{\mathbf{e}}^{i_r}; \vec{\mathbf{e}}_{j_1}, \ldots, \vec{\mathbf{e}}_{j_s}), \tag{2.46}$$

then it can be proved that

$${}^r{}_s\mathbf{T} = \tau^{i_1\ldots i_r}_{j_1\ldots j_s}\vec{\mathbf{e}}_{i_1} \otimes \cdots \otimes \vec{\mathbf{e}}_{i_r} \otimes \tilde{\mathbf{e}}^{j_1} \otimes \cdots \otimes \tilde{\mathbf{e}}^{j_s}. \tag{2.47}$$

(Note that there are N^{r+s} terms *suppressed* in the summation of the right-hand side of (2.47)!)

Now, we shall discuss the transformation of the components under the usual change of basis set

$$\widehat{\vec{\mathbf{e}}}_a = \lambda^b_a\, \vec{\mathbf{e}}_b, \;\; \vec{\mathbf{e}}_a = \mu^b_a \widehat{\vec{\mathbf{e}}}_b.$$

The transformation of the N^{r+s} components $\tau^{i_1\cdots i_r}_{j_1\cdots j_s}$ of the $(r+s)$th-order tensor ${}^r{}_s\mathbf{T}$ can be proved to be

$$\widehat{\tau}^{a_1\cdots a_r}{}_{b_1\cdots b_s} = \mu^{a_1}{}_{c_1}\cdots\mu^{a_r}{}_{c_r}\lambda^{d_1}{}_{b_1}\cdots\lambda^{d_s}{}_{b_s}\,\tau^{c_1\ldots c_r}{}_{d_1\ldots d_s}, \tag{2.48}$$

$$\tau^{a_1\ldots a_r}{}_{b_1\ldots b_s} = \lambda^{a_1}{}_{c_1}\cdots\lambda^{a_r}{}_{c_r}\,\mu^{d_1}{}_{b_1}\cdots\mu^{d_s}{}_{b_s}\,\widehat{\tau}^{c_1\ldots c_r}{}_{d_1\ldots d_s}. \tag{2.49}$$

Exercise 2.2.8. Suppose that $\mathcal{V} = \mathbb{C}^2$ and $\tilde{\mathbf{v}}(\beta^1, \beta^2) := \beta^1 + \beta^2$. We want to evaluate the class of trilinear functions $(\alpha^1, \alpha^2) \otimes \tilde{\mathbf{v}} \otimes (\beta^1, \beta^2)$ at the subset of points $(\tilde{\mathbf{u}}, (-i, i), \tilde{\mathbf{w}})$ in the domain set $\tilde{\mathcal{V}} \times \mathcal{V} \times \tilde{\mathcal{V}}$. Using (2.39) several times, we get

$$[(\alpha^1, \alpha^2) \otimes \tilde{\mathbf{v}} \otimes (\beta^1, \beta^2)]\,(\tilde{\mathbf{u}}, (-i, i), \tilde{\mathbf{w}}) = [\tilde{\mathbf{u}}(\alpha^1, \alpha^2)]\,[\tilde{\mathbf{v}}(-i, i)]\,[\tilde{\mathbf{w}}(\beta^1, \beta^2)] \equiv 0.$$

□

Let ${}^r{}_s\mathbf{T}$ be a tensor such that $1 \le r$, $1 \le s$. For $1 \le i \le r$ and $1 \le j \le s$, the **contraction** or **trace operation** is defined by a function ${}^i{}_j\mathcal{C}$ from ${}^r{}_s\mathcal{T}(\mathcal{V})$ into ${}^{r-1}{}_{s-1}\mathcal{T}(\mathcal{V})$ such that

$$\begin{aligned}&[({}^i{}_j\mathcal{C})({}^r{}_s\mathbf{T})](\tilde{\mathbf{w}}^1, \ldots, \tilde{\mathbf{w}}^{i-1}, \tilde{\mathbf{w}}^{i+1}, \ldots, \tilde{\mathbf{w}}^r; \vec{\mathbf{a}}_1, \ldots, \vec{\mathbf{a}}_{j-1}, \vec{\mathbf{a}}_{j+1}, \ldots \vec{\mathbf{a}}_s)\\ &:= \sum_{c=1}^{N} {}^r{}_s\mathbf{T}(\tilde{\mathbf{w}}^1, \ldots, \tilde{\mathbf{w}}^{i-1}, \tilde{\mathbf{e}}^c, \tilde{\mathbf{w}}^{i+1}, \ldots, \tilde{\mathbf{w}}^r; \vec{\mathbf{a}}_1, \ldots, \vec{\mathbf{a}}_{j-1}, \vec{\mathbf{e}}_c, \vec{\mathbf{a}}_{j+1}, \ldots \vec{\mathbf{a}}_s)\end{aligned} \tag{2.50}$$

for all $\vec{\mathbf{a}}_1, \ldots, \vec{\mathbf{a}}_s$ in $\mathcal{V}$ and all $\tilde{\mathbf{w}}^1, \ldots, \tilde{\mathbf{w}}^r$ in $\mathcal{V}$.

Example 2.2.9

$$\begin{aligned}&[({}^1{}_1\mathcal{C})\mathbf{T}^{\cdot\cdot}{}_{\cdot\cdot}]\,(\tilde{\mathbf{e}}^a; \vec{\mathbf{e}}_b)\\ &= \sum_{c=1}^{N} \mathbf{T}^{\cdot\cdot}{}_{\cdot\cdot}\,(\tilde{\mathbf{e}}^c, \tilde{\mathbf{e}}^a; \vec{\mathbf{e}}_c, \vec{\mathbf{e}}_b)\\ &= \sum_{c=1}^{N} \tau^{ca}{}_{cb} \equiv \tau^{ca}{}_{cb}.\end{aligned}$$

□

Example 2.2.10

$$[({}^1{}_1\mathcal{C})\mathbf{T}^{\cdot}{}_{\cdot}] = \tau^c{}_c \in \mathcal{F}.$$

□

Example 2.2.11

$$
\begin{aligned}
&[({}^{i}{}_{j}\mathcal{C})({}^{r}{}_{s}\mathbf{T})] = \\
&\tau^{a_1 \ldots a_{i-1} c a_{i+1} \ldots a_r}{}_{b_1 \ldots b_{j-1} c b_{j+1} \ldots b_s} \vec{\mathbf{e}}_{a_1} \otimes \\
&\ldots \vec{\mathbf{e}}_{a_{i-1}} \otimes \vec{\mathbf{e}}_{a_{i+1}} \otimes \ldots \vec{\mathbf{e}}_{a_r} \otimes \tilde{\mathbf{e}}^{b_1} \otimes \cdots \otimes \tilde{\mathbf{e}}^{b_{j-1}} \otimes \tilde{\mathbf{e}}^{b_{j+1}} \otimes \cdots \otimes \tilde{\mathbf{e}}^{b_s}.
\end{aligned}
\tag{2.51}
$$

□

The relevant properties of the contraction are listed below.

Theorem 2.2.12 *The contraction of $(r+s)$th-order tensors must satisfy the following properties:*

(i)

$$
({}^{i}{}_{j}\mathcal{C})(\lambda\ {}^{r}{}_{s}\mathbf{A} + \mu\ {}^{r}{}_{s}\mathbf{B}) = \lambda[({}^{i}{}_{j}\mathcal{C})({}^{r}{}_{s}\mathbf{A})] + \mu[{}^{i}{}_{j}\mathcal{C}({}^{r}{}_{s}\mathbf{B})], \tag{2.52}
$$

(ii) $({}^{i}{}_{j}\mathcal{C})({}^{r}{}_{s}\mathbf{T})$ *is independent of the choice of the basis set in* $\mathcal{V}$ *and* $\tilde{\mathcal{V}}$.

Proof of part (ii). Using (2.48), (2.49), (2.34), (2.35), and (1.9), we obtain

$$
\begin{aligned}
&\sum_{c=1}^{N} {}^{r}{}_{s}\mathbf{T}(\tilde{\mathbf{w}}^1, \ldots, \tilde{\mathbf{w}}^{i-1}, \widehat{\tilde{\mathbf{e}}}^c, \tilde{\mathbf{w}}^{i+1}, \ldots, \tilde{\mathbf{w}}^r ; \vec{\mathbf{q}}_1, \ldots, \vec{\mathbf{a}}_{j-1}, \widehat{\vec{\mathbf{e}}}_c, \vec{\mathbf{a}}_{j+1}, \ldots, \vec{\mathbf{a}}_a) \\
&= \sum_{c=1}^{N} \sum_{g=1}^{N} \sum_{h=1}^{N} (\mu^c_g\, \lambda^h_c) \\
&\quad {}^{r}{}_{s}\mathbf{T}\,(\tilde{\mathbf{w}}^1, \ldots, \tilde{\mathbf{w}}^{i-1}, \tilde{\mathbf{e}}^g, \tilde{\mathbf{w}}^{i+1}, \ldots, \tilde{\mathbf{w}}^r; \vec{\mathbf{a}}_1, \ldots, \vec{\mathbf{a}}_{j-1}, \vec{\mathbf{e}}_h, \vec{\mathbf{a}}_{j+1}, \ldots, \vec{\mathbf{a}}_s) \\
&= \sum_{k=1}^{N} {}^{r}{}_{s}\mathbf{T}\,(\tilde{\mathbf{w}}^1, \ldots, \tilde{\mathbf{w}}^{i-1}, \tilde{\mathbf{e}}^k, \tilde{\mathbf{w}}^{i+1}, \ldots, \tilde{\mathbf{w}}^r; \vec{\mathbf{a}}_1, \ldots, \vec{\mathbf{a}}_{j-1}, \vec{\mathbf{e}}_k, \vec{\mathbf{a}}_{j+1}, \ldots, \vec{\mathbf{a}}_s).
\end{aligned}
$$

■

One has to exercise *caution* in calculating the multiple contractions of a tensor.

Example 2.2.13

$$
\begin{aligned}
({}^{1}{}_{1}\mathcal{C})[({}^{1}{}_{1}\mathcal{C})(\mathbf{T}^{\cdot\cdot}{}_{\cdot\cdot})] &= ({}^{1}{}_{1}\mathcal{C})[\tau^{ca}{}_{cb}\vec{\mathbf{e}}_a \otimes \tilde{\mathbf{e}}^b] \\
&= \tau^{ck}{}_{ck}, \\
({}^{1}{}_{1}\mathcal{C})[({}^{2}{}_{1}\mathcal{C})(\mathbf{T}^{\cdot\cdot}{}_{\cdot\cdot})] &= \tau^{kc}{}_{ck} \not\equiv \tau^{ck}{}_{ck}.
\end{aligned}
$$

□

Consider the infinite-dimensional vector space $\mathcal{T}(\mathcal{V})$ obtained by the (weak) **direct sum** $\sum_{r=0}^{\infty}\sum_{s=0}^{\infty} {}^r{}_s\mathcal{T}(\mathcal{V})$. The vector space $\mathcal{T}(\mathcal{V})$, together with the tensor product rule $\otimes$, constitutes the **tensor algebra**.

There exist some special tensors for which the values of the components remain *unchanged* under a change of basis sets. These are called **numerical tensors.** We shall list some numerical tensors below.

(i) The $(r+s)$th-order zero tensor ${}^r{}_s\mathbf{O}$, which has for its components $O^{a_1\ldots a_r}{}_{b_1\ldots b_s} \equiv 0$, is numerical. (These values obviously remain intact under the transformation rules (2.48).)

(ii) The identity tensors $\mathbf{I}^{\cdot}{}_{\cdot}$ and $\mathbf{I}_{\cdot}{}^{\cdot}$ in (2.28) and (2.29) are also numerical. To prove this claim, consider the transformation rules (2.32) for the components of $\delta^a{}_b$ of $\mathbf{I}^{\cdot}{}_{\cdot}$. We have

$$\widehat{\delta}^a{}_b = \mu^a{}_c\,\lambda^d{}_b\,\delta^c{}_d = \mu^a{}_c\,\lambda^c{}_b = \delta^a{}_b\,.$$

(iii) Various tensor products $\mathbf{I}^{\cdot}{}_{\cdot} \otimes \cdots \otimes \mathbf{I}_{\cdot}{}^{\cdot}$ are also numerical tensors.

In the next section, we shall discuss other generalizations of the numerical tensors.

Exercises 2.2

1. Let $\{\vec{\mathbf{e}}_1, \vec{\mathbf{e}}_2\}$ be the standard basis set for $\mathcal{V} = \mathbb{R}^2$. Let $\vec{\mathbf{a}} := (-1,1)$, $\vec{\mathbf{b}} := (1,2)$, and $\vec{\mathbf{c}} := (2,1)$. Obtain the components of the $(3+0)$th-order tensor $\vec{\mathbf{a}} \otimes \vec{\mathbf{b}} \otimes \vec{\mathbf{c}}$ relative to the basis set $\{\vec{\mathbf{e}}_i \otimes \vec{\mathbf{e}}_j \otimes \vec{\mathbf{e}}_k\}$.

2. Let $\mathbf{S}_{..}$ and $\mathbf{G}_{..}$ be two non-zero symmetric covariant tensors in a four-dimensional vector space. Furthermore, let $\mathbf{S}_{..}$ and $\mathbf{G}_{..}$ satisfy the identity

$$\begin{aligned}&[\mathbf{G}_{..} \otimes \mathbf{S}_{..}]\,(\vec{\mathbf{a}}, \vec{\mathbf{b}}, \vec{\mathbf{c}}, \vec{\mathbf{d}}) - [\mathbf{G}_{..} \otimes \mathbf{S}_{..}](\vec{\mathbf{a}}, \vec{\mathbf{d}}, \vec{\mathbf{b}}, \vec{\mathbf{c}}) + [\mathbf{G}_{..} \otimes \mathbf{S}_{..}](\vec{\mathbf{b}}, \vec{\mathbf{c}}, \vec{\mathbf{a}}, \vec{\mathbf{d}})\\ &\quad -[\mathbf{G}_{..} \otimes \mathbf{S}_{..}](\vec{\mathbf{c}}, \vec{\mathbf{d}}, \vec{\mathbf{a}}, \vec{\mathbf{b}}) \equiv 0\end{aligned}$$

for all $\vec{\mathbf{a}}, \vec{\mathbf{b}}, \vec{\mathbf{c}}, \vec{\mathbf{d}}$ in $\mathcal{V}$. Prove that there must exist a scalar $\lambda \neq 0$ such that

$$\mathbf{G}_{..} = \lambda\,\mathbf{S}_{..}\,.$$

3. Consider the identity tensor $\mathbf{I}^{\cdot}_{\ \cdot}$. Prove that

$$\left[{}^1{}_1\mathcal{C}\left(\underbrace{{}^1{}_2\mathcal{C}\ldots{}^1{}_2\mathcal{C}}_{N-1}\right)\right]\left(\underbrace{\mathbf{I}^{\cdot}_{\ \cdot}\otimes\cdots\otimes\mathbf{I}^{\cdot}_{\ \cdot}}_{N}\right)=\dim(\mathcal{V}).$$

2.3 Exterior or Grassmann Algebra

Let an integer p satisfy $2 \leq p \leq \dim(\mathcal{V})$. Consider a $(0+p)$th-order covariant tensor ${}_p\mathbf{W} := {}^0{}_p\mathbf{W}$ such that

$$ {}_p\mathbf{W}(\vec{\mathbf{a}}_1,\ldots,\vec{\mathbf{a}}_i,\ldots,\vec{\mathbf{a}}_j,\ldots,\vec{\mathbf{a}}_p) = -{}_p\mathbf{W}(\vec{\mathbf{a}}_1,\ldots,\vec{\mathbf{a}}_j,\ldots,\vec{\mathbf{a}}_i,\ldots,\vec{\mathbf{a}}_p) \tag{2.53}$$

for all $\vec{\mathbf{a}}_1,\ldots,\vec{\mathbf{a}}_i,\ldots,\vec{\mathbf{a}}_j,\ldots,\vec{\mathbf{a}}_p$ in $\mathcal{V}$ and all i,j in $\{1,\ldots,p\}$. Such a tensor is called a **totally antisymmetric** (or **skew**, or **alternating**) **covariant tensor**.

The components satisfy

$$\begin{aligned} W_{a_1\ldots a_i\ldots a_j\ldots a_p} &:= {}_p\mathbf{W}(\vec{\mathbf{e}}_{a_1},\ldots,\vec{\mathbf{e}}_{a_i},\ldots,\vec{\mathbf{e}}_{a_j},\ldots,\vec{\mathbf{e}}_{a_p}) \\ &= -W_{a_1\ldots a_j\ldots a_i\ldots a_p} \end{aligned}$$

for all indices $a_1,\ldots,a_i,\ldots,a_j,\ldots,a_p$.

Note that, by (2.53),

$$ {}_p\mathbf{W}(\vec{\mathbf{a}}_1,\ldots,\vec{\mathbf{b}},\ldots,\vec{\mathbf{b}},\ldots,\vec{\mathbf{a}}_p) \equiv 0. \tag{2.54}$$

The subset of all totally antisymmetric covariant tensors of order $(0+p)$ is denoted by $\Lambda^p\,(\tilde{\mathcal{V}})$. It turns out to be a vector subspace of ${}^0{}_p\mathcal{T}(\mathcal{V})$.

Now, let us discuss the permutations of the positive integers $(1,2,\ldots,p)$. The identity permutation takes $(1,2,\ldots,p)$ into itself. A single (special) **transposition** (or switch, or interchange) takes $(1,2,3,\ldots,p)$ into $(2,1,3,\ldots,p)$. A typical permutation, σ takes $(1,2,\ldots,p)$ into $(\sigma(1),\sigma(2),\ldots,\sigma(p))$. The sign of the permutation σ is denoted by $\text{sgn}(\sigma)$. For an *odd* permutation (which involves odd numbers of transpositions), $\text{sgn}(\sigma) := -1$. For an *even* permutation, $\text{sgn}(\sigma) := +1$.

Example 2.3.1 Consider the integers $(1,2,3)$. If $\sigma(1,2,3) \equiv (\sigma(1),\sigma(2),\sigma(3)) := (1,3,2)$, $\text{sgn}(\sigma) = -1$. If $\sigma(1,2,3) \equiv (\sigma(1),\sigma(2),\sigma(3)) := (3,1,2)$, $\text{sgn}(\sigma) = 1$. This is because it involves two transpositions $(1,2,3) \xrightarrow{\sigma_1} (1,3,2) \xrightarrow{\sigma_2} (3,1,2)$. □

Theorem 2.3.2 *If ${}_p\mathbf{W}$ is a totally antisymmetric covariant tensor and σ is a permutation of $(1,2,\ldots,p)$, then*

$$ {}_p\mathbf{W}(\vec{\mathbf{a}}_{\sigma(1)},\ldots,\vec{\mathbf{a}}_{\sigma(p)}) \equiv [\text{sgn}\,(\sigma)]\ {}_p\mathbf{W}(\vec{\mathbf{a}}_1,\ldots,\vec{\mathbf{a}}_p). \tag{2.55}$$

The proof follows from the fact that every permutation σ is obtainable by compositions of transpositions (or switches).

The set of all $p!$ permutations of $(1, 2, \ldots, p)$ is denoted by S_p and constitutes a finite group (called the **symmetric group**).

Now we shall generate a new antisymmetric tensor out of an ordinary tensor. The process is called **alternating operation** and is defined by

$$[\mathrm{Alt}\,({}_p\mathbf{T})]\,(\vec{\mathbf{a}}_1, \ldots, \mathbf{a}_p) := (1/p!) \sum_{\sigma \in S_p} [\mathrm{sgn}(\sigma)]\; {}_p\mathbf{T}\,(\vec{\mathbf{a}}_{\sigma(1)}, \ldots, \vec{\mathbf{a}}_{\sigma(p)}) \tag{2.56}$$

for all $\vec{\mathbf{a}}_1, \ldots, \vec{\mathbf{a}}_p$ in $\mathcal{V}$.

Similarly, we can define the **symmetrization operation**

$$[\mathrm{Symm}\,({}_p\mathbf{T})]\,(\vec{\mathbf{a}}_1, \ldots, \mathbf{a}_p) := (1/p!) \sum_{\sigma \in S_p} {}_p\mathbf{T}(\vec{\mathbf{a}}_{\sigma(1)}, \ldots, \vec{\mathbf{a}}_{\sigma(p)}). \tag{2.57}$$

Example 2.3.3 Consider permutations of $(1, 2)$. There are just two elements in S_2. One is the identity (which is considered to be an even permutation). The other one is the single transposition $\sigma(1, 2) := (2, 1)$. Therefore, by (2.56),

$$\begin{aligned}
[\mathrm{Alt}\,(\tilde{\mathbf{u}} \otimes \tilde{\mathbf{v}})]\,(\vec{\mathbf{a}}_1, \mathbf{a}_2) &= (1/2!) \sum_{\sigma \in S_2} [\mathrm{sgn}(\sigma)]\,[\tilde{\mathbf{u}} \otimes \tilde{\mathbf{v}}](\vec{\mathbf{a}}_{\sigma(1)}, \vec{\mathbf{a}}_{\sigma(2)}) \\
&= (1/2)\{[\tilde{\mathbf{u}} \otimes \tilde{\mathbf{v}}](\vec{\mathbf{a}}_1, \vec{\mathbf{a}}_2) - [\tilde{\mathbf{u}} \otimes \tilde{\mathbf{v}}](\vec{\mathbf{a}}_2, \vec{\mathbf{a}}_1)\} \\
&= (1/2)\{[\tilde{\mathbf{u}}(\vec{\mathbf{a}}_1)]\,[\tilde{\mathbf{v}}(\vec{\mathbf{a}}_2)] - [\tilde{\mathbf{u}}(\vec{\mathbf{a}}_2)]\,[\tilde{\mathbf{v}}(\vec{\mathbf{a}}_1)]\} \\
&= (1/2)\{[\tilde{\mathbf{u}} \otimes \tilde{\mathbf{v}}] - [\tilde{\mathbf{v}} \otimes \tilde{\mathbf{u}}]\}(\vec{\mathbf{a}}_1, \vec{\mathbf{a}}_2)
\end{aligned}$$

for all $\vec{\mathbf{a}}_1, \vec{\mathbf{a}}_2$ in $\mathcal{V}$. Since the domains of the functions $\mathrm{Alt}\,(\tilde{\mathbf{u}} \otimes \tilde{\mathbf{v}})$ and $(1/2)\{[\tilde{\mathbf{u}} \otimes \tilde{\mathbf{v}}] - [\tilde{\mathbf{v}} \otimes \tilde{\mathbf{u}}]\}$ are identical, we conclude that

$$\mathrm{Alt}\,(\tilde{\mathbf{u}} \otimes \tilde{\mathbf{v}}) = (1/2)\{[\tilde{\mathbf{u}} \otimes \tilde{\mathbf{v}}] - [\tilde{\mathbf{v}} \otimes \tilde{\mathbf{u}}]\}. \tag{2.58}$$

(Compare the equation above with (2.20).) We can work out the symmetrization operation (2.57) to obtain

$$\mathrm{Symm}\,(\tilde{\mathbf{u}} \otimes \tilde{\mathbf{v}}) = (1/2)\{[\tilde{\mathbf{u}} \otimes \tilde{\mathbf{v}}] + [\tilde{\mathbf{v}} \otimes \tilde{\mathbf{u}}]\}. \tag{2.59}$$

We can generalize the examples above to the third-order tensor $\tilde{\mathbf{u}} \otimes \tilde{\mathbf{v}} \otimes \tilde{\mathbf{w}}$, etc. □

Theorem 2.3.4 *(i) If ${}_p\mathbf{T}$ belongs to ${}^0{}_p\mathcal{T}(\mathcal{V})$, then* $\mathrm{Alt}\,({}_p\mathbf{T})$ *belongs to* $\Lambda^p(\tilde{\mathcal{V}})$.

(ii) If ${}_p\mathbf{W}$ belongs to $\Lambda^p(\tilde{\mathcal{V}})$, then

$$\mathrm{Alt}\,({}_p\mathbf{W}) = {}_p\mathbf{W}. \tag{2.60}$$

(iii)

$$\mathrm{Alt}\,(\lambda\,{}_p\mathbf{W} + \mu\,{}_p\mathbf{U}) = \lambda[\mathrm{Alt}\,{}_p\mathbf{W}] + \mu\,[\mathrm{Alt}\,{}_p\mathbf{U}]. \tag{2.61}$$

The proof is left to the reader.

Now we are ready to define a new product called the **exterior** or **wedge product**. Let ${}_p\mathbf{W}$ belong to $\Lambda^p(\tilde{\mathcal{V}})$ and ${}_q\mathbf{U}$ belong to $\Lambda^q(\tilde{\mathcal{V}})$. Then the wedge product is defined by

$$ {}_p\mathbf{W} \wedge {}_q\mathbf{U} := [(p+q)!\,/(p!)(q!)]\,\mathrm{Alt}\,({}_p\mathbf{W} \otimes {}_q\mathbf{U}), \tag{2.62} $$

$$ \begin{aligned} &[{}_p\mathbf{W} \wedge {}_q\mathbf{U}](\vec{\mathbf{a}}_1, \ldots, \vec{\mathbf{a}}_p; \vec{\mathbf{a}}_{p+1}, \ldots, \vec{\mathbf{a}}_{p+q}) \\ &\quad = [1\,/(p!)(q!)] \sum_{\sigma \in S_{p+q}} [\mathrm{sgn}(\sigma)]\,[{}_p\mathbf{W} \otimes {}_q\mathbf{U}] \\ &\qquad (\vec{\mathbf{a}}_{\sigma(1)}, \ldots, \vec{\mathbf{a}}_{\sigma(p)}; \vec{\mathbf{a}}_{\sigma(p+1)}, \ldots, \vec{\mathbf{a}}_{\sigma(p+q)}) \end{aligned} \tag{2.63} $$

for all $\vec{\mathbf{a}}_1, \ldots, \vec{\mathbf{a}}_{p+q}$ in $\mathcal{V}$.

Example 2.3.5

$$ \begin{aligned} \tilde{\mathbf{W}} \wedge \tilde{\mathbf{U}} &= (2!)\mathrm{Alt}\,(\tilde{\mathbf{W}} \otimes \tilde{\mathbf{U}}) \\ &= [\tilde{\mathbf{W}} \otimes \tilde{\mathbf{U}}] - [\tilde{\mathbf{U}} \otimes \tilde{\mathbf{W}}]. \end{aligned} $$ □

Example 2.3.6 There are six distinct permutations of the integers $(1,2,3)$. These can be listed as follows: $I(1,2,3) := (1,2,3)$, $\sigma_2(1,2,3) := (2,1,3)$, $\sigma_3(1,2,3) := (1,3,2)$, $\sigma_4(1,2,3) := (3,2,1)$, $\sigma_5(1,2,3) := (2,3,1)$, $\sigma_6(1,2,3) := (3,1,2)$. The permutations $\sigma_2, \sigma_3, \sigma_4$ are odd and permutations I, σ_5, σ_6 are even. Let us consider the wedge product between $\mathbf{w}..$ and $\tilde{\mathbf{u}}$. By (2.63), it is given by

$$ \begin{aligned} &[\mathbf{w}.. \wedge \tilde{\mathbf{u}}]\,(\vec{\mathbf{a}}_1, \vec{\mathbf{a}}_2; \vec{\mathbf{a}}_3) \\ &\quad = (1/2)\,[\mathbf{w}..\,(\vec{\mathbf{a}}_1, \vec{\mathbf{a}}_2)\,\tilde{\mathbf{u}}(\vec{\mathbf{a}}_3) - \mathbf{w}..(\vec{\mathbf{a}}_2, \vec{\mathbf{a}}_1)\tilde{\mathbf{u}}(\vec{\mathbf{a}}_3) - \mathbf{w}..\,(\vec{\mathbf{a}}_1, \vec{\mathbf{a}}_3)\tilde{\mathbf{u}}(\vec{\mathbf{a}}_2) \\ &\qquad - \mathbf{w}..\,(\vec{\mathbf{a}}_3, \vec{\mathbf{a}}_2)\tilde{\mathbf{u}}(\vec{\mathbf{a}}_1) + \mathbf{w}..(\vec{\mathbf{a}}_2, \vec{\mathbf{a}}_3)\tilde{\mathbf{u}}(\vec{\mathbf{a}}_1) + \mathbf{w}..(\vec{\mathbf{a}}_3, \vec{\mathbf{a}}_1)\tilde{\mathbf{u}}(\vec{\mathbf{a}}_2)] \\ &\quad = [\tilde{\mathbf{u}} \wedge \mathbf{w}..](\vec{\mathbf{a}}_1, \vec{\mathbf{a}}_2; \vec{\mathbf{a}}_3) = (-1)^{(2\cdot 1)}[\tilde{\mathbf{u}} \wedge \mathbf{w}..]\,(\vec{\mathbf{a}}_1, \vec{\mathbf{a}}_2; \vec{\mathbf{a}}_3) \end{aligned} $$

for all $\vec{\mathbf{a}}_1, \vec{\mathbf{a}}_2; \vec{\mathbf{a}}_3$ in $\mathcal{V}$. □

Now we shall discuss the relevant properties of the wedge product.

Theorem 2.3.7 *The wedge product satisfies the following rules:*

(*i*)
$$ ({}_p\mathbf{A} + {}_p\mathbf{B}) \wedge {}_q\mathbf{C} = ({}_p\mathbf{A} \wedge {}_q\mathbf{C}) + ({}_p\mathbf{B} \wedge {}_q\mathbf{C}), \tag{2.64} $$

(*ii*)
$$ {}_p\mathbf{A} \wedge ({}_q\mathbf{B} + {}_q\mathbf{C}) = ({}_p\mathbf{A} \wedge {}_q\mathbf{B}) + ({}_p\mathbf{A} \wedge {}_q\mathbf{C}), \tag{2.65} $$

(iii)
$$(\lambda\, {}_p\mathbf{U}) \wedge (\mu\, {}_q\mathbf{W}) = (\lambda\mu)({}_p\mathbf{U} \wedge {}_q\mathbf{W}), \tag{2.66}$$

(iv)
$$({}_p\mathbf{A} \wedge {}_q\mathbf{B}) \wedge {}_r\mathbf{C} = {}_p\mathbf{A} \wedge ({}_q\mathbf{B} \wedge {}_r\mathbf{C}) =:\ {}_p\mathbf{A} \wedge {}_q\mathbf{B} \wedge {}_r\mathbf{C}, \tag{2.67}$$

(v)
$${}_q\mathbf{U} \wedge {}_p\mathbf{W} = (-1)^{pq}({}_p\mathbf{W} \wedge {}_q\mathbf{U}). \tag{2.68}$$

Proof of part (v). By (2.63) and (2.39),

$$\begin{aligned}
&[{}_q\mathbf{U} \wedge {}_p\mathbf{W}](\vec{\mathbf{a}}_1, \ldots, \vec{\mathbf{a}}_q;\ \vec{\mathbf{a}}_{q+1}, \ldots, \vec{\mathbf{a}}_{q+p}) \\
&\quad = (1/q!p!) \sum_{\sigma \in S_{q+p}} [\mathrm{sgn}(\sigma)]\, [{}_q\mathbf{U} \otimes {}_p\mathbf{W}]\, (\vec{\mathbf{a}}_{\sigma(1)}, \ldots, \vec{\mathbf{a}}_{\sigma(q+p)}) \\
&\quad = (1/q!p!) \sum_{\sigma \in S_{q+p}} [\mathrm{sgn}(\sigma)]\, [{}_q\mathbf{U}(\vec{\mathbf{a}}_{\sigma(1)}, \ldots, \vec{\mathbf{a}}_{\sigma(q)})]\, [{}_p\mathbf{W}(\vec{\mathbf{a}}_{\sigma(q+1)}, \ldots, \vec{\mathbf{a}}_{\sigma(q+p)})] \\
&\quad = (1/p!q!) \sum_{\sigma \in S_{p+q}} [\mathrm{sgn}(\sigma)]\, [{}_p\mathbf{W}(\vec{\mathbf{a}}_{\sigma(q+1)}, \ldots, \vec{\mathbf{a}}_{\sigma(q+p)})]\, [{}_q\mathbf{U}(\vec{\mathbf{a}}_{\sigma(1)}, \ldots, \vec{\mathbf{a}}_{\sigma(q)})] \\
&\quad = [{}_p\mathbf{W} \wedge {}_q\mathbf{U}]\, (\vec{\mathbf{a}}_{q+1}, \ldots, \vec{\mathbf{a}}_{q+p};\ \vec{\mathbf{a}}_1, \ldots, \vec{\mathbf{a}}_q).
\end{aligned}$$

Now, to permute $q+p$ vectors $(\vec{\mathbf{a}}_{q+1}, \ldots, \vec{\mathbf{a}}_{q+p}; \vec{\mathbf{a}}_1, \ldots, \vec{\mathbf{a}}_q)$ into $(\vec{\mathbf{a}}_1, \ldots, \vec{\mathbf{a}}_q;\ \vec{\mathbf{a}}_{q+1}, \ldots, \vec{\mathbf{a}}_{q+p})$, we need qp transposition of vector arguments. But each of the transpositions introduces a factor of (-1) by (2.53). Thus (2.68) is established. ■

Example 2.3.8

$$\begin{aligned}
\tilde{\mathbf{a}} \wedge \tilde{\mathbf{b}} &= (\alpha_c\, \tilde{\mathbf{e}}^c) \wedge (\beta_d\, \tilde{\mathbf{e}}^d) \\
&= \alpha_c\, \beta_d\, (\tilde{\mathbf{e}}^c \wedge \tilde{\mathbf{e}}^d) \\
&= (1/2!)\, (\alpha_c\, \beta_d - \beta_c\, \alpha_d)\, (\tilde{\mathbf{e}}^c \wedge \tilde{\mathbf{e}}^d).
\end{aligned}$$
□

Example 2.3.9

$$\omega_{a_1 \ldots a_p}(\tilde{\mathbf{e}}^{a_1} \wedge \cdots \wedge \tilde{\mathbf{e}}^{a_p}) = (p!) \sum_{1 \le a_1 < \cdots < a_p \le N} \omega_{a_1 \ldots a_p}(\tilde{\mathbf{e}}^{a_1} \wedge \cdots \wedge \tilde{\mathbf{e}}^{a_p}). \tag{2.69}$$

(On the right-hand side of (2.69), the summation is carried out with the restriction $1 \le a_1 < \cdots < a_p \le N$.) □

Next we shall introduce a $(p+p)$th-order $(2 \leq p \leq N)$ totally antisymmetric tensor. It is called the **generalized Kronecker delta** ${}^{p}{}_{p}\boldsymbol{\delta}$. Its components are furnished by the determinant of a $p \times p$ matrix as

$$\delta^{a_1 \ldots a_p}{}_{b_1 \ldots b_p} := \begin{vmatrix} \delta^{a_1}{}_{b_1} & \delta^{a_1}{}_{b_2} & \cdots & \delta^{a_1}{}_{b_p} \\ \vdots & \vdots & \vdots & \vdots \\ \delta^{a_p}{}_{b_1} & \delta^{a_p}{}_{b_2} & \cdots & \delta^{a_p}{}_{b_p} \end{vmatrix}. \tag{2.70}$$

Example 2.3.10 $\boldsymbol{\delta}^{\cdot}{}_{\cdot} = \mathbf{I}^{\cdot}{}_{\cdot}$, the identity tensor in (2.28). □

Example 2.3.11 $\delta^{a_1 a_2}{}_{b_1 b_2} = \delta^{a_1}{}_{b_1}\delta^{a_2}{}_{b_2} - \delta^{a_1}{}_{b_2}\delta^{a_2}{}_{b_1}$ are the components of $\boldsymbol{\delta}^{\cdot\cdot}{}_{\cdot\cdot}$. □

The important properties of the components of ${}^{p}{}_{p}\boldsymbol{\delta}$ are listed below.

Theorem 2.3.12 *The components of* ${}^{p}{}_{p}\boldsymbol{\delta}$ *satisfy the following rules:*

(i)
$$\delta^{a_1 \ldots a_p}{}_{b_1 \ldots b_p} = \begin{cases} 1 \text{ if } (b_1, \ldots, b_p) \text{ is an even permutation of } (a_1, \ldots, a_p), \\ -1 \text{ if } (b_1, \ldots, b_p) \text{ is an odd permutation of } (a_1, \ldots, a_p), \\ 0 \text{ otherwise}\,; \end{cases} \tag{2.71}$$

(ii)
$$\delta^{a_1 \ldots a_p}{}_{b_1 \ldots b_p} \equiv 0 \quad \text{for} \quad p > N; \tag{2.72}$$

(iii)
$$\delta^{a_1 \ldots a_p}{}_{a_1 \ldots a_p} = (N!/(N-p)!); \tag{2.73}$$

(iv)
$$\begin{aligned} &\delta^{a_1 \ldots a_q\, a_{q+1} \ldots a_p}{}_{b_1 \ldots b_q\, a_{q+1} \ldots a_p} \\ &= [(N-q)!/(N-p)!]\, \delta^{a_1 \ldots a_q}{}_{b_1 \ldots b_q} \quad \text{for} \quad 1 \leq q < p \leq N; \end{aligned} \tag{2.74}$$

(v)
$$\det[\lambda^a{}_b] = (1/N!)\delta^{a_1 \ldots a_N}{}_{b_1 \ldots b_N}\, \lambda^{b_1}{}_{a_1} \ldots \lambda^{b_N}{}_{a_N}; \tag{2.75}$$

(vi)
$$[\tilde{\mathbf{e}}^{a_1} \wedge \cdots \wedge \tilde{\mathbf{e}}^{a_p}](\vec{\mathbf{e}}_{b_1}, \ldots, \vec{\mathbf{e}}_{b_p}) = \delta^{a_1 \ldots a_p}{}_{b_1 \ldots b_p}. \tag{2.76}$$

Proof of part (v). Recall the following properties of a determinant:

$$\det[\lambda^a{}_b] := \sum_{\sigma \in S_N} \operatorname{sgn}(\sigma)\, \lambda^1{}_{\sigma(1)} \ldots \lambda^N{}_{\sigma(N)}, \tag{2.77}$$

$$\begin{vmatrix} \lambda^{a_1}{}_{b_1} & \ldots & \lambda^{a_1}{}_{b_p} \\ \vdots & & \vdots \\ \lambda^{a_p}{}_{b_1} & \ldots & \lambda^{a_p}{}_{b_p} \end{vmatrix} = \sum_{\sigma \in S_p} \operatorname{sgn}(\sigma)\, \lambda^{a_1}{}_{b_{\sigma(1)}} \ldots \lambda^{a_p}{}_{b_{\sigma(p)}}. \tag{2.78}$$

By (2.71), (2.69), and (2.78) we obtain

$$(1/N!)\delta^{a_1\ldots a_N}{}_{b_1\ldots b_N}\lambda^{b_1}{}_{a_1}\cdots\lambda^{b_N}{}_{a_N} = \sum_{1\le b_1<\cdots<b_N}^{N} \delta^{a_1\ldots a_N}{}_{b_1\ldots b_N}\lambda^{b_1}{}_{a_1}\cdots\lambda^{b_N}{}_{a_N}$$

$$= \delta^{a_1\ldots a_N}{}_{1\ldots N}\lambda^1_{a_1}\cdots\lambda^N_{a_N} = \sum_{\sigma\in S_N}\mathrm{sgn}(\sigma)\delta^{a_1}{}_{\sigma(1)}\cdots\delta^{a_N}{}_{\sigma(N)}\lambda^1_{a_1}\cdots\lambda^N_{a_N}$$

$$= \sum_{\sigma\in S_N}\mathrm{sgn}(\sigma)\lambda^1_{\sigma(1)}\cdots\lambda^N_{\sigma(N)} = \det[\lambda^a{}_b].$$ ■

Example 2.3.13 Let $\mathcal{V}$ be a real four-dimensional vector space. By (2.73) and (2.74), we obtain

$$\delta^{a_1\ldots a_4}{}_{a_1\ldots a_4} = 4!\,, \tag{2.79}$$

$$\delta^{a_1a_2a_3a_4}{}_{b_1a_2a_3a_4} = (3!)\,\delta^{a_1}{}_{b_1}, \tag{2.80}$$

$$\delta^{abcd}{}_{efcd} = 2\,\delta^{ab}{}_{ef}, \tag{2.81}$$

$$\delta^{abcd}{}_{efgd} = \delta^{abc}{}_{efg}. \tag{2.82}$$

These equations are applicable in the theory of relativity. □

Now we shall construct a basis set for $\Lambda^p(\tilde{\mathcal{V}})$. Consider a totally antisymmetric covariant tensor $\tilde{\mathbf{e}}^{a_1}\wedge\tilde{\mathbf{e}}^{a_2}\wedge\cdots\wedge\tilde{\mathbf{e}}^{a_p}$ $(2\le p\le N)$. If any two superscripts of $a_1,\ldots,a_p$ coincide, the tensor is zero by (2.54). The number of choices for p distinct integers out of N is $\binom{N}{p} = [N!/p!\,(N-p)!]$.

Theorem 2.3.14 *The set of all totally antisymmetric covariant tensors* $\{(\tilde{\mathbf{e}}^{a_1}\wedge\cdots\wedge\tilde{\mathbf{e}}^{a_p}) : 1\le a_1<a_2<\cdots<a_p\le N\}$ *is a basis set for* $\Lambda^p(\tilde{\mathcal{V}})$.

The proof is left to the reader.

Corollary 2.3.15

$$\dim[\Lambda^p(\tilde{\mathcal{V}})] = [N!/p!(N-p)!]. \tag{2.83}$$

Example 2.3.16 $\dim[\Lambda^N(\tilde{\mathcal{V}})] = 1$. Therefore, every non-zero totally antisymmetric covariant tensor ${}_N\mathbf{W}$ is a basis tensor for $\Lambda^N(\tilde{\mathcal{V}})$. □

Example 2.3.17 Consider a real four-dimensional vector space $\mathcal{V}$ and a basis set $\{\vec{\mathbf{e}}_1,\vec{\mathbf{e}}_2,\vec{\mathbf{e}}_3,\vec{\mathbf{e}}_4\}$. The corresponding basis sets for $\Lambda^2(\tilde{\mathcal{V}})$, $\Lambda^3(\tilde{\mathcal{V}})$, and $\Lambda^4(\tilde{\mathcal{V}})$ are respectively: $\{\tilde{\mathbf{e}}^1\wedge\tilde{\mathbf{e}}^2, \tilde{\mathbf{e}}^1\wedge\tilde{\mathbf{e}}^3, \tilde{\mathbf{e}}^1\wedge\tilde{\mathbf{e}}^4, \tilde{\mathbf{e}}^2\wedge\tilde{\mathbf{e}}^3, \tilde{\mathbf{e}}^2\wedge\tilde{\mathbf{e}}^4, \tilde{\mathbf{e}}^3\wedge\tilde{\mathbf{e}}^4\}$, $\{\tilde{\mathbf{e}}^1\wedge\tilde{\mathbf{e}}^2\wedge\tilde{\mathbf{e}}^3, \tilde{\mathbf{e}}^1\wedge\tilde{\mathbf{e}}^2\wedge\tilde{\mathbf{e}}^4, \tilde{\mathbf{e}}^1\wedge\tilde{\mathbf{e}}^3\wedge\tilde{\mathbf{e}}^4, \tilde{\mathbf{e}}^2\wedge\tilde{\mathbf{e}}^3\wedge\tilde{\mathbf{e}}^4\}$, and $\{\tilde{\mathbf{e}}^1\wedge\tilde{\mathbf{e}}^2\wedge\tilde{\mathbf{e}}^3\wedge\tilde{\mathbf{e}}^4\}$. These basis sets are relevant in the theory of relativity. □

The components $w_{a_1 \ldots a_p}$ for $1 \le a_1 < a_2 < \cdots < a_p \le N$ of a totally antisymmetric tensor ${}_p\mathbf{W}$ are called the **strict components**. These are denoted by $w_{a_1 < \cdots < a_p}$. Maintaining the summation convention, we can express

$$\begin{aligned} {}_p\mathbf{W} &= w_{a_1 \ldots a_p}(\tilde{\mathbf{e}}^{a_1} \otimes \cdots \otimes \tilde{\mathbf{e}}^{a_p}) \\ &= (1/p!) w_{a_1 \ldots a_p}(\tilde{\mathbf{e}}^{a_1} \wedge \cdots \wedge \tilde{\mathbf{e}}^{a_p}) \\ &= w_{a_1 < \cdots < a_p}(\tilde{\mathbf{e}}^{a_1} \wedge \cdots \wedge \tilde{\mathbf{e}}^{a_p}). \end{aligned} \tag{2.84}$$

Example 2.3.18 Choose $N = 3$ and $p = 2$. An antisymmetric tensor of order $(0+2)$ is provided by

$$\begin{aligned} \mathbf{w}_{..} &= w_{ab}\tilde{\mathbf{e}}^a \otimes \tilde{\mathbf{e}}^b = (1/2) w_{ab}\tilde{\mathbf{e}}^a \wedge \tilde{\mathbf{e}}^b \\ &= w_{12}\tilde{\mathbf{e}}^1 \wedge \tilde{\mathbf{e}}^2 + w_{13}\tilde{\mathbf{e}}^1 \wedge \tilde{\mathbf{e}}^3 + w_{23}\tilde{\mathbf{e}}^2 \wedge \tilde{\mathbf{e}}^3 \\ &= w_{a<b}\tilde{\mathbf{e}}^a \wedge \tilde{\mathbf{e}}^b. \end{aligned}$$

□

Theorem 2.3.19 *Prove that under the change of basis set in* (1.8)*, the strict components $w_{a_1 < \cdots < a_p}$ of a totally antisymmetric tensor ${}_p\mathbf{W}$ undergo the following transformation rules:*

$$\widehat{w}_{a_1 < \cdots < a_p} = \begin{vmatrix} \lambda^{b_1}{}_{a_1} & \cdots & \lambda^{b_p}{}_{a_1} \\ \vdots & & \vdots \\ \lambda^{b_1}{}_{a_p} & \cdots & \lambda^{b_p}{}_{a_p} \end{vmatrix} w_{b_1 < \cdots < b_p}, \tag{2.85}$$

$$w_{a_1 < \cdots < a_p} = \begin{vmatrix} \mu^{b_1}{}_{a_1} & \cdots & \mu^{b_p}{}_{a_1} \\ \vdots & & \vdots \\ \mu^{b_1}{}_{a_p} & \cdots & \mu^{b_p}{}_{a_p} \end{vmatrix} \widehat{w}_{b_1 < \cdots < b_p}. \tag{2.86}$$

The proof is left as an exercise. (Note that the summation convention is maintained in (2.85) and (2.86).)

Now we shall introduce another totally antisymmetric **permutation** (or **numerical**) **symbol** of Levi-Civita. It is denoted by ${}_N\mathbf{E}$, and its contravariant analogue is indicated by ${}^N\mathbf{E}$. The corresponding components are defined respectively by

$$\varepsilon_{a_1 \ldots a_N} := \delta^{1 \ldots N}{}_{a_1 \ldots a_N}, \tag{2.87}$$

$$\varepsilon^{b_1 \ldots b_N} := \delta^{b_1 \ldots b_N}{}_{1 \ldots N}. \tag{2.88}$$

These N^N components are mostly zero. Only $N!$ among the components are ± 1.

Example 2.3.20

$$\varepsilon_{123 \ldots N} = -\varepsilon_{213 \ldots N} = \delta^{12 \ldots N}{}_{12 \ldots N} = 1.$$

□

The main properties of the permutation symbols are summarized below.

Theorem 2.3.21 *The components of permutation symbols satisfy the following:*

(i) $\varepsilon_{a_1 \ldots a_N} = \varepsilon^{a_1 \ldots a_N} =$

$$\begin{cases} 1 & \text{if } (a_1, \ldots, a_N) \text{ is an even permutation of } (1, \ldots, N), \\ -1 & \text{if } (a_1, \ldots, a_N) \text{ is an odd permutation of } (1, \ldots, N), \\ 0 & \text{otherwise}; \end{cases} \tag{2.89}$$

(ii)
$$\varepsilon^{a_1 \ldots a_N} \varepsilon_{b_1 \ldots b_N} = \delta^{a_1 \ldots a_N}{}_{b_1 \ldots b_N}; \tag{2.90}$$

(iii)
$$\varepsilon^{a_1 \ldots a_p \; a_{p+1} \ldots a_N} \varepsilon_{b_1 \ldots b_p \; a_{p+1} \ldots a_N} = (N-p)! \delta^{a_1 \ldots a_p}{}_{b_1 \ldots b_p}; \tag{2.91}$$

(iv)
$$\lambda^{b_1}{}_{a_1} \ldots \lambda^{b_N}{}_{a_N} \varepsilon_{b_1 \ldots b_N} = \{\det[\lambda^c{}_d]\} \varepsilon_{a_1 \ldots a_N}; \tag{2.92}$$

(v)
$$\mu^{a_1}{}_{b_1} \ldots \mu^{a_N}{}_{b_N} \varepsilon^{b_1 \ldots b_N} = \{\det[\mu^c{}_d]\} \varepsilon^{a_1 \ldots a_N}. \tag{2.93}$$

The proof will be left as an exercise.

Example 2.3.22 Consider the two-dimensional vector space $\mathcal{V}$, the **spinor space in an arbitrary field** $\mathcal{F}$. In this case, the permutation symbols satisfy

$$\varepsilon_{11} = \varepsilon^{11} = \varepsilon_{22} = \varepsilon^{22} = 0, \;\; \varepsilon_{12} = \varepsilon^{12} = -\varepsilon_{21} = -\varepsilon^{21} = 1;$$

$$\varepsilon^{ab} \varepsilon_{cd} = \delta^{ab}{}_{cd}, \;\; \varepsilon^{ab} \varepsilon_{ab} = 2;$$

$$\varepsilon_{ab} \psi^a \chi^b = \det \begin{bmatrix} \psi^1 & \psi^2 \\ \chi^1 & \chi^2 \end{bmatrix}.$$

This example is relevant in the theory of spin-1/2 particles in nature. □

Example 2.3.23 Consider a three-dimensional vector space. In such a case,

$$\varepsilon^{abc} \varepsilon_{abc} = 6, \tag{2.94}$$

$$\varepsilon^{abc} \varepsilon_{dbc} = 2\delta^a{}_d, \tag{2.95}$$

$$\varepsilon^{abc} \varepsilon_{edc} = \delta^a{}_e \delta^b{}_d - \delta^a{}_d \delta^b{}_e. \tag{2.96}$$

This example is pertinent in the non-relativistic mechanics of particles and deformable bodies. □

Example 2.3.24 In the case of a four-dimensional real vector space, we have

$$\varepsilon^{abcd}\varepsilon_{abcd} = 4!\,, \tag{2.97}$$

$$\varepsilon^{abcd}\varepsilon_{ebcd} = 6\delta^a{}_e, \tag{2.98}$$

$$\varepsilon^{abcd}\varepsilon_{efcd} = 2(\delta^a{}_e\delta^b{}_f - \delta^a{}_f\delta^b{}_e), \tag{2.99}$$

$$\begin{aligned}\varepsilon^{abcd}\varepsilon_{efgd} &= \delta^a{}_e(\delta^b{}_f\delta^c{}_g - \delta^b{}_g\delta^c{}_f) + \delta^a{}_f(\delta^b{}_g\delta^c{}_e - \delta^b{}_e\delta^c{}_g)\\ &\quad+\delta^a{}_g(\delta^b{}_e\delta^c{}_f - \delta^b{}_f\delta^c{}_e).\end{aligned} \tag{2.100}$$

This example is useful in the theory of relativity. □

Now we shall discuss the transformation properties of the components $\varepsilon_{a_1\ldots a_N}$ and $\varepsilon^{a_1\ldots a_N}$ under the change of basis set in (1.8). It is natural to demand that the permutation symbols be *numerical symbols*; that is,

$$\widehat{\varepsilon}_{a_1\ldots a_N} = \varepsilon_{a_1\ldots a_N}, \; \widehat{\varepsilon}^{a_1\ldots a_N} = \varepsilon^{a_1\ldots a_N}.$$

To accomplish the equalities above, we must adopt the following transformation rules:

$$\widehat{\varepsilon}_{a_1\ldots a_N} = \{\det[\lambda^c{}_d]\}^{-1}\lambda^{b_1}_{a_1}\ldots\lambda^{b_N}{}_{a_N}\varepsilon_{b_1\ldots b_N}, \tag{2.101}$$

$$\widehat{\varepsilon}^{a_1\ldots a_N} = \{\det[\lambda^c{}_d]\}\mu^{a_1}_{b_1}\ldots\mu^{a_N}{}_{b_N}\varepsilon^{b_1\ldots b_N}. \tag{2.102}$$

(By (2.92) and (2.93), it is evident that transformation rules (2.101) and (2.102) render permutation symbols to behave as numerical objects.) Comparing (2.101) and (2.102) with the tensor transformation rules in (2.48), we conclude that the permutation symbols are generalizations of tensors. This fact prompts us to invent the concept of a **relative tensor** ${}^r_s\Theta_\omega$ **of weight** ω (an integer) with transformation rules for its components as

$$\widehat{\theta}^{a_1\ldots a_r}{}_{b_1\ldots b_s} = \{\det[\lambda^e{}_f]\}^\omega\mu^{a_1}{}_{c_1}\ldots\mu^{a_r}{}_{c_r}\lambda^{d_1}{}_{b_1}\ldots\lambda^{d_s}{}_{b_s}\theta^{c_1\ldots c_r}{}_{d_1\ldots d_s}. \tag{2.103}$$

For a positive weight $\omega > 0$, the relative tensor ${}^r_s\Theta_w$ belongs to the set $\underbrace{\Lambda^N(\tilde{\mathcal{V}})\otimes\cdots\otimes\Lambda^N(\tilde{\mathcal{V}})}_{\omega}\otimes\,{}^r_s\mathcal{T}(\mathcal{V})$, whereas for a negative weight it belongs to the set $\underbrace{\Lambda^N(\mathcal{V})\otimes\cdots\otimes\Lambda^N(\mathcal{V})}_{\omega}\otimes\,{}^r_s\mathcal{T}(\mathcal{V})$.

A relative tensor weight $\omega = 1$ is called a **tensor density**.

Example 2.3.25 Consider an antisymmetric tensor ${}_N\mathcal{A}$ belonging to $\Lambda^N(\tilde{\mathcal{V}})$. Let its strict components be denoted by

$$\begin{aligned}\alpha &:= \alpha_{a_1<\cdots<a_N} = \alpha_{1\ldots N},\\ \alpha_{a_1\ldots a_N} &= \alpha\varepsilon_{a_1\ldots a_N}.\end{aligned}$$

By the rules

$$\begin{aligned} {}_N\mathcal{A} &= (1/N!)\,\alpha_{a_1\ldots a_N}\tilde{\mathbf{e}}^{a_1}\wedge\cdots\wedge\tilde{\mathbf{e}}^{a_N} \\ &= (\alpha_{1\ldots N})(\tilde{\mathbf{e}}^1\wedge\cdots\wedge\mathbf{e}^N), \\ \widehat{\alpha}_{1\ldots N} &= \{\det[\lambda^c{}_d]\}\,\alpha_{1\ldots N}, \end{aligned}$$

we have

$$\widehat{\alpha} = \{\det[\lambda^c{}_d]\}\,\alpha.$$

Thus ${}_N\mathcal{A}$ is a **scalar density**. □

Example 2.3.26 The permutation symbol ${}^N\mathbf{E}$ is a contravariant Nth-order tensor density according to the transformation rules (2.102) and (2.103). □

We denote the set of all pth-order, antisymmetric, relative covariant tensors of weight ω by $\Lambda^p{}_\omega(\tilde{\mathcal{V}})$ and for similar contravariant tensors by $\Lambda^p{}_\omega(\mathcal{V})$. We observe that the dimension of $\Lambda^p{}_\omega(\tilde{\mathcal{V}})$ or $\Lambda^p{}_\omega(\mathcal{V})$ for $p \leq N$ is $\binom{N}{p}$. This is precisely the dimension of $\Lambda^{N-p}{}_\omega(\tilde{\mathcal{V}})$ or $\Lambda^{N-p}{}_\omega(\mathcal{V})$. Therefore, it is possible to set up a linear and one-to-one mapping from $\Lambda^p{}_\omega(\tilde{\mathcal{V}})$ into $\Lambda^{N-p}{}_{\omega+1}(\mathcal{V})$. This is called the **up-star** (*) **duality mapping** and is defined by the following rules on the components:

$$ {}^*\omega^{b_1\ldots b_{N-p}} := (1/p!)\,\varepsilon^{a_1\ldots a_p\,b_1\ldots b_{N-p}}\,\omega_{a_1\ldots a_p}. \tag{2.104}$$

The corresponding inverse transformation is furnished by

$$\omega_{a_1\ldots a_p} = [1/(N-p)!]\,\varepsilon_{a_1\ldots a_p\,b_1\ldots b_{N-p}}\,{}^*\omega^{b_1\ldots b_{N-p}}. \tag{2.105}$$

We define the **down-star** ($_*$) **duality mapping** from $\Lambda^p{}_{w+1}(\mathcal{V})$ into $\Lambda^{N-p}{}_w(\tilde{\mathcal{V}})$ by the equations

$$ {}_*\zeta_{a_1\ldots a_{N-p}} := (1/p!)\,\varepsilon_{a_1\ldots a_{N-p}\,b_1\ldots b_p}\zeta^{b_1\ldots b_p}. \tag{2.106}$$

$$\zeta^{b_1\ldots b_p} = [1/(N-p)!]\,\varepsilon^{a_1\ldots a_{N-p}\,b_1\ldots b_p}\,{}_*\zeta_{a_1\ldots a_{N-p}}. \tag{2.107}$$

Example 2.3.27 Consider the spinor space. It is a two-dimensional complex vector space. Moreover, we restrict the change of basis sets so that

$$\det[\lambda^c{}_d] = 1.$$

(*Remark:* The set of such unimodular transformations constitutes the group $SGL(2,\mathbb{C})$.)

For such transformations, there are no distinctions between tensors (or spinors) and relative tensors (or relative spinors). The star dual mappings in (2.104) and (2.106) are

$$
\begin{aligned}
{}^*\omega^b &= \varepsilon^{ab}\,\omega_a, \quad \omega_a = \varepsilon_{ab}\,{}^*\omega^b,\\
{}_*\zeta_a &= \varepsilon_{ab}\,\zeta^b, \quad \zeta^b = \varepsilon^{ab}\,{}_*\zeta_b,\\
{}^*\omega^b\,\omega_b &= {}_*\zeta_a\zeta^a \equiv 0.
\end{aligned}
$$

Here, ε^{ab} and ε_{cd} serve the purposes of raising and lowering indices. (Nevertheless, these should *not* be called metric tensor components!) □

We consider now composite mappings ${}_*\circ{}^*$ and ${}^*\circ{}_*$. Using (2.106), (2.104), and (2.91), we obtain

$$
\begin{aligned}
{}_*({}^*\omega_{a_1\ldots a_{N-p}}) &= (1/p!)\varepsilon_{a_1\ldots a_{N-p}b_1\ldots b_p}\,{}^*\omega^{b_1\ldots b_p}\\
&= (1/p!)\varepsilon_{a_1\ldots a_{N-p}b_1\ldots b_p}\{[1/(N-p)!]\varepsilon^{c_1\ldots c_{N-p}b_1\ldots b_p}\omega_{c_1\ldots c_{N-p}}\}\\
&= (1/p!)\delta^{c_1\ldots c_{N-p}}{}_{a_1\ldots a_{N-p}}\omega_{c_1\ldots c_{N-p}} = \omega_{a_1\ldots a_{N-p}}.
\end{aligned}
$$

Thus, the mappings ${}_*\circ{}^*$, and ${}^*\circ{}_*$ are *identity* mappings.

The basis set $\{\vec{\mathbf{e}}_1, \ldots, \vec{\mathbf{e}}_N\}$ for $\mathcal{V}$ can have one of two possible orientations. In the case ${}_N\mathbf{E}(\vec{\mathbf{e}}_1, \ldots, \vec{\mathbf{e}}_N) > 0$, the basis set is said to be **positively oriented**. Otherwise, it is **negatively oriented**. For a change of basis sets given by $\widehat{\vec{\mathbf{e}}}^a = \lambda^a{}_b\vec{\mathbf{e}}_b$, we have $[\tilde{\mathbf{e}}^1 \wedge \cdots \wedge \tilde{\mathbf{e}}^N](\widehat{\vec{\mathbf{e}}}_1, \ldots, \widehat{\vec{\mathbf{e}}}_N) = \det[\lambda^a{}_b] \neq 0$. Therefore, the sign of the $\det[\lambda^a{}_b]$ divides the transformations into two distinct classes. In the case $\det[\lambda^a{}_b] > 0$, the basis $\{\widehat{\vec{\mathbf{e}}}_1, \ldots, \widehat{\vec{\mathbf{e}}}_N\}$ is positively oriented relative to $\{\vec{\mathbf{e}}_1, \ldots, \vec{\mathbf{e}}_N\}$. For $\det[\lambda^a{}_b] < 0$, the basis is negatively oriented with respect to $\{\vec{\mathbf{e}}_1, \ldots, \vec{\mathbf{e}}_N\}$. We can introduce a generalization of relative tensors as given in (2.103). The components of an **oriented** (or **pseudo**) **relative tensor** transform as

$$
\begin{aligned}
&\widehat{\theta}^{a_1\ldots a_r}{}_{b_1\ldots b_s}\\
&\quad = \{\mathrm{sgn}[\det[\lambda^e{}_f]]\}\cdot\{\det[\lambda^i{}_j]\}^w\cdot\mu^{a_1}{}_{c_1}\cdots\mu^{a_r}{}_{c_r}\lambda^{d_1}{}_{b_1}\cdots\lambda^{d_s}{}_{b_s}\theta^{c_1\ldots c_r}{}_{d_1\ldots d_s}.
\end{aligned}
\tag{2.108}
$$

Example 2.3.28 The transformation of the components of a second-order covariant tensor is given by (see (2.13))

$$
\widehat{\tau}_{ab} = \lambda^c{}_a\,\lambda^d{}_b\,\tau_{cd} = \lambda^c{}_a\,\tau_{cd}\,\lambda^d{}_b.
$$

Taking the determinant of both sides, we have

$$
\det[\widehat{\tau}_{ab}] = \{\det[\lambda^a{}_c]^T\}\,\{\det[\tau_{cd}]\}\,\{\det[\lambda^d{}_b]\}.
$$

Now, taking positive square roots of the absolute values of both sides, we obtain that

$$
\sqrt{|\det[\widehat{\tau}_{ab}]|} = \{\mathrm{sgn}[\det[\lambda^e{}_f]]\}\{\det[\lambda^i{}_j]\}\sqrt{|\det[\tau_{cd}]|}. \tag{2.109}
$$

Therefore, $\sqrt{|\det[\tau_{ab}]|}$ *transforms as an oriented (or pseudo) scalar density.* □

Exercises 2.3

1. Let $\tilde{\mathbf{u}}^1, \ldots, \tilde{\mathbf{u}}^p$ be p covariant vectors. Prove that if $\tilde{\mathbf{u}}^1, \ldots, \tilde{\mathbf{u}}^p$ are linearly dependent, then $\tilde{\mathbf{u}}^1 \wedge \cdots \wedge \tilde{\mathbf{u}}^p = {}_p\mathbf{O}$.

2. Prove that

$$[\tilde{\mathbf{u}}^1 \wedge \cdots \wedge \tilde{\mathbf{u}}^p](\vec{\mathbf{a}}_1, \ldots, \tilde{\mathbf{a}}_p) = \det[\tilde{\mathbf{u}}^a(\vec{\mathbf{a}}_b)].$$

3. Prove the following equations.

(i) $\varepsilon^{j_1 \cdots j_r j_{r+1} \cdots j_N} \delta^{h_{r+1} \ldots h_N}{}_{j_{r+1} \cdots j_N} \equiv (N-r)! \varepsilon^{j_1 \cdots j_r h_{r+1} \ldots h_N}$.

(ii) $\det[a_{ij}] = \varepsilon^{j_1 \ldots j_N} a_{1 j_1} \ldots a_{N j_N}$.

(iii) The characteristic determinant

$$\det[\alpha^h{}_j - \lambda \delta^h{}_j] =$$
$$(-\lambda)^N + \sum_{k=1}^{N} (1/k!)(-\lambda)^{N-k} \delta^{j_1 \ldots j_k}{}_{h_1 \ldots h_k} \alpha^{h_1}{}_{j_1} \ldots \alpha^{h_k}{}_{j_k}.$$

4. Prove theorems 2.3.19 and 2.3.21.

2.4 Inner Product Vector Spaces and the Metric Tensor

In the preceding sections, *we have assumed a general field* $\mathcal{F}$. In the sequel, we shall use mostly the real field $\mathbb{R}$ (and occasionally $\mathbb{C}$). The vector spaces can be made more interesting by allowing additional structures. For example, we may introduce the notion of the **norm** or **length** of a real vector. The norm or length, denoted by $||\cdot||$, maps a vector into a non-negative real number. The axioms regarding a norm are listed below.

$N1.$ $$||\vec{\mathbf{a}}|| \geq 0; \quad ||\vec{\mathbf{a}}|| = 0 \text{ if and only if } \vec{\mathbf{a}} = \vec{\mathbf{o}}. \tag{2.110}$$

$N2.$ $$||\lambda \vec{\mathbf{a}}|| = |\lambda| \; ||\vec{\mathbf{a}}||. \tag{2.111}$$

$N3.$ $$||\vec{\mathbf{a}} + \vec{\mathbf{b}}|| \leq ||\vec{\mathbf{a}}|| + ||\vec{\mathbf{b}}|| \text{ (the triangle inequality)}. \tag{2.112}$$

Example 2.4.1 The familiar notion of the Euclidean norm or length of a vector $\vec{\mathbf{a}} = \alpha^b \vec{\mathbf{e}}_b$ is given by

$$||\vec{\mathbf{a}}|| := +\sqrt{\delta_{bc}\alpha^b\alpha^c} = \sqrt{(\alpha^1)^2 + \cdots + (\alpha^N)^2}. \qquad \square$$

A vector space can have a more elaborate structure, known as the **inner products** (or **scalar product**). Inner products can exist *independently* of the norm. The inner product is a function $\mathbf{g}..$ with the following axioms:

I1. $$\mathbf{g}..\,(\vec{\mathbf{a}}, \vec{\mathbf{b}}) \in \mathbb{R}.$$

I2. $$\mathbf{g}..\,(\vec{\mathbf{b}}, \vec{\mathbf{a}}) \equiv \mathbf{g}..(\vec{\mathbf{a}}, \vec{\mathbf{b}}). \tag{2.113}$$

I3. $$\mathbf{g}..\,(\lambda\vec{\mathbf{a}} + \mu\vec{\mathbf{b}}, \vec{\mathbf{c}}) \equiv [\lambda\mathbf{g}..\,(\vec{\mathbf{a}}, \vec{\mathbf{c}})] + [\mu\mathbf{g}..\,(\vec{\mathbf{b}}, \vec{\mathbf{c}})]. \tag{2.114}$$

I4. $$\mathbf{g}..\,(\vec{\mathbf{a}}, \vec{\mathbf{x}}) = 0 \text{ for all } \vec{\mathbf{x}} \text{ in } \mathcal{V} \text{ if and only if } \vec{\mathbf{a}} = \vec{\mathbf{o}}.$$

It is clear that $\mathbf{g}..$ is a symmetric second order covariant tensor. It is also called the **metric tensor**. The axiom I4 is known as the **non-degeneracy axiom**. (An alternate notation for an inner product is $(\vec{\mathbf{a}}, \vec{\mathbf{b}})_g \equiv \mathbf{g}..\,(\vec{\mathbf{a}}, \vec{\mathbf{b}})$.)

Example 2.4.2 The **Euclidean metric** is defined by

$$\begin{aligned}
\mathbf{g}..\,(\vec{\mathbf{a}}, \vec{\mathbf{b}}) &\equiv \mathbf{I}..(\vec{\mathbf{a}}, \vec{\mathbf{b}}) = \mathbf{I}..\,(\alpha^a\vec{\mathbf{e}}_a, \beta^b\vec{\mathbf{e}}_b) \\
&:= \delta_{ab}\alpha^a\alpha^b = (\alpha^1\beta^1) + \cdots + (\alpha^N\beta^N); \\
\mathbf{g}.. &\equiv \mathbf{I}.. = \delta_{ab}(\tilde{\mathbf{e}}^a \otimes \tilde{\mathbf{e}}^b); \\
\|\vec{\mathbf{a}}\|^2 &:= \mathbf{I}..\,(\vec{\mathbf{a}}, \vec{\mathbf{a}}).
\end{aligned} \tag{2.115}$$

$\square$

Example 2.4.3 We can define a **Lorentz metric** by

$$\begin{aligned}
\mathbf{g}..\,(\vec{\mathbf{a}}, \vec{\mathbf{b}}) \equiv \mathbf{d}..(\vec{\mathbf{a}}, \vec{\mathbf{b}}) &= d_{ab}\alpha^a\beta^a \\
&:= (\alpha^1\beta^1) + \cdots + (\alpha^{N-1}\beta^{N-1}) - (\alpha^N\beta^N).
\end{aligned} \tag{2.116}$$

$$\mathbf{g}.. = d_{ab}(\tilde{\mathbf{e}}^a \otimes \tilde{\mathbf{e}}^b) = (\tilde{\mathbf{e}}^1 \otimes \tilde{\mathbf{e}}^1) + \cdots + (\tilde{\mathbf{e}}^{N-1} \otimes \tilde{\mathbf{e}}^{N-1}) - (\tilde{\mathbf{e}}^N \otimes \tilde{\mathbf{e}}^N). \qquad \square$$

The metric tensor components $g_{ij} := \mathbf{g}..\,(\vec{\mathbf{e}}_i, \vec{\mathbf{e}}_j) = g_{ji}$. Therefore, the eigenvalues $\lambda_1, \ldots, \lambda_N$ of the symmetric matrix $[g_{ij}]$ are all *real*. Moreover, by the non-degeneracy axiom I4, none of the eigenvalues can be zero. Therefore, we can introduce the notion of the **signature of a metric** by

$$\begin{aligned} \mathrm{sgn}(\mathbf{g}..) &:= \sum_{k=1}^{N} \mathrm{sgn}(\lambda_k) = p - n; \\ N &= p + n. \end{aligned} \tag{2.117}$$

Here, p is the number of positive eigenvalues and n is the number of negative eigenvalues. For the Lorentz metric in (2.116), $\mathrm{sgn}(\mathbf{d}..) = N - 2$. In the theory of relativity, $N = 4$ and $\mathrm{sgn}(\mathbf{d}..) = 2$. (Caution: Some relativists denote the metric by $\widehat{\mathbf{d}}.. := -\mathbf{d}..$, $\mathrm{sgn}(\widehat{\mathbf{d}}..) = -2$!)

There is a stronger concept of inner product. It is called a **positive-definite inner product**. In this case, one of the axioms I4 in (2.113) and (2.114) is replaced by a stronger axiom,

I4+. $\quad \mathbf{g}..(\vec{\mathbf{a}}, \vec{\mathbf{a}}) \geq 0; \quad \mathbf{g}..(\vec{\mathbf{a}}, \vec{\mathbf{a}}) = 0 \text{ iff } \vec{\mathbf{a}} = \vec{\mathbf{o}}.$

In the case of a positive-definite inner product, we can define in a natural way the norm or length of a vector $\vec{\mathbf{a}}$ as

$$\|\vec{\mathbf{a}}\| := +\sqrt{\mathbf{g}..(\vec{\mathbf{a}}, \vec{\mathbf{a}})}. \tag{2.118}$$

We can define the "**angle**" θ between two non-zero vectors $\vec{\mathbf{a}}$ and $\vec{\mathbf{b}}$ by

$$\cos\theta := \left[\mathbf{g}..(\vec{\mathbf{a}}, \vec{\mathbf{b}}) / \sqrt{\mathbf{g}..(\vec{\mathbf{a}}, \vec{\mathbf{a}}) \cdot \mathbf{g}..(\vec{\mathbf{b}}, \vec{\mathbf{b}})}\,\right]. \tag{2.119}$$

By the Cauchy-Schwartz inequality, $|\cos\theta| \leq 1$.

If the inner product is not positive-definite, there is a problem of defining the norm. The best we can do is to define the **separation** σ of a vector by

$$\sigma(\vec{\mathbf{a}}) := +\sqrt{|\mathbf{g}..(\vec{\mathbf{a}}, \vec{\mathbf{a}})|}. \tag{2.120}$$

(The pair of vertical lines indicate the *absolute value*!) The separation defined in (2.120) satisfies the following two conditions:

$$\begin{aligned} &\sigma(\vec{\mathbf{a}}) \geq 0, \\ &\sigma(\lambda\vec{\mathbf{a}}) = |\lambda|\sigma(\vec{\mathbf{a}}). \end{aligned} \tag{2.121}$$

The rules above definitely *differ* from the rules of a norm in (2.110)–(2.112). For a Lorentz metric in (2.116) and a vector $\vec{\mathbf{a}} \neq \vec{\mathbf{o}}$ with components $\alpha^1 = \alpha^N = 1$, $\alpha^2 = \alpha^3 = \cdots = \alpha^{N-1} = 0$, the separation $\sigma(\vec{\mathbf{a}}) = 0$. This *clearly violates* the axiom (2.110) for a norm or length.

For a vector space with a Lorentz metric $\mathbf{g}..$, we define the following subsets of vectors.

(i) A **spacelike vector** $\vec{\mathbf{s}}$ must satisfy

$$\mathbf{g}..(\vec{\mathbf{s}}, \vec{\mathbf{s}}) > 0. \tag{2.122}$$

(ii) A **timelike vector** $\vec{\mathbf{t}}$ must satisfy

$$\mathbf{g}..(\vec{\mathbf{t}}, \vec{\mathbf{t}}) < 0. \tag{2.123}$$

(iii) A **null vector** $\vec{\mathbf{n}}$ must satisfy

$$\mathbf{g}..(\vec{\mathbf{n}}, \vec{\mathbf{n}}) = 0. \tag{2.124}$$

Example 2.4.4 Consider the four-dimensional real vector space with the Lorentz metric

$$\mathbf{d}.. = (\tilde{\mathbf{e}}^1 \otimes \tilde{\mathbf{e}}^1) + (\tilde{\mathbf{e}}^2 \otimes \tilde{\mathbf{e}}^2) + (\tilde{\mathbf{e}}^3 \otimes \tilde{\mathbf{e}}^3) - (\tilde{\mathbf{e}}^4 \otimes \tilde{\mathbf{e}}^4). \tag{2.125}$$

Here, the basis $\{\vec{\mathbf{e}}_1, \ldots, \vec{\mathbf{e}}_4\}$ is the standard one. For the vector $\vec{\mathbf{s}} := \vec{\mathbf{e}}_1 + \vec{\mathbf{e}}_2 + \vec{\mathbf{e}}_3$, $\sigma(\vec{\mathbf{s}}) = \mathbf{d}..(\vec{\mathbf{s}}, \vec{\mathbf{s}}) = 3$. Therefore, $\vec{\mathbf{s}}$ is spacelike. For the vector $\vec{\mathbf{t}} := \vec{\mathbf{e}}_1 - \vec{\mathbf{e}}_2 + \vec{\mathbf{e}}_3 - 2\vec{\mathbf{e}}_4$, $\mathbf{d}..(\vec{\mathbf{t}}, \vec{\mathbf{t}}) = -1$. Thus, $\vec{\mathbf{t}}$ is timelike. For the vector $\vec{\mathbf{n}} := \vec{\mathbf{e}}_1 + \vec{\mathbf{e}}_2 - \vec{\mathbf{e}}_3 + \sqrt{3}\vec{\mathbf{e}}_4$, $\mathbf{d}..(\vec{\mathbf{n}}, \vec{\mathbf{n}}) = 0$. Therefore, $\vec{\mathbf{n}}$ is a (non-zero) null vector. □

Now let us discuss the "angle" between two vectors in the Lorentz metric. The natural generalization of (2.119) for two non-null vectors is provided by

$$\cos\theta := \left[\mathbf{d}..(\vec{\mathbf{a}}, \vec{\mathbf{b}}) \Big/ \sqrt{\left|\mathbf{d}..(\vec{\mathbf{a}}, \vec{\mathbf{a}})\mathbf{d}..(\vec{\mathbf{b}}, \vec{\mathbf{b}})\right|}\,\right].$$

If we choose the standard vectors $\vec{\mathbf{e}}_1$ and $\vec{\mathbf{e}}_N$ and define the sequence of vectors $\vec{\mathbf{b}}_m := \vec{\mathbf{e}}_1 + (m - 1/m)\vec{\mathbf{e}}_N$, $m \in \{1, 2, \cdots\}$, then $\mathbf{d}..(\vec{\mathbf{e}}_1, \vec{\mathbf{e}}_1) = 1$, $\mathbf{d}..(\vec{\mathbf{e}}_N, \vec{\mathbf{e}}_N) = -1$, $\mathbf{d}..(\vec{\mathbf{b}}_m, \vec{\mathbf{b}}_m) = 1 - (m - 1/m)^2$, $\mathbf{d}..(\vec{\mathbf{e}}_1, \vec{\mathbf{b}}_m) \equiv 1$. Therefore, the sequence of "angles" between $\vec{\mathbf{e}}_1$ and $\vec{\mathbf{b}}_m$ are given by

$$\begin{aligned}\cos(\theta_m) &= \left[\mathbf{d}..(\vec{\mathbf{e}}_1, \vec{\mathbf{b}}_m) \Big/ \sqrt{\left|\mathbf{d}..(\vec{\mathbf{e}}_1, \vec{\mathbf{e}}_1)\mathbf{d}..(\vec{\mathbf{b}}_m, \vec{\mathbf{b}}_m)\right|}\,\right] \\ &= (m/\sqrt{2m-1}).\end{aligned}$$

Thus $1 \leq \cos\theta_m$ and $\lim_{m\to\infty} \cos\theta_m \longrightarrow \infty$. This displays the violation of the Cauchy-Schwartz inequality and shows that the use of a trigonometric function, "cosine," is meaningless for a non–positive-definite metric.

From these discussions, *we conclude that there is neither a length nor an angle in a vector space with a Lorentz metric.*

A **"unit" vector** $\vec{\mathbf{u}}$ is defined in general by the condition

$$\sigma(\vec{\mathbf{u}}) = 1. \tag{2.126}$$

Two "**orthogonal vectors**" $\vec{\mathbf{a}}$ and $\vec{\mathbf{b}}$ must satisfy the condition

$$\mathbf{g}_{..}(\vec{\mathbf{a}}, \vec{\mathbf{b}}) = 0. \tag{2.127}$$

In the case where $\{\vec{\mathbf{e}}_1, \ldots, \vec{\mathbf{e}}_N\}$ is an "**orthonormal basis**" set, the vectors must satisfy

$$\mathbf{g}_{..}(\vec{\mathbf{e}}_k, \vec{\mathbf{e}}_j) = \begin{cases} \pm 1 & \text{for } k = j, \\ 0 & \text{for } k \neq j. \end{cases} \tag{2.128}$$

(Subsequently, we shall *drop* " " from the orthonormality.)

The non-degeneracy axiom I4 implies that the $N \times N$ matrix $[g_{ab}]$ is non-singular. Therefore, the matrix $[g_{ab}]$ possesses a unique and symmetric inverse matrix. Denoting the entries of the inverse matrix by g^{ab}, we obtain

$$g^{ba} = g^{ab}, \; g^{ab} g_{bc} = g_{cb} g^{ba} = \delta^a{}_c. \tag{2.129}$$

The **contravariant** (or **conjugate**) **metric tensor** is defined by

$$\mathbf{g}^{..} := \mathbf{g}^{ab}(\vec{\mathbf{e}}_a \otimes \vec{\mathbf{e}}_b). \tag{2.130}$$

The contravariant metric tensor induces an inner product in $\tilde{\mathcal{V}}$ by the rules

$$\mathbf{g}^{..}(\tilde{\mathbf{e}}^a, \tilde{\mathbf{e}}^b) = g^{ab}, \tag{2.131}$$

$$\mathbf{g}^{..}(\tilde{\mathbf{x}}, \tilde{\mathbf{y}}) = \mathbf{g}^{..}(\xi_a \tilde{\mathbf{e}}^a, \eta_b \tilde{\mathbf{e}}^b) = g^{ab} \xi_a \eta_b. \tag{2.132}$$

The metric tensor $\mathbf{g}_{..}$ induces an *isomorphism* between $\mathcal{V}$ and $\tilde{\mathcal{V}}$. The following theorem establishes the isomorphism.

Theorem 2.4.5 *Let $\mathcal{V}$ be a finite-dimensional real vector space with a metric $\mathbf{g}_{..}$. Then, to every covariant vector $\tilde{\mathbf{u}}$ in $\tilde{\mathcal{V}}$ there corresponds a unique vector $\vec{\mathbf{u}}$ in $\mathcal{V}$ such that $\tilde{\mathbf{u}}(\vec{\mathbf{a}}) = \mathbf{g}_{..}(\vec{\mathbf{a}}, \vec{\mathbf{u}})$ for all $\vec{\mathbf{a}}$ in $\mathcal{V}$. Moreover, the function $\mathcal{I}_g$ from $\tilde{\mathcal{V}}$ into $\mathcal{V}$, defined by $\vec{\mathbf{u}} := \mathcal{I}_g(\tilde{\mathbf{u}})$, is a vector space isomorphism.*

The proof is left as an exercise. By the preceding theorem, the metric $\mathbf{g}_{..}$ makes $\mathcal{V}$ and $\tilde{\mathcal{V}}$ exactly similar. In terms of the components, the isomorphism can be expressed as

$$\begin{aligned} \xi^a &:= g^{ab} \xi_b, \\ \xi_c &= g_{ca} \xi^a. \end{aligned} \tag{2.133}$$

The procedures above are called **raising** and **lowering indices**, respectively.

The isomorphism due to the metric can be extended to tensor spaces with fixed $(r+s)$ order. Thus, the spaces ${}^{(r+s)}_{0}\mathcal{T}(\mathcal{V})$, ${}^{r}_{s}\mathcal{T}(\mathcal{V})$, ${}^{0}_{(r+s)}\mathcal{T}(\mathcal{V})$, etc., are exactly similar due to the metric. In terms of the components, the isomorphism can be obtained by the following (and similar) rules for raising and lowering indices:

$$\begin{aligned}\mathcal{T}^{a_1\ldots a_r}{}_{b_1\ldots b_s} &:= g_{b_1 d_1}\cdots g_{b_s d_s}\mathcal{T}^{a_1\ldots a_r d_1\ldots d_s}\\ &= g^{a_1 c_1}\ldots g^{a_r c_r}\mathcal{T}_{c_1\ldots c_r b_1\ldots b_s}\\ &= g^{a_r k}\mathcal{T}^{a_1\ldots a_{r-1}}{}_{k b_1\ldots b_s}.\end{aligned} \tag{2.134}$$

Example 2.4.6 For a Euclidean metric $g_{ab} \equiv \delta_{ab}$, (2.134) implies that

$$\mathcal{T}^{a_1\ldots a_r}{}_{b_1\ldots b_s} = \mathcal{T}^{a_1\ldots a_r b_1\ldots b_s} = \mathcal{T}_{a_1\ldots a_r b_1\ldots b_s}.$$

□

Example 2.4.7 Let us work out the raising and lowering of second-order tensor components from (2.134):

$$\begin{aligned}\mathcal{T}^{ab} &= g^{bc}\mathcal{T}^{a}{}_{c} = g^{ac}g^{bd}\mathcal{T}_{cd} = g^{ac}\mathcal{T}_{c}{}^{b},\\ \mathcal{T}_{ab} &= g_{ac}\mathcal{T}^{c}{}_{b} = g_{ac}g_{bd}\mathcal{T}^{cd} = g_{bd}\mathcal{T}_{a}{}^{d},\\ \mathcal{T}^{a}{}_{b} &= g^{ac}\mathcal{T}_{cb} = g_{bd}\mathcal{T}^{ad} = g^{ac}g_{bd}\mathcal{T}_{c}{}^{d},\\ \mathcal{T}_{a}{}^{b} &= g_{ac}\mathcal{T}^{cb} = g^{bd}\mathcal{T}_{ad} = g_{ac}g^{bd}\mathcal{T}^{c}{}_{d}.\end{aligned} \tag{2.135}$$

□

Example 2.4.8 For the metric tensor components, we obtain from (2.129) and (2.134)

$$\begin{aligned}g^{a}{}_{b} &= g^{ac}g_{cb} = \delta^{a}{}_{b},\\ g_{a}{}^{b} &= g_{ac}g^{cb} = \delta_{a}{}^{b}.\end{aligned} \tag{2.136}$$

□

The metric $\mathbf{g}_{..}$ in $\mathcal{V}$ induces a generalized metric ${}^{2s}{}_{2r}\mathbf{g}$ for the N^{r+s}-dimensional vector space ${}^{r}{}_{s}\mathcal{T}(\mathcal{V})$. Therefore, the function ${}^{2s}{}_{2r}\mathbf{g}$ should map a pair of ordered tensors ${}^{r}{}_{s}\mathbf{A} = \alpha^{a_1\ldots a_r}{}_{b_1\ldots b_s}\vec{\mathbf{e}}_{a_1}\otimes\cdots\otimes\tilde{\mathbf{e}}_{b_s}$ and ${}^{r}{}_{s}\mathbf{B} = \beta^{c_1\ldots c_r}{}_{d_1\ldots d_s}\vec{\mathbf{e}}_{c_1}\otimes\cdots\otimes\tilde{\mathbf{e}}_{d_s}$ into a real number. We define this function by the following rule:

$$\begin{aligned}&{}^{2s}{}_{2r}\mathbf{g}({}^{r}{}_{s}\mathbf{A}, {}^{r}{}_{s}\mathbf{B}) :=\\ &g^{b_1 d_1}\ldots g^{b_s d_s}g_{a_1 c_1}\cdots g_{a_r c_r}\alpha^{a_1\ldots a_r}{}_{b_1\ldots b_s}\beta^{c_1\ldots c_r}{}_{d_1\ldots d_s}.\end{aligned} \tag{2.137}$$

Theorem 2.4.9 *The induced metric* ${}^{2s}{}_{2r}\mathbf{g}$ *in* (2.137) *satisfies the rules of inner product* I1 – I4 *in* (2.113) *and* (2.114) *for the* N^{r+s}*-dimensional vector space* ${}^{r}{}_{s}\mathcal{T}(\mathcal{V})$.

The proof is left as an exercise.

We can define the separation function $\tilde{\sigma}$ for the tensor space ${}^{r}{}_{s}\mathcal{T}(\mathcal{V})$. It is given by

$$\tilde{\sigma}({}^{r}{}_{s}\mathbf{T}) := +\sqrt{|{}^{2s}{}_{2r}\mathbf{g}({}^{r}{}_{s}\mathbf{T}, {}^{r}{}_{s}\mathbf{T})|}. \tag{2.138}$$

Example 2.4.10 Let us consider the induced metric $\wedge\mathbf{g}_{....}$ for the space $\Lambda^2(\mathcal{V})$. (It is also called the **bivector space**.) Choose two contravariant antisymmetric tensors $\mathbf{w}^{\cdot\cdot}$ and $\mathbf{A}^{\cdot\cdot}$. By (2.84), we can express

$$\mathbf{w}^{\cdot\cdot} = \omega^{ab}\vec{\mathbf{e}}_a \otimes \vec{\mathbf{e}}_b =: \tfrac{1}{2}\omega^{ab}\vec{\mathbf{e}}_a \wedge \vec{\mathbf{e}}_b,$$
$$\mathbf{A}^{\cdot\cdot} = \alpha^{cd}\vec{\mathbf{e}}_c \otimes \vec{\mathbf{e}}_d =: \tfrac{1}{2}\alpha^{cd}\vec{\mathbf{e}}_c \wedge \vec{\mathbf{e}}_d.$$

By the definition in (2.137), we have

$$\wedge\mathbf{g}_{....}(\mathbf{w}^{\cdot\cdot}, \mathbf{A}^{\cdot\cdot}) = g_{ac}g_{bd}\omega^{ab}\alpha^{cd}.$$

The components of $\vec{\mathbf{e}}_e \wedge \vec{\mathbf{e}}_f$ relative to their own basis set are furnished by

$$\vec{\mathbf{e}}_e \wedge \vec{\mathbf{e}}_f = \frac{1}{2}\delta^{ab}{}_{ef}\vec{\mathbf{e}}_a \wedge \vec{\mathbf{e}}_b.$$

Therefore, we obtain for the components

$$\begin{aligned}\wedge g_{efgh} &= \wedge\mathbf{g}_{....}(\vec{\mathbf{e}}_e \wedge \vec{\mathbf{e}}_f, \vec{\mathbf{e}}_g \wedge \vec{\mathbf{e}}_h) \\ &= \left(\frac{1}{4}\right) g_{ac}g_{bd}\delta^{ab}{}_{ef}\delta^{cd}{}_{gh} \\ &= \left(\frac{1}{4}\right)\begin{vmatrix} g_{eg} & g_{eh} \\ g_{fg} & g_{fh}\end{vmatrix}.\end{aligned}$$

Moreover,

$$\wedge g_{fegh} = -\wedge g_{efgh} = \wedge g_{efhg} = \wedge g_{hgef}.$$

□

Now we shall *modify* the permutation symbols of the preceding section. We introduce totally antisymmetric pseudo (or oriented) tensors ${}_N\boldsymbol{\eta}$ and ${}^N\boldsymbol{\eta}$ of Levi-Civita:

$$\eta_{a_1\ldots a_N} := +\sqrt{|\det[g_{cd}]|}\,\varepsilon_{a_1\ldots a_N} \equiv \sqrt{|g|}\,\varepsilon_{a_1\ldots a_N}, \tag{2.139}$$

$$\eta^{a_1\ldots a_N} := \frac{\sqrt{|g|}}{g}\varepsilon^{a_1\ldots a_N} = \frac{\mathrm{sgn}(g)}{\sqrt{|g|}}\varepsilon^{a_1\ldots a_N}. \tag{2.140}$$

The second choice is made to *accommodate the raising and lowering* rules in (2.134).

Now we shall examine the transformation rules of the components of antisymmetric pseudo-tensors under the usual change of basis sets

$$\widehat{\vec{\mathbf{e}}}_a = \lambda^b{}_a \vec{\mathbf{e}}_b, \vec{\mathbf{e}}_a = \mu^b{}_a \widehat{\vec{\mathbf{e}}}_b.$$

Theorem 2.4.11 *Under the change of basis sets in $\mathcal{V}$, the components of antisymmetric pseudo-tensors $\eta_{b_1\ldots b_N}$ and $\eta^{b_1\ldots b_N}$ transform as*

$$\widehat{\eta}_{a_1\ldots a_N} = \text{sgn}\{\det[\lambda^c{}_d]\} \cdot \lambda^{b_1}{}_{a_1} \ldots \lambda^{b_N}{}_{a_N} \eta_{b_1\ldots b_N}, \tag{2.141}$$

$$\widehat{\eta}^{a_1\ldots a_N} = \text{sgn}\{\det[\lambda^c{}_d]\} \cdot \mu^{a_1}{}_{b_1} \ldots \mu^{a_N}{}_{b_N} \eta^{b_1\ldots b_N}. \tag{2.142}$$

The proof is left as an exercise. The transformation rules (2.141) and (2.142) are slightly simpler compared with those of the permutation symbols in (2.101) and (2.102).

There is a geometrical significance to these pseudo-tensors or oriented tensors. Consider a set of vectors $\{\vec{\mathbf{a}}_1, \ldots, \vec{\mathbf{a}}_N\}$ in $\mathcal{V}$. Imagine these vectors all start from the origin. These can generate an N-dimensional *hyper-parallelepiped.* The N-dimensional **oriented volume** of such a hyper-parallelepiped is defined by

$$\begin{aligned} V(\vec{\mathbf{a}}_1, \ldots, \vec{\mathbf{a}}_N) &:= {}_N\boldsymbol{\eta}(\vec{\mathbf{a}}_1, \ldots, \vec{\mathbf{a}}_N) \\ &= {}_N\boldsymbol{\eta}(\alpha_1^{a_1}\vec{\mathbf{e}}_{a_1}, \ldots, \alpha_N^{a_N}\vec{\mathbf{e}}_{a_N}) \\ &= \sqrt{|g|}\det[\alpha_d^c]. \end{aligned} \tag{2.143}$$

Therefore, the volume is non-zero if and only if the set of vectors $\{\vec{\mathbf{a}}_1, \ldots, \vec{\mathbf{a}}_N\}$ is a basis set. Furthermore, the oriented volume is positive or else negative according to whether $\{\vec{\mathbf{a}}_1, \ldots, \vec{\mathbf{a}}_N\}$ is positively or negatively oriented relative to the basis $\{\vec{\mathbf{e}}_1, \ldots, \vec{\mathbf{e}}_N\}$.

The pseudo-tensors can be used to modify slightly the star operations in (2.104) and (2.106). We call this duality operation the **Hodge star operation.** The Hodge $*$ function maps an antisymmetric tensor (or an antisymmetric pseudo-tensor) ${}_p\mathbf{W}$ into an antisymmetric pseudo-tensor (or an antisymmetric tensor) ${}^*_{N-p}\mathbf{W}$. (*The star is neither up nor down!*) This one-to-one and linear function is defined by

$$*\omega_{b_1\ldots b_{N-p}} := (1/p!)\eta^{a_1\ldots a_p}{}_{b_1\ldots b_{N-p}}\omega_{a_1\ldots a_p}. \tag{2.144}$$

We can also write that

$$*\zeta_{c_1\ldots c_p} = [1/(N-p)!]\eta_{c_1\ldots c_p}^{b_1\ldots b_{N-p}}\zeta_{b_1\ldots b_{N-p}}. \tag{2.145}$$

Example 2.4.12 Consider a four-dimensional real vector space and an antisymmetric second-order covariant tensor $\mathbf{F}..$ given by

$$\mathbf{F}.. = \phi_{ab}\tilde{\mathbf{e}}^a \otimes \tilde{\mathbf{e}}^b =: \frac{1}{2}\phi_{[ab]}\tilde{\mathbf{e}}^a \otimes \tilde{\mathbf{e}}^b = \phi_{<a<b}\tilde{\mathbf{e}}^a \wedge \tilde{\mathbf{e}}^b. \tag{2.146}$$

The Hodge star mapping will generate the components

$$\begin{aligned} *\phi^{cd} &= (1/2)\eta^{abcd}\phi_{ab} \\ &= (1/2)[\operatorname{sgn}(g)/\sqrt{|g|}]\varepsilon^{abcd}\phi_{ab}. \end{aligned} \tag{2.147}$$

Thus, we have for the Lorentz metric

$$*\phi^{12} = -\left(1/\sqrt{|g|}\right)\phi_{34}, \quad *\phi^{23} = -\left(\frac{1}{\sqrt{|g|}}\right)\phi_{14},$$

$$*\phi^{31} = -\left(1/\sqrt{|g|}\right)\phi_{24}, \quad *\phi^{14} = -\left(\frac{1}{\sqrt{|g|}}\right)\phi_{23},$$

$$*\phi^{24} = -\left(1/\sqrt{|g|}\right)\phi_{31}, \quad *\phi^{34} = -\left(\frac{1}{\sqrt{|g|}}\right)\phi_{12}.$$

This example is relevant in the relativistic electromagnetic field theory. □

Exercises 2.4

1. Let a metric $\mathbf{g}..$ in a three-dimensional vector space be defined by

$$\mathbf{g}..\,(\vec{\mathbf{a}}, \vec{\mathbf{b}}) = \mathbf{g}..(\alpha^a\vec{\mathbf{e}}_a, \beta^b\vec{\mathbf{e}}_b) := \alpha^1\beta^1 + 3(\alpha^1\beta^2 + \alpha^2\beta^1) + 4\alpha^2\beta^2 + \alpha^3\beta^3.$$

Determine whether or not the metric $\mathbf{g}..$ is positive-definite.

2. Let an N-dimensional vector space be endowed with a *positive-definite* metric $\mathbf{g}..$.

(i) Prove the Cauchy-Schwartz inequality:

$$\mathbf{g}..\,(\vec{\mathbf{a}}, \vec{\mathbf{b}}) \leq +\sqrt{\mathbf{g}..(\vec{\mathbf{a}}, \vec{\mathbf{a}})\mathbf{g}..(\vec{\mathbf{b}}, \vec{\mathbf{b}})}.$$

(ii) Prove the polarization identity:

$$\mathbf{g}..\,(\vec{\mathbf{a}}, \vec{\mathbf{b}}) \equiv (1/4)[\mathbf{g}..(\vec{\mathbf{a}} + \vec{\mathbf{b}}, \vec{\mathbf{a}} + \vec{\mathbf{b}}) - \mathbf{g}..\,(\vec{\mathbf{a}} - \vec{\mathbf{b}}, \vec{\mathbf{a}} - \vec{\mathbf{b}})].$$

3. Prove theorems 2.4.5 and 2.4.9.

4. Prove the following equations:

(i)
$$\eta^{a_1 \cdots a_N} = g^{a_1 b_1} \ldots g^{a_N b_N} \eta_{b_1 \ldots b_N}.$$

(ii)
$$\eta^{a_1 \ldots a_p b_1 \ldots b_{N-p}} \eta_{c_1 \ldots c_p b_1 \ldots b_{N-p}} = \mathrm{sgn}(g)(N-p)!\delta^{a_1 \ldots a_p}{}_{c_1 \ldots c_p},$$

for $p \leq N$.

5. Show that, for the Hodge star operation,

$$*(*\omega_{c_1 \ldots c_p}) = (-1)^{p(N-p)} \mathrm{sgn}(g) \omega_{c_1 \ldots c_p}.$$

Chapter 3

Tensor Analysis on a Differentiable Manifold

3.1 Differentiable Manifolds

We shall go *briefly* through the definition of an N-dimensional differentiable manifold M. There are few assumptions in this definition. A set with a **topology** is one in which open subsets are known. Furthermore, if for every two distinct elements (or points) p and q, there exist open and disjoint subsets containing p and q, respectively, then the topology is called **Hausdorff**. A connected Hausdorff manifold is **paracompact** if and only if it has a countable basis of open sets. (See the book by Hocking & young [20].)

(1) The first assumption we make about an applicable differentiable manifold M is that it is endowed with a paracompact topology.

(*Remark:* This assumption is necessary for the purpose of integration in any domain.)

We also consider only a **connected set** M for physical reasons. Moreover, we deal mostly with an open set M.

Now we shall introduce local coordinates for M. A **chart** (χ, U) or a **local coordinate system** is a pair consisting of an open subset $U \subset M$ together with a continuous one-to-one mapping (**homeomorphism**) χ from U into (codomain) $D \subset \mathbb{R}^N$. Here D is an open subset of $\mathbb{R}^N$ with the usual Euclidean topology. For a point $p \in M$, we have $x \equiv (x^1, x^2, \ldots, x^N) = \chi(p) \in D$. The coordinates $(x^1, x^2, \ldots, x^N)$ are the coordinates of the point p in the chart (χ, U).

Each of the N coordinates is obtained by the projection mappings $\pi^k : D \subset \mathbb{R}^n \longrightarrow \mathbb{R}$, $k \in \{1, \ldots, N\}$. These are defined by $\pi^k(x) \equiv \pi^k(x^1, \ldots, x^N) := x^k \in \mathbb{R}$. (See fig. 3.1.) (See the references [19], [28], [30] and [35].)

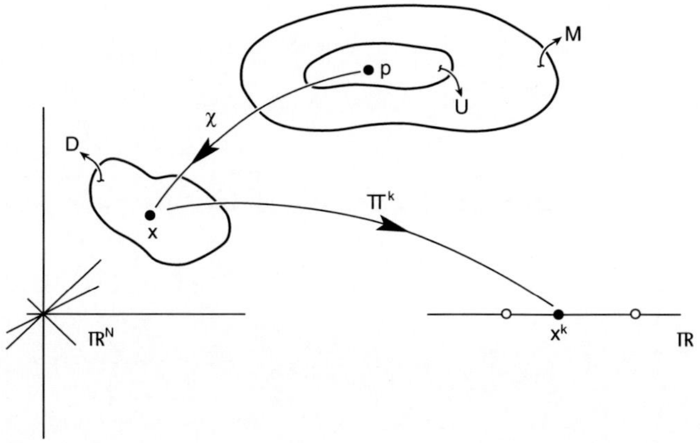

Figure 3.1: A chart (χ, U) and projection mappings.

Consider two coordinate systems or charts (χ, U) and $(\widehat{\chi}, \widehat{U})$ such that the point p is in the non-empty intersection of U and $\widehat{U}$. From fig. 3.2 we conclude that

$$\begin{aligned} \widehat{x} &= (\widehat{\chi} \circ \chi^{-1})(x), \\ x &= (\chi \circ \widehat{\chi}^{-1})(\widehat{x}), \end{aligned} \tag{3.1}$$

where $x \in D_s \subset D$ and $\widehat{x} \in \widehat{D}_s \subset \widehat{D}$.

From the preceding considerations, the mappings $\widehat{\chi} \circ \chi^{-1}$ and $\chi \circ \widehat{\chi}^{-1}$ are continuous and one-to-one. By projection of these points, we get

$$\begin{aligned} \widehat{x}^k &= [\pi^k \circ \widehat{\chi} \circ \chi^{-1}](x) := \widehat{X}^k(x) \equiv \widehat{X}^k(x^1, \ldots, x^N), \\ x^k &= [\pi^k \circ \chi \circ \widehat{\chi}^{-1}](\widehat{x}) := X^k(\widehat{x}) \equiv X^k(\widehat{x}^1, \ldots, \widehat{x}^N). \end{aligned} \tag{3.2}$$

The $2N$ functions $\widehat{X}^k$ and X^k are continuous. However, for a differentiable manifold, we assume that $\widehat{X}^k$ and X^k are differentiable functions of N real variables. In this context, we introduce some new notations. In the case where a function $f : D \subset \mathbb{R}^N \longrightarrow \mathbb{R}^M$ can be continuously differentiated r times with respect to every variable, we define the function f as belonging to the **class** $C^r(D \subset \mathbb{R}^N; \mathbb{R}^M)$, $r \in \{0, 1, 2, \ldots\}$. In the case where the function is Taylor-expandable (or **real-analytic**), the symbol $C^w(D \subset \mathbb{R}^N; \mathbb{R}^M)$ is used for the class.

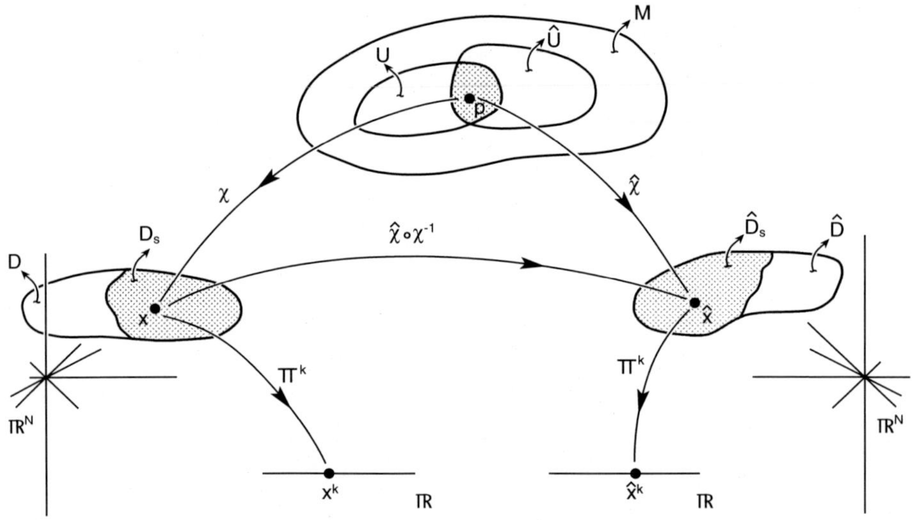

Figure 3.2: Two charts in M and a coordinate transformation.

By the first assumption of paracompactness, we can conclude that open subsets U_h exist such that $M = \bigcup_h U_h$. The second assumption about M is the following.

(2) There exist countable charts (χ_h, U_h) for M. Moreover, wherever there is a non-empty intersection between charts, coordinate transformations as in (3.2) of class C^r can be found. Such a basis of charts for M is called a C^r**-atlas**. A maximal collection of C^r-related atlases is called a maximal C^r**-atlas**. (It is also called the complete atlas.)

Finally, we are in a position to define a differentiable manifold.

(3) An N-dimensional C^r-differentiable manifold is a set M with a maximal C^r-atlas.

(*Remark:* For $r = 0$, the set M is called a **topological manifold**.)

A differentiable manifold is said to be **orientable** if there exists an atlas (χ_h, U_h) such that the **Jacobian** $\det\left[\frac{\partial \widehat{X}^k(x)}{\partial x^j}\right]$ is non-zero of one sign. We shall furnish some examples now.

Example 3.1.1 Consider $\mathbb{E}_1$, the one-dimensional Euclidean manifold. This is the simplest example of a differentiable manifold. A **global chart** $(\chi, \mathbb{E}_1)$

exists such that

$$x = \chi(p) \text{ for all } p \text{ in } \mathbb{E}_1 \text{ with } D = \mathbb{R}.$$

Another global chart $(\widehat{\chi}, \mathbb{E}_1)$ exists such that

$$\widehat{x} = \widehat{\chi}(p) \in \mathbb{R}.$$

Let $\widehat{X} := \widehat{\chi} \circ \chi^{-1}$ be given by

$$\widehat{x} = \widehat{X}(x) := (x)^3, \widehat{D} = \mathbb{R}.$$

Here, $D_s = D = \mathbb{R}$ and $\widehat{D}_s = \widehat{D} = \mathbb{R}$. we have the one-to-one mapping $\widehat{X} \in C^\omega(\mathbb{R}; \mathbb{R})$ since $\widehat{X}$ is a polynomial function. The Jacobian is given by the derivative $\frac{d\widehat{X}(x)}{dx} = 3(x)^2 > 0$ for $x \neq 0$. However, the vanishing of $\frac{d\widehat{X}(x)}{dx}$ at $x = 0$ does *not* violate the condition of one-to-one mapping over $\mathbb{R}$. □

Example 3.1.2 Consider the three-dimensional Euclidean manifold $\mathbb{E}_3$. One global chart $(\chi, \mathbb{E}_3)$ is furnished by

$$x = (x^1, x^2, x^3) = \chi(p), \quad p \in \mathbb{E}_3; \quad D = \mathbb{R}^3.$$

Let $(\chi, \mathbb{E}_3)$ be one of the infinitely many **Cartesian coordinate systems**.

Another chart $(\widehat{\chi}, \widehat{U})$ is given by

$$\begin{aligned} &\widehat{x} \equiv (\widehat{x}^1, \widehat{x}^2, \widehat{x}^3) \equiv (r, \theta, \phi) = \widehat{\chi}(p); \\ &\widehat{D} := \left\{ (\widehat{x}^1, \widehat{x}^2, \widehat{x}^3) \in \mathbb{R}^3 : \widehat{x}^1 > 0,\ 0 < \widehat{x}^2 < \pi,\ -\pi < \widehat{x}^3 < \pi \right\}. \end{aligned}$$

This is a **spherical polar coordinate system**. It is not a global chart. The coordinate transformation between the two coordinate systems is characterized by the equations

$$\begin{aligned} x^1 &= X^1(\widehat{x}) := \widehat{x}^1 \sin \widehat{x}^2 \cos \widehat{x}^3, \\ x^2 &= X^2(\widehat{x}) := \widehat{x}^1 \sin \widehat{x}^2 \sin \widehat{x}^3, \\ x^3 &= X^3(\widehat{x}) := \widehat{x}^1 \cos \widehat{x}^2. \end{aligned} \tag{3.3}$$

It is slightly more complicated to obtain the hatted functions $\widehat{X}^k$ explicitly. For that purpose, recall that principal branch Arctan and Arccos functions take values from $(-\pi/2, \pi/2)$ and $(0, \pi)$, respectively. Now, with these notations, we can provide the explicit functions

$$\begin{aligned} \widehat{x}^1 &= \widehat{X}^1(x) := +\sqrt{(x^1)^2 + (x^2)^2 + (x^3)^2}, \\ \widehat{x}^2 &= \widehat{X}^2(x) := \arccos\left[x^3 / \sqrt{(x^1)^2 + (x^2)^2 + (x^3)^2} \right], \\ \widehat{x}^3 &= \widehat{X}^3(x) := \operatorname{arc}(x^1, x^2) \\ &:= \begin{cases} \arctan(x^2/x^1) \text{ for } x^1 > 0, \\ (\pi/2)\operatorname{sgn}(x^2) \text{ for } x^1 = 0 \text{ and } x^2 \neq 0, \\ \arctan(x^2/x^1) + \pi \operatorname{sgn}(x^2) \text{ for } x^1 < 0 \text{ and } x^2 \neq 0. \end{cases} \end{aligned} \tag{3.4}$$

In this transformation, $\widehat{X}^k \in C^\infty(D_s \subset \mathbb{R}^3; \mathbb{R})$. Moreover, the domains are

$$D_s = \{(x^1, x^2, x^3) \in \mathbb{R}^3 : x^3 \in \mathbb{R}; x^1 > 0 \text{ or } x^2 \neq 0\} \subset D = \mathbb{R}^3,$$

$$\widehat{D}_s = \{(\widehat{x}^1, \widehat{x}^2, \widehat{x}^3) \in \mathbb{R}^3 : \widehat{x}^1 > 0, 0 < \widehat{x}^2 < \pi, -\pi < \widehat{x}^3 < \pi\} = \widehat{D} \subset \mathbb{R}^3,$$

$$\frac{\partial(x^1, x^2, x^3)}{\partial(\widehat{x}^1, \widehat{x}^2, \widehat{x}^3)} := \det\left[\frac{\partial X^k(\widehat{x})}{\partial \widehat{x}^j}\right] = (\widehat{x}^1)^2 \sin \widehat{x}^2 > 0. \qquad \square$$

Example 3.1.3 The two-dimensional (boundary) surface S^2 of the unit solid sphere can be constructed in the Euclidean space $\mathbb{E}_3$ by one constraint expressible in terms of the Cartesian coordinates as $(x^1)^2 + (x^2)^2 + (x^3)^2 = 1$. Equivalently, the same constraint is expressible in the spherical polar coordinates as $\widehat{x}^1 = 1$. (We used the preceding example.) The remaining spherical polar coordinates $(\widehat{x}^2, \widehat{x}^3) =: (\theta, \phi)$ can be used as a possible coordinate system over an open subset of S^2, which is a two-dimensional differentiable manifold in its own right. This coordinate chart is characterized by

$$x = (\theta, \phi) = \chi(p), \quad p \in U \subset S^2;$$

$$D := \{(\theta, \phi) \in \mathbb{R}^2 : 0 < \theta < \pi, -\pi < \phi < \pi\} \subset \mathbb{R}^2.$$

Another *distinct* spherical polar chart is furnished by

$$\begin{aligned}
\widehat{\theta} &= \widehat{\Theta}(\theta, \phi) := \text{Arccos}\,(-\sin\theta \sin\phi), \\
\widehat{\phi} &= \widehat{\Phi}(\theta, \phi) := arc\,(-\sin\theta\cos\phi, \cos\theta), \\
\widehat{D} &:= \left\{(\widehat{\theta}, \widehat{\phi}) \in \mathbb{R}^2 : 0 < \widehat{\theta} < \pi, -\pi < \widehat{\phi} < \pi\right\} \subset \mathbb{R}^2.
\end{aligned}$$

Neither of the two charts is global. However, the union $U \cup \widehat{U} = S^2$. Thus, the collection of two charts (χ, U) and $(\widehat{\chi}, \widehat{U})$ constitutes an atlas for S^2. (The minimum number of charts for an atlas of S^2 is two.) See fig. 3.3 for the illustration. $\square$

Now, let us discuss a function F from $U \subset M$ into $\mathbb{R}$. (See fig. 3.4.) This function induces another function $f : F \circ \chi^{-1}$ from $D \subset \mathbb{R}^N$ into $\mathbb{R}$. We can represent it by the usual notation

$$y = F(p) = \left[F \circ \chi^{-1}\right](x) =: f(x) \equiv f(x^1, \ldots, x^N). \qquad (3.5)$$

The function f may be of differentiability class $C^k(D \subset \mathbb{R}^N; \mathbb{R})$, $k \in \{0, 1, \ldots\}$, a totally differentiable function, a function obeying the Lipschitz condition, or a discontinuous function.

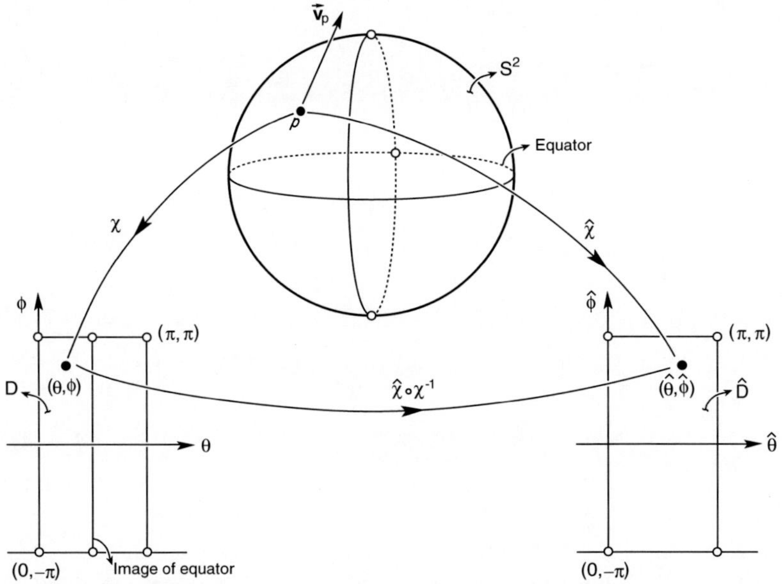

Figure 3.3: Spherical polar coordinates.

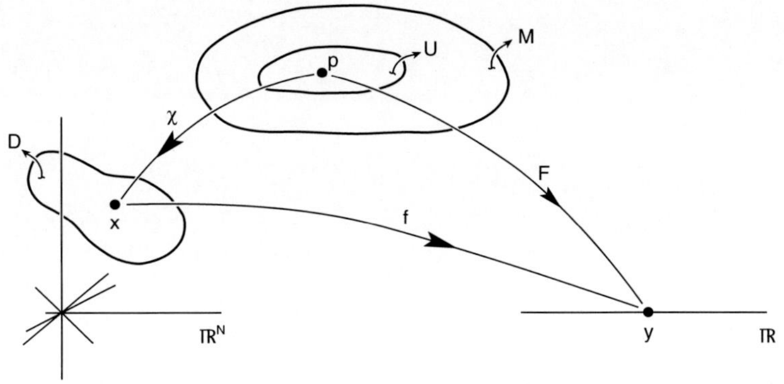

Figure 3.4: A function from $U \subset M$ into $\mathbb{R}$.

Example 3.1.4 Let us consider a function F over a Euclidean plane $\mathbb{E}_2$. We adopt a global Cartesian chart $(\chi, \mathbb{E}_2)$. Let the corresponding f over $\mathbb{R}^2$ be defined by

$$f(x) \equiv f(x^1, x^2) := \begin{cases} 0 \ \text{ for } \ (x^1, x^2) = (0,0), \\ \left[\dfrac{(x^1)^2 - (x^2)^2}{(x^1)^2 + (x^2)^2}\right] \ \text{ for } \ (x^1, x^2) \neq (0,0). \end{cases}$$

We can deduce that

$$\lim_{x^1 \to 0} \lim_{x^2 \to 0} f(x^1, x^2) = - \lim_{x^2 \to 0} \lim_{x^1 \to 0} f(x^1, x^2) = 1.$$

Obviously, the function is discontinuous at the origin. We denote the function by $f \in \mathrm{Map}\,(\mathbb{R}^2; \mathbb{R})$. □

A function F from M into $\widehat{M}$ is a C^k**-diffeomorphism** ($k \geq 1$) provided F is one-to-one, onto, and of class C^k. Moreover, F^{-1} must also be one-to-one, onto, and of class C^k.

(*Remark:* In the case where $k = 0$, such a mapping is a **homeomorphism**.)

A diffeomorphism is elaborated in fig. 3.5.

In terms of coordinate charts (χ, U) and $(\widehat{\chi}, \widehat{U})$, the diffeomorphism F yields the following transformations:

$$\begin{aligned} y^i &= \left[\pi^i \circ \widehat{\chi} \circ F \circ \chi^{-1}\right](x) =: Y^i(x^1, \ldots, x^N), \\ \det\left[\frac{\partial y^i}{\partial x^j}\right] &=: \partial(y^1, \ldots, y^N)/\partial(x^1, \ldots, x^N) \neq 0. \end{aligned} \tag{3.6}$$

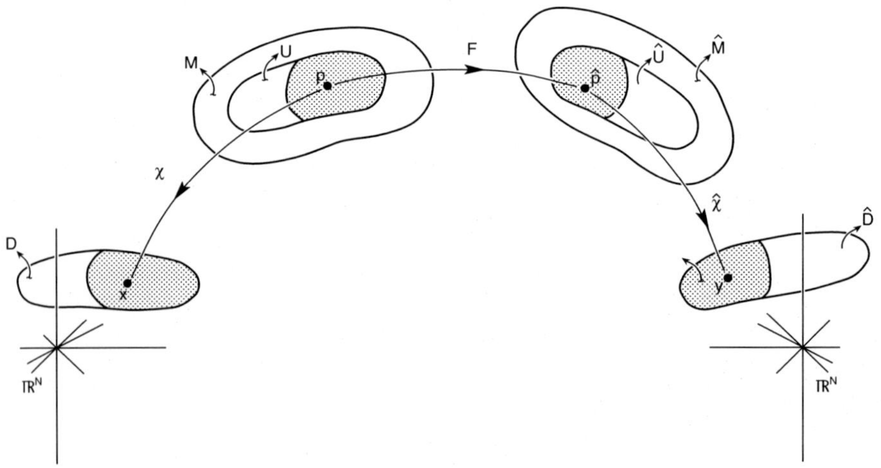

Figure 3.5: A C^k-diffeomorphism F.

Example 3.1.5 Consider $M = \mathbb{E}_1$. Let the mapping F be specified by

$$y = \widehat{\chi} \circ F(p) = f(x) := \tanh x, \quad x \in \mathbb{R}.$$
$$\frac{df(x)}{dx} = \sec h^2 x > 0.$$

Since $|\tanh x| < 1$, F is a C^w-diffeomorphism from $\mathbb{E}_1$ into a *proper* connected subset of $\widehat{M}$ (which is coordinatized as $(-1,1)$). □

Example 3.1.6 Consider $M = \widehat{M} = \mathbb{E}_1$. Let the mapping F be characterized by

$$y = \widehat{\chi} \circ F(p) := (x)^5, \; x \in \mathbb{R};$$
$$x = \sqrt[5]{y}, \; \frac{dx}{dy} = \frac{1}{5y^{4/5}}.$$

The derivative of the inverse mapping F^{-1} is not defined to correspond to the point $y = 0$. However, it is a continuous and one-to-one mapping. Therefore, we conclude that F is a homeomorphism that is *not* a diffeomorphism. □

Exercises 3.1

1. Consider the four-dimensional (flat space-time) differentiable manifold M of special relativity. The Poincaré (or inhomogeneous Lorentz) transformations between two C^w-related global charts (χ, M) and $(\widehat{\chi}, M)$ are characterized by

$$\widehat{x}^k = \widehat{X}^k(x) := c^k + l^k_j x^j; \;\; D_s = D = \mathbb{R}^4.$$

Here, c^k's are arbitrary constants or parameters. Moreover, the other parameters l^i_js satisfy

$$d_{ij} l^i{}_a l^j{}_b = d_{ab}.$$

(The metric tensor components d_{ab} are defined in (2.116).) Prove that the Jacobian of the Poincaré transformations satisfies

$$\frac{\partial(\widehat{x}^1, \widehat{x}^2, \widehat{x}^3, \widehat{x}^4)}{\partial(x^1, x^2, x^3, x^4)} := \det \left[\frac{\partial \widehat{X}^i(x)}{\partial x^j} \right] = \pm 1.$$

(*Remark:* The set of all Poincaré transformations constitutes the group denoted by $\mathcal{I}0(3,1;\mathbb{R})$.)

2. Consider the following function over $\mathbb{R}^2$:

$$f(x^1, x^2) := \begin{cases} (x^1 x^2) \left[((x^1)^2 - (x^2)^2)/((x^1)^2 + (x^2)^2) \right] \text{ for } (x^1, x^2) \neq (0,0), \\ 0 \text{ for } (x^1, x^2) = (0,0). \end{cases}$$

Prove that the second partials exist at $(0, 0)$ but

$$\left.\frac{\partial^2 f(x^1, x^2)}{\partial x^1 \partial x^2}\right|_{(0,0)} \neq \left.\frac{\partial^2 f(x^1, x^2)}{\partial x^2 \partial x^1}\right|_{(0,0)}.$$

(See Gelbaum & Olmsted [15].)

3.2 Tangent Vectors, Cotangent Vectors, and Parametrized Curves

Let us consider the space in the arena of Newtonian physics. It is mathematically represented by the three-dimensional Euclidean space $\mathbb{E}_3$. This space admits infinitely many global Cartesian charts. Vectors in the physical space describe velocities, accelerations, forces, etc. Each of these vectors has a (homeomorphic) image in $\mathbb{R}^3$ of a Cartesian chart. To obtain an intuitive definition of a **tangent vector** $\vec{\mathbf{v}}_p$ or its image $\vec{\mathbf{v}}_x$ in $\mathbb{R}^3$, we shall visualize an "arrow" in $\mathbb{R}^3$ with the starting point $x \equiv (x^1, x^2, x^3)$ in $\mathbb{R}^3$ and the directed displacement $\vec{\mathbf{v}} \equiv (v^1, v^2, v^3)$ necessary to reach the point $(x^1 + v^1, x^2 + v^2, x^3 + v^3)$ in $\mathbb{R}^3$. (See fig. 3.6.)

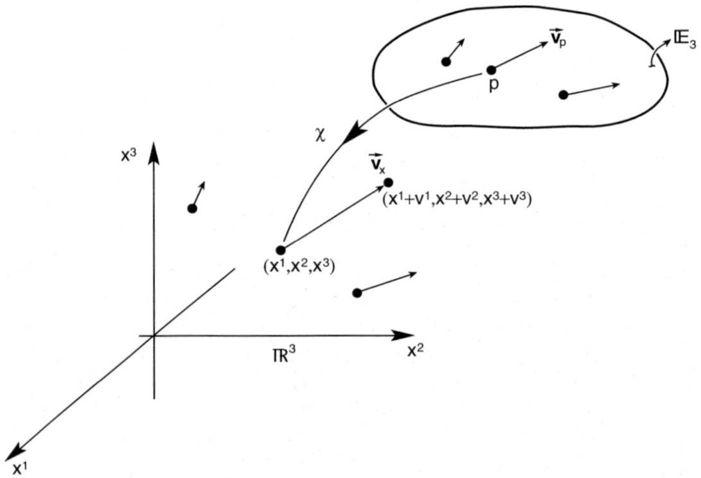

Figure 3.6: Tangent vector in $\mathbb{E}_3$ and $\mathbb{R}^3$.

Example 3.2.1 Let us choose as standard (Cartesian) basis vectors at x in $\mathbb{R}^3$

$$\vec{\mathbf{i}}_x \equiv \vec{\mathbf{e}}_{1x} := (1,0,0)_x, \quad \vec{\mathbf{j}}_x \equiv \vec{\mathbf{e}}_{2x} := (0,1,0)_x,$$
$$\vec{\mathbf{k}}_x \equiv \vec{\mathbf{e}}_{3x} := (0,0,1)_x.$$

Let a three-dimensional vector be given by $\vec{\mathbf{v}} := 3\vec{\mathbf{e}}_1 + 2\vec{\mathbf{e}}_2 + \vec{\mathbf{e}}_3 = (3,2,1)$. Let us choose a point $x \equiv (x^1, x^2, x^3) = (1,2,3)$. Therefore, the tangent vector $\vec{\mathbf{v}}_x = (3,2,1)_{(1,2,3)}$ starts from the point $(1,2,3)$ and terminates at $(4,4,4)$. □

But such a simple definition runs into problems in a curved manifold. A simple curved manifold is the spherical surface S^2 in fig. 3.3. We have drawn an intuitive picture of a tangent vector $\vec{\mathbf{v}}_p$ on S^2. The starting point p of $\vec{\mathbf{v}}_p$ is on S^2. However, the end point of $\vec{\mathbf{v}}_p$ is *not* on S^2. The problem is how to define a tangent vector *intrinsically* on S^2 without going out of the spherical surface. One logical possibility is to introduce *directional derivatives* of a smooth function F defined at p in a subset of S^2. Such a definition involves only the point p and its neighboring points, all on S^2. Thus, we shall represent tangent vectors by the directional derivatives. This concept appears to be very abstract at the beginning. (But recall that the position coordinates and the momentum variables are represented by linear operators in quantum mechanics!)

Now, let us define a **generalized directional derivative**. We shall use a coordinate chart (χ, U) of the abstract manifold M. Let $x \equiv (x^1, \ldots, x^N) = \chi(p)$. Let an N-tuple of vector components be given by $(v^1, \ldots, v^N)$. Then, the **tangent vector** $\vec{\mathbf{v}}_x$ in $D \subset \mathbb{R}^N$ is defined by the generalized directional derivative

$$\begin{aligned} \vec{\mathbf{v}}_x &:= \sum_{a=1}^{N} v^a \frac{\partial}{\partial x^a} \equiv v^a \frac{\partial}{\partial x^a}, \\ \vec{\mathbf{v}}_x[f] &:= v^a \frac{\partial f(x)}{\partial x^a}. \end{aligned} \tag{3.7}$$

Here, f belongs to $C^1(D \subset \mathbb{R}^N; \mathbb{R})$. (Note that in the notation of the usual calculus, $\vec{\mathbf{v}}_x[f] = \vec{\mathbf{v}}_x \cdot \mathbf{grad}\, f$.)

The set of all tangent vectors $\vec{\mathbf{v}}_x$ constitutes the N-dimensional **tangent vector space** $T_x(\mathbb{R}^N)$ in $D \subset \mathbb{R}^N$. It is an isomorphic image of the tangent vector space $T_p(M)$. The standard basis set $\{\vec{\mathbf{e}}_{1x}, \ldots, \vec{\mathbf{e}}_{Nx}\}$ for $T_x(\mathbb{R}^N)$ is defined by the differential operators

$$\begin{aligned} \vec{\mathbf{e}}_{1x} &:= \frac{\partial}{\partial x^1}, \ldots, \vec{\mathbf{e}}_{Nx} := \frac{\partial}{\partial x^N}; \\ \vec{\mathbf{e}}_{kx}[f] &:= \frac{\partial f(x)}{\partial x^k}. \end{aligned} \tag{3.8}$$

Now, we shall define a tangent vector field $\vec{\mathbf{v}}(p)$ in $U \subset M$ or equivalently the tangent vector field $\vec{\mathbf{v}}(x)$ in $D \subset \mathbb{R}^N$. It involves N real-valued functions

$v^1(x), \ldots, v^N(x)$. The tangent vector field is defined by the operator

$$\begin{aligned} \vec{\mathbf{v}}(x) &:= v^j(x)\frac{\partial}{\partial x^j}, \\ \vec{\mathbf{v}}(x)[f] &:= v^j(x)\frac{\partial f(x)}{\partial x^j}. \end{aligned} \tag{3.9}$$

Here, $f \in C^1(D \subset \mathbb{R}^N; \mathbb{R})$.

Example 3.2.2 Let $D := \{(x^1, x^2) \in \mathbb{R}^2 \ : \ x^1 \in \mathbb{R}, 1 < x^2\}$. Moreover, let $f(x) := x^1 \left[(x^2)^{x^2}\right]$ and

$$\vec{\mathbf{v}}(x^1, x^2) := e^{x^2}\frac{\partial}{\partial x^1} - (2\cosh x^1)\frac{\partial}{\partial x^2}.$$

Therefore, by (2.37) we get

$$\begin{aligned} &\vec{\mathbf{v}}(x^1, x^2)[f] = [(x^2)^{x^2}][e^{x^2} - 2(x^1\cosh x^1)(1 + \ln x^2)], \\ &\lim_{x^1 \to 1}\ \lim_{x^2 \to +1} \{\vec{\mathbf{v}}(x^1, x^2)[f]\} = -e^{-1}. \end{aligned}$$

□

Example 3.2.3 The standard basis vectors from (3.8) can be expressed as

$$\vec{\mathbf{e}}_j(x) = \frac{\partial}{\partial x^j} = \delta^k{}_j\frac{\partial}{\partial x^k}.$$

Therefore, for the function $f(x) := x^i$,

$$\begin{aligned} \vec{\mathbf{e}}_j(x)[x^i] &= \delta^k{}_j\frac{\partial}{\partial x^k}(x^i) = \delta^k{}_j\delta^i{}_k = \delta^i{}_j, \\ \vec{\mathbf{v}}(x)[x^i] &= v^j(x)\frac{\partial}{\partial x^j}(x^i) = v^j(x)\delta^i{}_j = v^i(x). \end{aligned} \tag{3.10}$$

□

The main properties of tangent vector fields can be summarized in the following theorem.

Theorem 3.2.4 *If $\vec{\mathbf{v}}$ and $\vec{\mathbf{w}}$ are tangent vector fields in $D \subset \mathbb{R}^N$, and $f, g, h \in C^1(D \subset \mathbb{R}^N; \mathbb{R})$, then*

(i)
$$[f(x)\vec{\mathbf{v}}(x) + g(x)\vec{\mathbf{w}}(x)][h] = f(x)(\vec{\mathbf{v}}(x)[h]) + g(x)(\vec{\mathbf{w}}(x)[h]); \tag{3.11}$$

(ii)
$$\vec{\mathbf{v}}(x)[cf + kg] = c(\vec{\mathbf{v}}(x)[f]) + k(\vec{\mathbf{v}}(x)[g]) \tag{3.12}$$

for all constants c and k;

(iii)
$$\vec{\mathbf{v}}(x)[fg] = (\vec{\mathbf{v}}(x)[f])g(x) + f(x)(\vec{\mathbf{v}}(x)[g]). \tag{3.13}$$

The proof is left as an exercise.

Now we shall discuss the **cotangent** (or **covariant**) **vector field** $\tilde{\mathbf{v}}(p)$ in $U \subset N$ or the isomorphic image $\tilde{\mathbf{v}}(x)$ in $D \subset \mathbb{R}^N$. But first we need to define a (totally) **differentiable function** f in $D \subset \mathbb{R}^N$. In the case where f satisfies the criterion

$$\lim_{(h^1,\ldots h^N)\to(0,\ldots,0)} \left\{ \left[f(x^1+h^1,\ldots,x^N+h^N) - f(x^1,\ldots,x^N) - h^j \frac{\partial f(x)}{\partial x^j} \right] \Big/ \sqrt{(h^1)^2+\cdots+(h^N)^2} \right\} = 0, \tag{3.14}$$

for arbitrary $(h^1,\ldots,h^N)$, we call the function f **totally differentiable** at x. The usual condensed notation for (3.14) is to write the total differential as

$$df(x) = \frac{\partial f(x)}{\partial x^j} dx^j. \tag{3.15}$$

The existence of the first partials is a necessity for (total) differentiability. Moreover, continuities of the first partials are sufficient conditions for (total) differentiability. Clearly, (3.14) or (3.15) can provide the rate of change of the function f in every direction. The equation (3.15) vaguely resembles the second equation in (3.9). Moreover, by (1.17), we can write $\tilde{\mathbf{f}}(\vec{\mathbf{v}}) := (\vec{\mathbf{v}})[f] = v^j f_j$. These facts motivate us to identify $\tilde{\mathbf{f}}(x)$ with $df(x)$. Thus, we are prompted to define abstractly

$$df(x)[\vec{\mathbf{v}}(x)] := \vec{\mathbf{v}}(x)[f] = v^j(x)\frac{\partial f(x)}{\partial x^j}. \tag{3.16}$$

We can prove by the preceding definition and (3.11) that

$$df(x)[g(x)\vec{\mathbf{v}}(x) + h(x)\vec{\mathbf{w}}(x)] = g(x)(\vec{\mathbf{v}}(x)[f]) + h(x)(\vec{\mathbf{w}}(x)[h]). \tag{3.17}$$

Thus, $df(x)$ is a linear mapping. By the comparison of (3.16) and (3.17) with (1.14), we conclude that $df(x)$ is indeed a **covariant vector field**. It is also called a **cotangent vector field** or a **1-form**. Analogously to (1.15), we define the linear combination as

$$[\Lambda(x)df(x)+\Omega(x)dg(x)]\ [\vec{\mathbf{v}}(x)] := \Lambda(x)(df(x)[\vec{\mathbf{v}}(x)])+\Omega(x)(dg(x)[\vec{\mathbf{v}}(x)]). \tag{3.18}$$

Here $\Lambda(x)$ and $\Omega(x)$ are arbitrary scalar fields in $D \subset \mathbb{R}^N$.

Thus, the set of all covariant vector fields at $x \in D \subset \mathbb{R}^N$ constitutes the N-dimensional **cotangent** (or **covariant vector**) **space** denoted by $\tilde{T}_x(\mathbb{R}^N)$. It has the isomorphic pre-image $\tilde{T}_p(M)$.

Example 3.2.5 Let $f(x) \equiv f(x^1,x^2) := (1/2)[(x^1)^2 - 2e^{x^2}]$, $(x^1,x^2) \in \mathbb{R}^2$. Therefore, by (3.15) and (3.16),

$$\begin{aligned} df(x) &= x^1 dx^1 - e^{x^2} dx^2, \\ df(x)[\vec{\mathbf{v}}(x)] &= x^1 v^1(x) - e^{x^2} v^2(x), \\ df(0,0)[\vec{\mathbf{v}}(0,0)] &= -v^2(0,0). \end{aligned}$$

□

Example 3.2.6 Consider the function $f(x) := x^i$. Then

$$df(x) = dx^i = \delta^i{}_j \, dx^j.$$

Furthermore, by (3.8) and (3.16),

$$dx^j \left[\vec{\mathbf{e}}_i(x)\right] = dx^j \left[\frac{\partial}{\partial x^i}\right] = dx^j \left[\delta^k{}_i \frac{\partial}{\partial x^k}\right] = \delta^k{}_i \frac{\partial(x^j)}{\partial x^k} = \delta^j{}_i.$$

Therefore, by the consequence of $\tilde{\mathbf{e}}^j[\vec{\mathbf{e}}_i] = \delta^j{}_i$ in (1.16), we identify the coordinate covariant basis field $\{\tilde{\mathbf{e}}^1(x), \ldots, \tilde{\mathbf{e}}^N(x)\}$ for $\tilde{T}_x(\mathbb{R}^N)$ as

$$\tilde{\mathbf{e}}^j(x) = dx^j, \quad j \in \{1, \ldots, N\}. \tag{3.19}$$

Thus, every covariant vector field $\tilde{\mathbf{W}}(x)$ admits the linear combination

$$\tilde{\mathbf{W}}(x) = W_j(x) dx^j \tag{3.20}$$

in terms of the basis covariant vectors dx^j's. □

Example 3.2.7 Let $f(x) := x^j$. Then, by (3.16),

$$df(x)\left[\vec{\mathbf{v}}(x)\right] = dx^j \left[v^k(x) \frac{\partial}{\partial x^k}\right] = v^k(x) \frac{\partial(x^j)}{\partial x^k} = v^j(x). \tag{3.21}$$

Thus, the value of the linear function dx^j at $\vec{\mathbf{v}}(x)$ is exactly the component $v^j(x)$. □

Now we shall discuss a different topic, namely a **parametrized curve** γ. Consider an interval $[a, b] \subset \mathbb{R}$.

(*Remark:* Open or semi-open intervals are also allowed. Moreover, unbounded intervals are permitted, too.)

The **parametrized curve** γ is a function from the interval $[a, b]$ into a differentiable manifold. (See fig. 3.7.) (Note that the *function* γ is called the parametrized curve.) The image of the composite function $\mathcal{X} := \chi \circ \gamma$ in $D \subset \mathbb{R}^n$ is denoted by the symbol Γ. The coordinates on Γ are furnished by

$$\begin{aligned} x &= [\chi \circ \gamma](t) = \mathcal{X}(t), \\ x^j &= [\pi^j \circ \chi \circ \gamma](t) = \left[\pi^j \circ \mathcal{X}\right](t) \equiv \mathcal{X}^j(t). \end{aligned} \tag{3.22}$$

Here, $t \in [a, b] \subset \mathbb{R}$. The functions χ^j are usually *assumed* to be differentiable and piecewise twice-differentiable. The condition of the **non-degeneracy** (or **regularity**) is

$$\sum_{j=1}^{N} \left[\frac{d\mathcal{X}^j(t)}{dt}\right]^2 > 0. \tag{3.23}$$

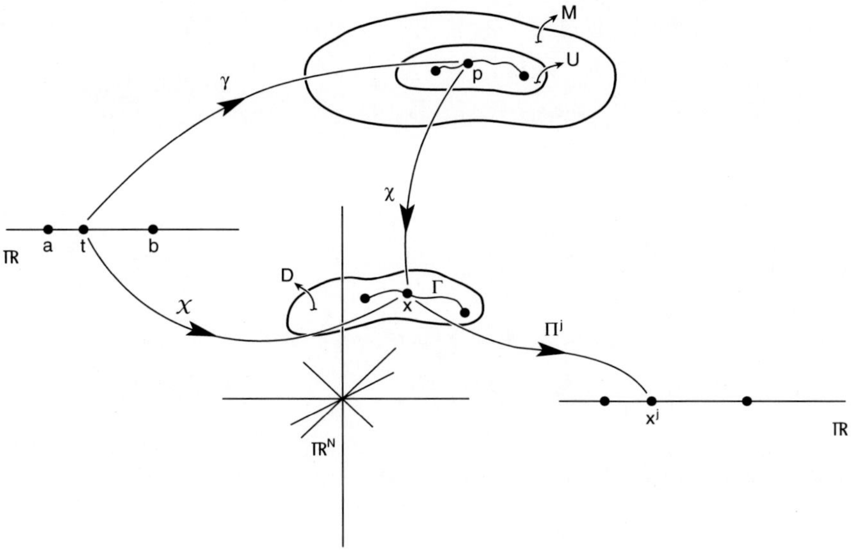

Figure 3.7: A curve γ into M.

In Newtonian physics, t is taken to be the time variable and $M = \mathbb{E}_3$, the physical space. Moreover, Γ is a particle trajectory relative to a Cartesian coordinate system. The non-degeneracy condition (3.23) implies that the speed of the motion is strictly positive.

Example 3.2.8 Let us consider a particle trajectory in the physical space in the absence of external forces. In this case, the acceleration is zero and Newton's second law of motion yields

$$\frac{d^2\mathcal{X}^j(t)}{dt^2} = 0, \quad 0 < t,$$

for $j \in \{1, 2, 3\}$. Solving these ordinary differential equations, we can obtain the general solution as

$$x^j = \mathcal{X}^j(t) = (b^j)t + c^j.$$

Here, b^j and c^j are six arbitrary constants of integration. The speed of this motion is $\sqrt{(b^1)^2 + (b^2)^2 + (b^3)^2} \geq 0$. If the speed is positive (or the curve γ is non-degenerate), Γ is a straight line in $\mathbb{R}^3$. If the speed is zero (or the curve γ is degenerate), the particle stays permanently at the point (c^1, c^2, c^3) in $\mathbb{R}^3$. □

Example 3.2.9 Let the image Γ in $\mathbb{R}^3$ be given by

$$x = \mathcal{X}(t) := (2\cos^2 t, \ \sin 2t, \ 2\sin t); \quad 0 < t < \pi/2.$$

The curve is non-degenerate. The coordinate functions are real-analytic. Consider a circular cylinder in $\mathbb{R}^3$ such that it intersects the $x^1 - x^2$ plane on the unit circle with the center at $(1, 0, 0)$. Now, consider a spherical surface in $\mathbb{R}^3$ given by the equation $(x^1)^2 + (x^2)^2 + (x^3)^2 = 4$. The image Γ lies in the intersection of the cylinder and the sphere. □

Let us consider the tangent vector $\vec{\mathbf{t}}_x$ of the image Γ at the point x in $\mathbb{R}^N$. (See fig. 3.7.) In calculus, the components of the tangent vector $\vec{\mathbf{t}}_{\mathcal{X}(t)}$ are taken to be $\left(\frac{d\mathcal{X}^1(t)}{dt}, \ldots, \frac{d\mathcal{X}^N(t)}{dt}\right)$. Therefore, the tangent vector $\vec{\mathbf{t}}_{\mathcal{X}(t)} \equiv \vec{\mathcal{X}}'(t)$ along Γ (according to (3.7)) must be defined as the generalized directional derivative

$$\vec{\mathbf{t}}_{\mathcal{X}(t)} \equiv \vec{\mathcal{X}}'(t) := \frac{d\mathcal{X}^j(t)}{dt} \left[\frac{\partial}{\partial x^j}\right]_{|\mathcal{X}(t)}. \tag{3.24}$$

The tangent vector field above belongs to the tangent vector space $T_{\mathcal{X}(t)}(\mathbb{R}^3)$ along Γ.

Example 3.2.10 Let a real-analytic, non-degenerate curve $\mathcal{X}$ into $\mathbb{R}^4$ be defined by

$$x = \mathcal{X}(t) := ((t)^2, t, e^t, \sinh t), \quad -\infty < t < \infty;$$
$$\mathcal{X}(0) = (0, 0, 1, 0).$$

The corresponding tangent vector field along Γ is furnished by the generalized directional derivative

$$\vec{\mathcal{X}}'(t) = (2t)\left[\frac{\partial}{\partial x^1}\right]_{|\mathcal{X}(t)} + \left[\frac{\partial}{\partial x^2}\right]_{|\mathcal{X}(t)} + (e^t)\left[\frac{\partial}{\partial x^3}\right]_{|\mathcal{X}(t)} + (\cosh t)\left[\frac{\partial}{\partial x^4}\right]_{|\mathcal{X}(t)},$$

$$\vec{\mathcal{X}}'(0) = \left[\frac{\partial}{\partial x^2}\right]_{|\mathcal{X}(0)} + \left[\frac{\partial}{\partial x^3}\right]_{|\mathcal{X}(0)} + \left[\frac{\partial}{\partial x^4}\right]_{|\mathcal{X}(0)}. \quad \square$$

The tangent vector $\vec{\mathcal{X}}'(t)$ can act on a differentiable function f (restricted on Γ) by the rule (3.7). On this topic, we state and prove the following theorem.

Theorem 3.2.11 *Let a parametrized curve $\mathcal{X} : [a, b] \subset \mathbb{R} \to \mathbb{R}^N$ be differentiable and non-degenerate. Let $f : D \subset \mathbb{R}^N \to \mathbb{R}$ be a (totally) differentiable function. Then*

$$\vec{\mathcal{X}}'(t)[f(\mathcal{X}(t))] = \frac{d}{dt}[f(\mathcal{X}(t))]. \tag{3.25}$$

Proof. By (3.9) and the chain rule of differentiation, the left-hand side of the equation above yields

$$\frac{d\mathcal{X}^j(t)}{dt}\left[\frac{\partial}{\partial x^j}\right]_{|\mathcal{X}(t)}[f(\mathcal{X}(t))] \equiv \frac{d\mathcal{X}^j(t)}{dt}\left[\frac{\partial f(x)}{\partial x^j}\right]_{|\mathcal{X}(t)} = \frac{d}{dt}[f(\mathcal{X}(t))]. \quad \blacksquare$$

Now, we shall discuss the **reparametrization** of a curve. Let h be a differentiable and one-to-one mapping from $[c,d] \subset \mathbb{R}$ into $\mathbb{R}$. (See fig. 3.8.)

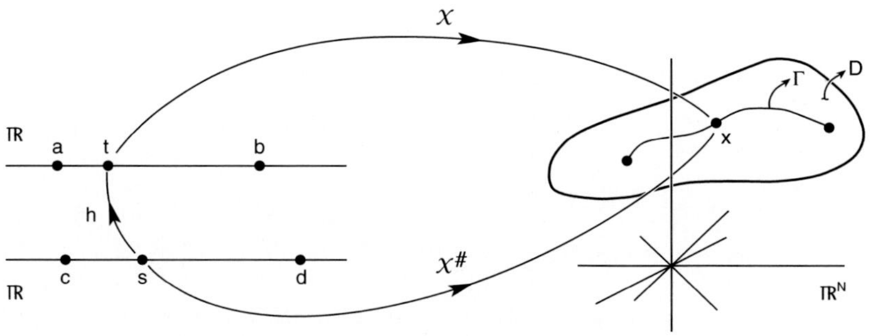

Figure 3.8: Reparametrization of a curve.

The image Γ in $\mathbb{R}^N$ is given by the points

$$\begin{aligned} t &= h(s), \\ x &= \mathcal{X}(t), \quad\quad a \leq t \leq b, \;\; a = h(c), \;\; b = h(d), \\ x &= \mathcal{X}^{\#}(s), \;\; c \leq s \leq d. \end{aligned}$$

Here,

$$\mathcal{X}^{\#} := \mathcal{X} \circ h = \chi \circ \gamma \circ h \tag{3.26}$$

is the **reparametrized curve** into $\mathbb{R}^N$.

Theorem 3.2.12 *If $\mathcal{X}^{\#}$ is the reparametrization of the differentiable curve $\mathcal{X}$ by the function h, then the tangent vector*

$$\vec{\mathcal{X}}^{\#\prime}(s) = \frac{dh(s)}{ds} \vec{\mathcal{X}}'(h(s)). \tag{3.27}$$

Proof. By (3.26), we get

$$\mathcal{X}^{\#j}(s) = \mathcal{X}^j(h(s)).$$

By the assumption of differentiabilities and the chain rule, we obtain

$$\frac{d\mathcal{X}^{\#j}(s)}{ds} = \frac{d\mathcal{X}^j(t)}{dt}\Big|_{t=h(s)} \cdot \frac{dh(s)}{ds}.$$

Therefore, the tangent vector

$$\begin{aligned} \vec{\mathcal{X}}^{\#\prime}(s) &= \frac{d\mathcal{X}^{\#j}(s)}{ds} \left[\frac{\partial}{\partial x^j}\right]_{|\mathcal{X}^{\#}(s)} = \frac{dh(s)}{ds} \left[\frac{d\mathcal{X}^j(t)}{dt}\right]_{|t=h(s)} \left[\frac{\partial}{\partial x^j}\right]_{|\mathcal{X}(h(s))} \\ &\equiv \frac{dh(s)}{ds} \vec{\mathcal{X}}'(h(s)). \end{aligned}$$

■

Example 3.2.13 Consider the Euclidean plane $\mathbb{E}_2$ and a parametrized (real-analytic) curve $\mathcal{X} := \chi \circ \gamma$ given by

$$\mathcal{X}(t) := (\cos t, \sin t), \quad -3\pi \leq t \leq 3\pi.$$

The image γ is a unit circle in $\mathbb{R}^2$ with self-intersections. The **winding number** of this circle is exactly three.

Let a (real-analytic) reparametrization be defined by

$$t = h(s) := 3s, \quad -\pi \leq s \leq \pi;$$
$$\frac{dh(s)}{ds} \equiv 3 > 0.$$

The reparametrized curve $\mathcal{X}^{\#}$ is furnished by

$$\mathcal{X}^{\#}(s) = \mathcal{X}(3s) = (\cos(3s), \sin(3s)).$$

The new tangent vector is given by

$$\begin{aligned} \vec{\mathcal{X}}^{\#\prime}(s) &= 3\left[-(\sin 3s)\frac{\partial}{\partial x^1} + (\cos 3s)\frac{\partial}{\partial x^2}\right]_{|\mathcal{X}^{\#}(s)} \\ &= 3\vec{\mathcal{X}}'(3s). \end{aligned}$$

□

Exercises 3.2

1. Prove theorem 3.2.4.

2. Let a tangent vector field in $\mathbb{R}^3$ be defined by

$$\vec{\mathbf{V}}(x) := x^j \frac{\partial}{\partial x^j}.$$

Let $P_n(x^1, x^2, x^3)$ be an nth-degree homogeneous polynomial of the variables x^1, x^2, x^3. Prove Euler's theorem

$$\vec{\mathbf{V}}(x)[P_n(x)] = nP_n(x^1, x^2, x^3).$$

3. Evaluate the 1-form $\tilde{\mathbf{w}}(x) := \delta_{ij} x^i dx^j$ on the tangent vector field on $\vec{\mathbf{V}}(x) := \sum_{j=1}^{N} (x^j)^3 \frac{\partial}{\partial x^j}$. Prove that $\tilde{\mathbf{w}}(x)[\vec{\mathbf{V}}(x)] \geq 0$.

4. Find the *unique* curve $\mathcal{X}$ into $\mathbb{R}^4$ such that $\mathcal{X}(0) = (0, 0, 0, 0)$ and $\vec{\mathcal{X}}'(t) = \left\{\sum_{\alpha=1}^{3} \left[\frac{\partial}{\partial x^\alpha}\right] + \sqrt{3}\frac{\partial}{\partial x^4}\right\}_{|\chi(t)}$; $t \in (0, \infty)$.

(*Remark:* The image of this curve can represent a possible trajectory of a photon.)

5. Consider the semi-cubical parabola in $\mathbb{R}^2$

$$\mathcal{X}(t) := (t^2, t^3), \quad t \in \mathbb{R}.$$

The curve is degenerate at $\mathcal{X}(0) = (0, 0)$. Prove that reparametrization of this curve *cannot* remove the degeneracy.

(*Remark:* The degenerate point is called the **cusp**.)

3.3 Tensor Fields over Differentiable Manifolds

Consider the tangent vector space $T_{x_0}(\mathbb{R}^N)$ at $x_0 \equiv (x_0^1, \ldots, x_0^N)$ in $\mathbb{R}^N$ (corresponding to $T_{p_0}(M)$ at p_0 in M). We can construct the cotangent space $\tilde{T}_{x_0}(\mathbb{R}^N)$ by the set of linear functions from $T_{x_0}(\mathbb{R}^N)$ into $\mathbb{R}$. We consider the change of basis set in $T_{x_0}(\mathbb{R}^N)$ and the corresponding change of basis in $\tilde{T}_{x_0}(\mathbb{R}^N)$. By (2.113) and (1.20), we write

$$\begin{aligned}
&\widehat{\vec{\mathbf{e}}}_{p|x_0} = \lambda^q{}_p \, \vec{\mathbf{e}}_{q|x_0}, \quad \vec{\mathbf{e}}_{p|x_0} = \mu^q{}_p \, \widehat{\vec{\mathbf{e}}}_{q|x_0}, \\
&\widehat{\tilde{\mathbf{e}}}^p{}_{|x_0} = \mu^p{}_q \, \tilde{\mathbf{e}}^q{}_{|x_0}, \quad \tilde{\mathbf{e}}^p{}_{|x_0} = \lambda^p{}_q \widehat{\tilde{\mathbf{e}}}^q{}_{|x_0}, \\
&\det[\lambda^p{}_q] \neq 0, \\
&[\mu^q{}_p] := [\lambda^p{}_q]^{-1}.
\end{aligned} \tag{3.28}$$

Following the discussions from section 2.1 to section 2.3, we introduce an $(r+s)$th-order (tangent) tensor

$${}^r{}_s\mathbf{T}_{|x_0} = \tau^{p_1 \ldots p_r}{}_{q_1 \ldots q_s} \vec{\mathbf{e}}_{p_1|x_0} \otimes \cdots \otimes \vec{\mathbf{e}}_{p_r|x_0} \otimes \tilde{\mathbf{e}}^{q_1}{}_{|x_0} \otimes \cdots \otimes \tilde{\mathbf{e}}^{q_s}{}_{|x_0}.$$

The tangent tensor ${}^r{}_s\mathbf{T}_{|x_0}$ belongs to the N^{r+s}-dimensional tangent tensor space ${}^r{}_s\mathcal{T}(T_{x_0}(\mathbb{R}^N))$. Under a change of basis sets in $T_{x_0}(\mathbb{R}^N)$, the tensor components $\tau^{p_1 \ldots p_r}_{q_1 \ldots q_s}$ undergo the transformation rules as in (2.48) and (2.49).

Now we shall introduce the concept of a **tensor field** in $D \subset \mathbb{R}^N$. A (tangent) **tensor field** ${}^r{}_s\mathbf{T}$ is a function that assigns a tensor ${}^r{}_s\mathbf{T}(x)$ in the (tangent) tensor space ${}^r{}_s\mathcal{T}(T_x(\mathbb{R}^N)) := T_x(\mathbb{R}^N) \otimes \cdots \otimes T_x(\mathbb{R}^N) \otimes \tilde{T}_x(\mathbb{R}^N) \otimes \cdots \otimes \tilde{T}_x(\mathbb{R}^N)$ for each point x in $D \subset \mathbb{R}^N$. (A more rigorous definition of a **tensor bundle** over the manifold M is provided in appendix 1.)

All the discussions of vectors, covariant vectors, and higher-order tensors from section 1.2 to section 2.3 are valid for tangent (or contravariant) vector fields, cotangent (or covariant) vector fields, and higher-order tensor fields at each point x in the domain D of $\mathbb{R}^N$.

If we choose a basis vector field $\{\vec{\mathbf{e}}_1(x), \ldots, \vec{\mathbf{e}}_N(x)\}$ and the corresponding covariant basis field $\{\tilde{\mathbf{e}}^1(x), \ldots, \tilde{\mathbf{e}}^N(x)\}$, then by (3.28) we can express the transformation rules

$$\begin{aligned}
&\widehat{\vec{\mathbf{e}}}_p(x) = \lambda^q{}_p(x)\vec{\mathbf{e}}_q(x), \qquad \vec{\mathbf{e}}_p(x) = \mu^q{}_p(x)\widehat{\vec{\mathbf{e}}}_q(x),\\
&\widehat{\tilde{\mathbf{e}}}^p(x) = \mu^p{}_q(x)\tilde{\mathbf{e}}^q(x), \qquad \tilde{\mathbf{e}}^p(x) = \lambda^p{}_q(x)\widehat{\tilde{\mathbf{e}}}^q(x),\\
&\det\left[\lambda^p{}_q(x)\right] \neq 0,\\
&\left[\mu^q{}_p(x)\right] := \left[\lambda^p{}_q(x)\right]^{-1},\\
&x \in D \subset \mathbb{R}^N.
\end{aligned} \tag{3.29}$$

The transformation of the tensor field components from (3.29) and from (2.48) and (2.49) are furnished by

$$\begin{aligned}
{}^r{}_s\mathbf{T}(x) &= T^{p_1\ldots p_r}{}_{q_1\ldots q_s}(x)\vec{\mathbf{e}}_{p_1}(x)\otimes\cdots\otimes\vec{\mathbf{e}}_{p_r}(x)\otimes\tilde{\mathbf{e}}^{q_1}(x)\otimes\cdots\otimes\tilde{\mathbf{e}}^{q_s}(x)\\
&= \widehat{T}^{p_1\ldots p_r}{}_{q_1\ldots q_s}(x)\widehat{\vec{\mathbf{e}}}_{p_1}(x)\otimes\cdots\otimes\widehat{\vec{\mathbf{e}}}_{p_r}(x)\otimes\widehat{\tilde{\mathbf{e}}}^{q_1}(x)\otimes\cdots\otimes\widehat{\tilde{\mathbf{e}}}^{q_s}(x),\\
\widehat{T}^{p_1\ldots p_r}{}_{q_1\ldots q_s}(x) &= \lambda^{p_1}{}_{u_1}(x)\ldots\lambda^{p_r}{}_{u_r}(x)\mu^{v_1}{}_{q_1}(x)\ldots\mu^{v_s}{}_{q_s}(x)T^{u_1\ldots u_r}{}_{v_1\ldots v_s}(x),\\
T^{p_1\ldots p_r}{}_{q_1\ldots q_s}(x) &= \mu^{p_1}{}_{u_1}(x)\ldots\mu^{p_r}{}_{u_r}(x)\lambda^{v_1}{}_{q_1}(x)\ldots\lambda^{v_s}{}_{q_s}(x)\widehat{T}^{u_1\ldots u_r}{}_{v_1\ldots v_s}(x).
\end{aligned} \tag{3.30}$$

Example 3.3.1 Consider $N = 2$ and $D = \mathbb{R}^2$. Let a transformation of basis sets be given by

$$\begin{aligned}
\widehat{\vec{\mathbf{e}}}_1(x) &= \left[\cosh(x^1+x^2)\right]\vec{\mathbf{e}}_1(x),\\
\widehat{\vec{\mathbf{e}}}_2(x) &= \left[\tanh(x^1-x^2)\right]\vec{\mathbf{e}}_1(x) + \left[\operatorname{sech}(x^1+x^2)\right]\vec{\mathbf{e}}_2(x).
\end{aligned}$$

By (3.29) and (3.30), the components of a tensor field ${}^1{}_1\mathbf{T}(x) \equiv \mathbf{T}^{\cdot}{}_{\cdot}(x)$ transform as

$$\begin{aligned}
\widehat{T}^1{}_1(x) &= T^1{}_1(x) + \left[\operatorname{sech}(x^1+x^2)\right]\left[\tanh(x^1-x^2)\right]T^2{}_1(x),\\
\widehat{T}^1{}_2(x) &= \left[\cosh(x^1+x^2)\right]\left[\tanh(x^1-x^2)\right]\left[T^1{}_2(x) - T^1{}_1(x)\right]\\
&\quad + \left[\cosh(x^1+x^2)\right]^2 T^1{}_2(x) - \left[\tanh(x^1-x^2)\right]^2 T^2{}_1(x),\\
\widehat{T}^2{}_1(x) &= \left[\operatorname{sech}(x^1+x^2)\right]^2 T^2{}_1(x),\\
\widehat{T}^2{}_2(x) &= T^2{}_2(x) - \left[\operatorname{sech}(x^1+x^2)\right]\left[\tanh(x^1-x^2)\right]T^2{}_1(x).
\end{aligned}$$

In the case where the tensor field is *restricted* to the straight line $x^1 = x^2 = t$,

$t \in \mathbb{R}$, the transformations above reduce to

$$\begin{aligned}
\widehat{T}^1{}_1(t,t) &= T^1{}_1(t,t), \\
\widehat{T}^1{}_2(t,t) &= [\cosh(2t)]^2\, T^1{}_2(t,t), \\
\widehat{T}^2{}_1(t,t) &= [\mathrm{sech}(2t)]^2\, T^2{}_1(t,t), \\
\widehat{T}^2{}_2(t,t) &= T^2{}_2(t,t), \\
\widehat{T}^p{}_q(0,0) &= T^p{}_q(0,0). \qquad \square
\end{aligned}$$

By the discussions of tensor algebra in chapter 2, we can derive some useful equations for tensor field components. For example, by (2.37) and (3.30), we obtain

$$\begin{aligned}
{}^r{}_s\mathbf{T}(x)+{}^r{}_s\mathbf{W}(x) = {}&[T^{p_1\ldots p_r}{}_{q_1\ldots q_s}(x) + W^{p_1\ldots p_r}{}_{q_1\ldots q_s}(x)]\,\vec{\mathbf{e}}_{p_1}(x)\otimes\cdots\otimes \\
&\vec{\mathbf{e}}_{p_r}(x)\otimes\tilde{\mathbf{e}}^{q_1}(x)\otimes\cdots\otimes\tilde{\mathbf{e}}^{q_s}(x),
\end{aligned} \tag{3.31}$$

$$\begin{aligned}
\lambda(x)\,[{}^r{}_s\mathbf{T}(x)] = {}&[\lambda(x)T^{p_1\ldots p_r}{}_{q_1\ldots q_s}(x)]\,\vec{\mathbf{e}}_{p_1}(x)\otimes\cdots\otimes \\
&\vec{\mathbf{e}}_{p_r}(x)\otimes\tilde{\mathbf{e}}^{q_1}(x)\otimes\cdots\otimes\tilde{\mathbf{e}}^{q_s}(x).
\end{aligned} \tag{3.32}$$

By (2.39) and (3.30), we can derive that

$$\begin{aligned}
{}^r{}_s\mathbf{T}(x)\otimes{}^t{}_w\mathbf{B}(x) = {}&[T^{p_1\ldots p_r}{}_{q_1\ldots q_s}(x)B^{u_1\ldots u_t}{}_{v_1\ldots v_w}(x)]\,\vec{\mathbf{e}}_{p_1}(x)\otimes\cdots\otimes \\
&\vec{\mathbf{e}}_{p_r}(x)\otimes\vec{\mathbf{e}}_{u_1}(x)\otimes\cdots\otimes\vec{\mathbf{e}}_{u_t}(x)\otimes\tilde{\mathbf{e}}^{q_1}(x)\otimes\cdots\otimes \\
&\tilde{\mathbf{e}}^{q_s}(x)\otimes\tilde{\mathbf{e}}^{v_1}(x)\otimes\cdots\otimes\tilde{\mathbf{e}}^{v_w}(x).
\end{aligned} \tag{3.33}$$

Finally, by (2.51) and (3.30), we can establish that

$$\begin{aligned}
{}^u{}_v\mathcal{C}\left({}^r{}_s\mathbf{T}(x)\right) = {}&\left[T^{p_1\ldots p_{u-1}cp_{u+1}\ldots p_r}{}_{q_1\ldots q_{v-1}cq_{v+1}\ldots q_s}(x)\right]\vec{\mathbf{e}}_{p_1}(x)\otimes\cdots\otimes \\
&\vec{\mathbf{e}}_{p_{u-1}}(x)\otimes\vec{\mathbf{e}}_{p_{u+1}}(x)\otimes\cdots\otimes\vec{\mathbf{e}}_{p_r}(x)\otimes\tilde{\mathbf{e}}^{q_1}(x)\otimes\cdots\otimes \\
&\tilde{\mathbf{e}}^{q_{v-1}}(x)\otimes\tilde{\mathbf{e}}^{q_{v+1}}(x)\otimes\cdots\otimes\tilde{\mathbf{e}}^{q_s}(x).
\end{aligned} \tag{3.34}$$

Example 3.3.2 Let us choose $N = 2$ and $D = \mathbb{R}^2$. Let a basis field and its conjugate covariant field be expressed as

$$\begin{aligned}
&\vec{\mathbf{e}}_p(x) := E^i{}_p(x)\frac{\partial}{\partial x^i}, \quad p \in \{1,2\}; \qquad\qquad i \in \{1,2\}; \\
&\tilde{\mathbf{e}}^q(x) = \tilde{E}^q{}_j(x)dx^j, \quad \tilde{E}^q_i(x)E^i{}_p(x) = \delta^q{}_p.
\end{aligned}$$

Let a scalar field be defined by

$$\lambda(x) := \exp(x^1 + x^2).$$

Furthermore, let two tensor fields be furnished by

$$\mathbf{T}^{\cdot}_{\cdot}(x) := (\delta_{qu} x^p x^u)\vec{\mathbf{e}}_p(x) \otimes \tilde{\mathbf{e}}^q(x),$$
$$\mathbf{W}^{\cdot}_{\cdot}(x) := \sum_{u=1}^{2}\sum_{v=1}^{2} \left\{ \left[\delta_{vt}(x^u)^3 (x^t)^3\right] \vec{\mathbf{e}}_u(x) \otimes \tilde{\mathbf{e}}^v(x) \right\}.$$

Then, by (3.31), (3.32), (3.33), and (3.34), we get

$$\begin{aligned}
\lambda(x) T^p{}_q(x) + W^p{}_q(x) &= \sum_{t=1}^{2} \delta_{qt}\left[\exp(x^1+x^2)x^p x^t + (x^p)^3 (x^t)^3\right], \\
\left[\mathbf{T}^{\cdot}_{\cdot}(x) \otimes \mathbf{W}^{\cdot}_{\cdot}(x)\right]^{pu}_{qv} &= T^p{}_q(x) W^u{}_v(x) \\
&= \sum_{s=1}^{2}\sum_{t=1}^{2} \left[\delta_{qt}\delta_{vs} x^p x^t (x^u)^3 (x^s)^3\right], \\
{}^1{}_1\mathcal{C}\left(\mathbf{T}^{\cdot}_{\cdot}(x)\right) = T^p{}_p(x) &= \delta_{pq} x^p x^q \geq 0, \\
\vec{\mathbf{e}}_u(x)\left[{}^1{}_1\mathcal{C}\left(\mathbf{T}^{\cdot}_{\cdot}(x)\right)\right] &= \delta_{pq}\left[x^p E^q{}_u(x) + x^q E^p{}_u(x)\right].
\end{aligned}$$

□

We shall now generalize transformation rules (3.30) of tensor field components to the components of an $(r+s)$th-order **relative tensor field** ${}^r{}_s\Theta_w(x)$ of weight w. This transformation, as furnished by (2.103), is given by

$$\begin{aligned}
&\widehat{\Theta}^{p_1\ldots p_r}{}_{q_1\ldots q_s}(x) \\
&= \left\{\det\left[\lambda^t{}_n(x)\right]\right\}^w \mu^{p_1}{}_{u_1}(x)\cdots\mu^{p_r}{}_{u_r}(x)\lambda^{v_1}{}_{q_1}(x)\cdots\lambda^{v_s}{}_{q_s}(x)\Theta^{u_1\ldots u_r}{}_{v_1\ldots v_s}(x).
\end{aligned} \tag{3.35}$$

A relative tensor of weight $w = 1$ is called a **tensor density field**.

Example 3.3.3 The constant-valued numerical $(0+N)$th antisymmetric components $\varepsilon_{q_1\ldots q_N}$ transform (by (2.101)) as

$$\widehat{\varepsilon}_{q_1\ldots q_N} = \left\{\det[\lambda^t{}_n(x)]\right\}^{-1} \lambda^{v_1}{}_{q_1}(x)\ldots\lambda^{v_N}{}_{q_N}(x)\varepsilon_{v_1\ldots v_N}. \tag{3.36}$$

Thus, these are the components of a numerical relative tensor field of weight -1. □

Example 3.3.4 The constant-valued numerical $(N+0)$th-order relative antisymmetric tensor components $\varepsilon^{p_1\cdots p_N}$ transform (by (2.102)) as

$$\widehat{\varepsilon}^{p_1\ldots p_N} = \left\{\det[\lambda^t{}_n(x)]\right\} \mu^{p_1}{}_{u_1}(x)\cdots\mu^{p_N}{}_{u_N}(x)\varepsilon^{u_1\ldots u_N}. \tag{3.37}$$

Therefore, these are the components of a relative tensor of weight 1 or a tensor density field. □

Now, we shall proceed to make another generalization. By (2.108), the components of an **oriented** (or **pseudo**) **relative** $(r+s)$**th-order tensor field** of weight w transform as

$$\begin{aligned}\widehat{\theta}^{p_1 \dots p_N}(x) = &\{\operatorname{sgn}[\det[\lambda^t{}_n(x)]]\}\{\det[\lambda^t{}_n(x)]\}^w \\ &\mu^{p_1}{}_{u_1}(x)\cdots\mu^{p_r}{}_{u_r}(x)\lambda^{v_1}{}_{q_1}(x)\dots\lambda^{v_s}{}_{q_s}(x)\theta^{u_1\dots u_r}{}_{v_1\dots v_s}(x).\end{aligned} \tag{3.38}$$

Example 3.3.5 Suppose that $T_{ab}(x)$ are the components of a $(0+2)$th-order tensor field. Then, by (2.109),

$$\sqrt{\left|\det\left[\widehat{T}_{pq}(x)\right]\right|} = \{\operatorname{sgn}[\det[\lambda^t{}_n(x)]]\}\{\det[\lambda^t{}_n(x)]\}\sqrt{|\det[T_{uv}(x)]|}. \tag{3.39}$$

Therefore, $\sqrt{|\det[T_{pq}(x)]|}$ transforms as an oriented (or pseudo) scalar density field. □

Now, we shall discuss the transformation of tensor components under a coordinate transformation introduced in (3.2) and illustrated in fig. 3.2. To reiterate briefly,

$$\begin{aligned}\widehat{x}^k &= \left[\pi^k\circ\widehat{X}\right](x) := \left[\pi^k\circ\widehat{\chi}\circ\chi^{-1}\right](x) \equiv \widehat{X}^k(x) \equiv \widehat{X}^k(x^1,\dots,x^N),\\ x^k &= \left[\pi^k\circ X\right](\widehat{x}) := \left[\pi^k\circ\chi\circ\widehat{\chi}^{-1}\right](\widehat{x}) \equiv X^k(\widehat{x}) \equiv X^k(\widehat{x}^1,\dots,\widehat{x}^N),\\ x &\in D_s\subset\mathbb{R}^N,\quad \widehat{x}\in\widehat{D}_s\subset\mathbb{R}^N.\end{aligned} \tag{3.40}$$

The functions $\widehat{X}^k$ and X^k are assumed to be one-to-one and of class $C^r(D_s\subset\mathbb{R}^N;\mathbb{R})$ and $C^r(\widehat{D}_s\subset\mathbb{R}^N;\mathbb{R})$, respectively. (For general relativity theory, $r=3$.) See fig. 3.9 for the simple case of $N=2$.
The coordinate transformation $\widehat{X}$ *induces* a one-to-one mapping $\widehat{\vec{\mathbf{X}}}'$ from $T_x(\mathbb{R}^N)$ into $T_{\widehat{x}}(\mathbb{R}^N)\equiv T_{\widehat{X}(x)}(\mathbb{R}^N)$. (See fig. 3.9.) This mapping is called the **Jacobian mapping** (or **derivative mapping**) and is furnished by

$$\begin{aligned}&\left[\widehat{\vec{\mathbf{X}}}'(\vec{\mathbf{t}}(x))\right][f(\widehat{x})] := \vec{\mathbf{t}}(x)\left[f\circ\widehat{X}(x)\right],\\ &\vec{\mathbf{X}}' = \left(\widehat{\vec{\mathbf{X}}}'\right)^{-1},\\ &\vec{\mathbf{t}}(x)\in T_x(\mathbb{R}^N),\quad \widehat{\vec{\mathbf{X}}}'\left(\vec{\mathbf{t}}(x)\right)\in T_{\widehat{x}}(\mathbb{R}^N),\\ &f\in C^1\left(\widehat{D}_s\subset\mathbb{R}^N;\mathbb{R}\right).\end{aligned} \tag{3.41}$$

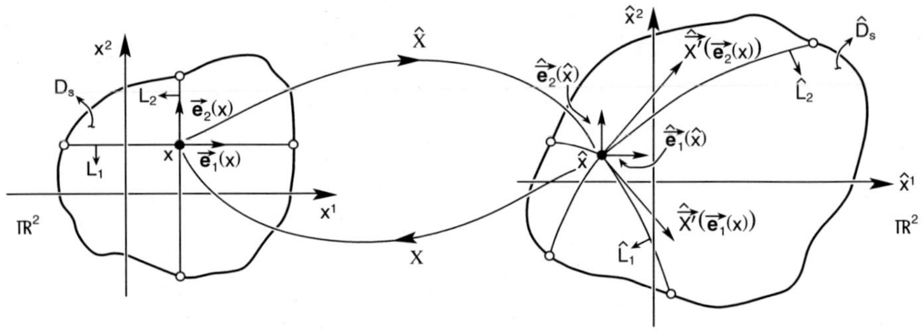

Figure 3.9: The Jacobian mapping of tangent vectors.

We can verify by (3.41) and (3.11) that

$$\begin{aligned}
&\left[\widehat{\vec{\mathbf{X}}}'\left(\lambda(x)\vec{\mathbf{a}}(x)+\mu(x)\vec{\mathbf{b}}(x)\right)\right][f(\widehat{x})] \\
&=[\lambda(x)\vec{\mathbf{a}}(x)+\mu(x)\vec{\mathbf{b}}(x)][f\circ\widehat{X}(x)] \\
&=\lambda(x)\left\{\vec{\mathbf{a}}(x)[f\circ\widehat{X}(x)]\right\}+\mu(x)\left\{\vec{\mathbf{b}}(x)[f\circ\widehat{X}(x)]\right\} \\
&=\lambda(x)\left\{\widehat{\vec{\mathbf{X}}}'(\vec{\mathbf{a}}(x))[f(\widehat{x})]\right\}+\mu(x)\left\{\widehat{\vec{\mathbf{X}}}'(\vec{\mathbf{b}}(x))[f(\widehat{x})]\right\}.
\end{aligned}$$

Thus, the Jacobian mapping $\widehat{\vec{\mathbf{X}}}'$ is *linear*. Moreover, we can prove that it is a vector space isomorphism.

Example 3.3.6 Let us choose $\vec{\mathbf{t}}(x)=\frac{\partial}{\partial x^i}$ in $T_x(\mathbb{R}^N)$. Also, we choose the projection mapping $\pi^j=f$. Then, by (3.41) and (3.40), we obtain

$$\left[\widehat{\vec{\mathbf{X}}}'\frac{\partial}{\partial x^i}\right][\pi^j(\widehat{x})]=\frac{\partial}{\partial x^i}\left[\pi^j\circ\widehat{X}(x)\right]=\frac{\partial\widehat{X}^j(x)}{\partial x^i}. \tag{3.42}$$

The $N\times N$ non-singular matrix $\left[\frac{\partial\widehat{X}^k(x)}{\partial x^j}\right]$ is called the **Jacobian matrix** (whose determinant is the usual Jacobian). We can express (3.42) as

$$\left[\widehat{\vec{\mathbf{X}}}'\left(\frac{\partial}{\partial x^i}\right)\right](\pi^j(\widehat{x}))=\left[\frac{\partial\widehat{X}^k(x)}{\partial x^i}\frac{\partial}{\partial\widehat{x}^k}\right](\pi^j(\widehat{x})).$$

Therefore, we have the transformation of basis vectors $\frac{\partial}{\partial\widehat{x}^k}$ into other basis vectors $\widehat{\vec{\mathbf{X}}}'\left(\frac{\partial}{\partial x^i}\right)$ by the rules

$$\widehat{\vec{\mathbf{X}}}'\left(\frac{\partial}{\partial x^i}\right)=\frac{\partial\widehat{X}^k(x)}{\partial x^i}\frac{\partial}{\partial\widehat{x}^k}. \tag{3.43}$$

Both of the vectors above belong to $T_{\widehat{X}(x)}(\mathbb{R}^N)$. □

The coordinate transformation (3.40) induces an isomorphism between the cotangent spaces $\tilde{T}_x(\mathbb{R}^N)$ and $\tilde{T}_{\widehat{X}(x)}(\mathbb{R}^N)$. It is explicitly furnished by

$$\left[\widehat{\tilde{\mathbf{X}}}'(\tilde{\boldsymbol{\omega}}(x))\right]\left[\widehat{\vec{\mathbf{V}}}(\widehat{x})\right] := \tilde{\boldsymbol{\omega}}(x)\left[\vec{\mathbf{X}}'\left(\widehat{\vec{\mathbf{V}}}(\widehat{x})\right)\right]. \tag{3.44}$$

Example 3.3.7 Consider the mappings of the basis 1-forms dx^i's. By (3.44) and (3.43), we have

$$\begin{aligned}
\left[\widehat{\tilde{\mathbf{X}}}'(dx^i)\right]\left[\frac{\partial}{\partial\widehat{x}^j}\right] &= dx^i\left[\vec{\mathbf{X}}'\left(\frac{\partial}{\partial\widehat{x}^j}\right)\right] \\
&= dx^i\left[\frac{\partial X^k(\widehat{x})}{\partial\widehat{x}^j}\frac{\partial}{\partial x^k}\right] = \frac{\partial X^k(\widehat{x})}{\partial\widehat{x}^j}\left[\frac{\partial x^i}{\partial x^k}\right] \\
&= \frac{\partial X^i(\widehat{x})}{\partial\widehat{x}^j}, \\
\left[\frac{\partial X^i(\widehat{x})}{\partial\widehat{x}^k}d\widehat{x}^k\right]\left[\frac{\partial}{\partial\widehat{x}^j}\right] &= \frac{\partial X^i(\widehat{x})}{\partial\widehat{x}^j}.
\end{aligned} \tag{3.45}$$

Thus, we conclude that

$$\widehat{\tilde{\mathbf{X}}}'(dx^i) = \frac{\partial X^i(\widehat{x})}{\partial\widehat{x}^k}d\widehat{x}^k. \tag{3.46}$$

Both of the covariant vectors above belong to $\tilde{T}_{\widehat{x}}(\mathbb{R}^N)$. □

The coordinate transformation (3.28) induces an isomorphism between the tensor spaces ${}^r{}_s\mathcal{T}_x(\mathbb{R}^N)$ and ${}^r{}_s\mathcal{T}_{\widehat{X}(x)}(\mathbb{R}^N)$. It is explicitly defined by

$$\begin{aligned}
&{}^r{}_s\widehat{\mathbf{T}}(\widehat{x}) := {}^r{}_s\widehat{\mathbf{X}}'\left[{}^r{}_s\mathbf{T}(x)\right], \\
&\left[{}^r{}_s\mathbf{T}(x)\right](\tilde{\mathbf{w}}_1(x), \ldots, \tilde{\mathbf{w}}_r(x), \vec{\mathbf{v}}_1(x), \ldots, \vec{\mathbf{v}}_s(x)) \\
&\quad =: \left[{}^r{}_s\widehat{\mathbf{X}}'({}^r{}_s\mathbf{T}(x))\right] \\
&\qquad \left(\widehat{\tilde{\mathbf{X}}}'(\tilde{\mathbf{w}}_1(x)), \ldots, \widehat{\tilde{\mathbf{X}}}'(\tilde{\mathbf{w}}_r(x)), \widehat{\vec{\mathbf{X}}}'(\vec{\mathbf{v}}_1(x)), \ldots, \widehat{\vec{\mathbf{X}}}'(\vec{\mathbf{v}}_s(x))\right),
\end{aligned} \tag{3.47}$$

for all $\vec{\mathbf{v}}_1(x), \ldots, \vec{\mathbf{v}}_s(x) \in T_x(\mathbb{R}^N)$ and all $\tilde{\mathbf{w}}_1(x), \ldots, \tilde{\mathbf{w}}_r(x) \in \tilde{T}_x(\mathbb{R}^N)$. We shall derive the transformation rules of tensor field components by (3.47). (*The tensor field components relative to a coordinate basis will be denoted by the indices* i, j, k, l, m, n, etc.)

Theorem 3.3.8 *The transformation rules of the $(r + s)$th tensor field $T^{i_1 \ldots i_r}{}_{j_1 \ldots j_s}(x)$ under a differentiable general coordinate transformation (3.40) are furnished by*

$$
\begin{aligned}
&\widehat{T}^{k_1 \ldots k_r}{}_{l_1 \ldots l_s}(\widehat{x}) \\
&= \frac{\partial \widehat{X}^{k_1}(x)}{\partial x^{i_1}} \cdots \frac{\partial \widehat{X}^{k_r}(x)}{\partial x^{i_r}} \frac{\partial X^{j_1}(\widehat{x})}{\partial \widehat{x}^{l_1}} \cdots \frac{\partial X^{j_s}(\widehat{x})}{\partial \widehat{x}^{l_s}} T^{i_1 \ldots i_r}{}_{j_1 \ldots j_s}(x),
\end{aligned}
\tag{3.48}
$$

where $x \in D_s \subset D \subset \mathbb{R}^N$ and $\widehat{x} \in \widehat{D}_s \subset \widehat{D} \subset \mathbb{R}^N$.

Proof. By the definition (3.47) and the equations (3.43) and (3.46), we obtain

$$
\begin{aligned}
&[{}^r{}_s\mathbf{T}(x)] \left(dx^{i_1}, \ldots, dx^{i_r}, \frac{\partial}{\partial x^{j_1}}, \ldots, \frac{\partial}{\partial x^{j_s}} \right) \\
&\qquad = \left[{}^r{}_s\widehat{\mathbf{T}}(\widehat{x})\right] \left(\frac{\partial X^{i_1}(\widehat{x})}{\partial \widehat{x}^{k_1}} d\widehat{x}^{k_1}, \ldots, \ldots, \frac{\partial \widehat{X}^{l_s}(x)}{\partial x^{j_s}} \frac{\partial}{\partial \widehat{x}^{l_s}} \right) \\
&\qquad = \frac{\partial X^{i_1}(\widehat{x})}{\partial \widehat{x}^{k_1}} \cdots \frac{\partial \widehat{X}^{l_s}(x)}{\partial x^{j_s}} \left\{ \left[{}^r{}_s\widehat{\mathbf{T}}(\widehat{x})\right] \left(d\widehat{x}^{k_1}, \ldots, \ldots, \frac{\partial}{\partial \widehat{x}^{l_s}} \right) \right\},
\end{aligned}
$$

$$
T^{i_1 \ldots i_r}{}_{j_1 \ldots j_s}(x) = \frac{\partial X^{i_1}(\widehat{x})}{\partial \widehat{x}^{k_1}} \cdots \frac{\partial \widehat{X}^{l_s}(x)}{\partial x^{j_s}} \widehat{T}^{k_1 \ldots k_r}{}_{l_1 \ldots l_s}(\widehat{x}). \tag{3.49}
$$

By inverting (3.49), we obtain the transformation rules (3.48). ■

The transformation rules for a relative tensor of weight w under (3.40) are prompted by (3.35) as

$$
\begin{aligned}
\widehat{\Theta}^{i_1 \ldots i_r}{}_{j_1 \ldots j_s}(\widehat{x}) = &\left\{ \det \left[\frac{\partial X^m(\widehat{x})}{\partial \widehat{x}^n} \right] \right\}^w \frac{\partial \widehat{X}^{i_1}(x)}{\partial x^{k_1}} \cdots \frac{\partial \widehat{X}^{i_r}(x)}{\partial x^{k_r}} \frac{\partial X^{l_1}(\widehat{x})}{\partial \widehat{x}^{j_1}} \cdots \\
&\frac{\partial X^{l_s}(\widehat{x})}{\partial \widehat{x}^{j_s}} \Theta^{k_1 \ldots k_r}{}_{l_1 \ldots l_s}(x).
\end{aligned}
\tag{3.50}
$$

The transformation rules for an oriented (or pseudo) relative tensor of weight w under (3.40) are taken to be (see (3.38))

$$
\begin{aligned}
\widehat{\Theta}^{i_1 \ldots i_r}{}_{j_1 \ldots j_s}(\widehat{x}) = &\left\{ \operatorname{sgn} \left[\det \left[\frac{\partial X^m(\widehat{x})}{\partial \widehat{x}^n} \right] \right] \right\} \left\{ \det \left[\frac{\partial X^m(\widehat{x})}{\partial \widehat{x}^n} \right] \right\}^w \frac{\partial \widehat{X}^{i_1}(x)}{\partial x^{k_1}} \cdots \\
&\frac{\partial \widehat{X}^{i_r}(x)}{\partial x^{k_r}} \frac{\partial X^{l_1}(\widehat{x})}{\partial \widehat{x}^{j_1}} \cdots \frac{\partial X^{l_s}(\widehat{x})}{\partial \widehat{x}^{j_s}} \Theta^{k_1 \ldots k_r}{}_{l_1 \ldots l_s}(x).
\end{aligned}
\tag{3.51}
$$

Example 3.3.9 Consider the simple case of $N = 2$ and $D = D_s = \mathbb{R}^2$. Let a (real) analytic coordinate transformation be provided by

$$\begin{aligned}
\widehat{x}^1 &= \widehat{X}^1(x) := x^1 + x^2, \\
\widehat{x}^2 &= \widehat{X}^2(x) := x^1 - x^2, \\
\frac{\partial(\widehat{x}^1, \widehat{x}^2)}{\partial(x^1, x^2)} &\equiv \det\left[\frac{\partial \widehat{X}^i(x)}{\partial x^j}\right] \equiv -2.
\end{aligned}$$

Let a special symmetric $(2+0)$th-order tensor field be furnished by

$$\begin{aligned}
&\mathbf{T}^{\cdot\cdot}(x) := (x^i x^j)\frac{\partial}{\partial x^i} \otimes \frac{\partial}{\partial x_j}, \\
&T^{ij}(x) = x^i x^j.
\end{aligned}$$

The transformed components $\widehat{T}^{kl}(\widehat{x})$ are given by (3.48) as

$$\begin{aligned}
\widehat{T}^{11}(\widehat{x}) &= \left[\frac{\partial \widehat{X}^1(x)}{\partial x^1}\right]^2 T^{11}(x) + 2\left[\frac{\partial \widehat{X}^1(x)}{\partial x^1} \cdot \frac{\partial \widehat{X}^1(x)}{\partial x^2}\right] T^{12}(x) \\
&\quad + \left[\frac{\partial \widehat{X}^1(x)}{\partial x^2}\right]^2 T^{22}(x) = (x^1 + x^2)^2 = (\widehat{x}^1)^2.
\end{aligned}$$

Similarly, we can deduce that

$$\widehat{T}^{kl}(\widehat{x}) = \widehat{x}^k \widehat{x}^l.$$

The values of $\widehat{T}^{kl}$ at the special point $(\widehat{x}^1, \widehat{x}^2) = (1/\sqrt{2}, -1/\sqrt{2})$ are given by

$$\widehat{T}^{11}(1/\sqrt{2}, -1/\sqrt{2}) = \widehat{T}^{22}(1/\sqrt{2}, -1/\sqrt{2}) = -\widehat{T}^{12}(1/\sqrt{2}, -1/\sqrt{2}) = 1/2. \qquad \square$$

Example 3.3.10 The constant-valued, numerical, totally antisymmetric relative tensors transform by (3.36) and (3.37) as

$$\begin{aligned}
\widehat{\varepsilon}_{j_1 \ldots j_N} &= \left\{\det\left[\frac{\partial X^m(\widehat{x})}{\partial \widehat{x}^n}\right]\right\}^{-1} \frac{\partial X^{i_1}(\widehat{x})}{\partial \widehat{x}^{j_1}} \cdots \frac{\partial X^{i_N}(\widehat{x})}{\partial \widehat{x}^{j_N}} \varepsilon_{i_1 \ldots i_N}, \\
\widehat{\varepsilon}^{i_1 \ldots i_N} &= \left\{\det\left[\frac{\partial X^m(\widehat{x})}{\partial \widehat{x}^n}\right]\right\} \frac{\partial \widehat{X}^{i_1}(x)}{\partial x^{j_1}} \cdots \frac{\partial \widehat{X}^{i_N}(x)}{\partial x^{j_N}} \varepsilon^{j_1 \ldots j_N}.
\end{aligned} \tag{3.52}$$

$\square$

Example 3.3.11 Suppose that $T_{ij}(x)$ are components of a differentiable tensor field $\mathbf{T}_{\cdot\cdot}(x)$. Then, by (3.39), we have the transformation rule

$$\begin{aligned}
&\sqrt{\left|\det\left[\widehat{T}_{ij}(\widehat{x})\right]\right|} \\
&= \left\{\operatorname{sgn}\det\left[\frac{\partial X^m}{\partial \widehat{x}^n}(\widehat{x})\right]\right\}\left\{\det\left[\frac{\partial X^m}{\partial \widehat{x}^n}(\widehat{x})\right]\right\}\sqrt{|\det\left[T_{kl}(x)\right]|}.
\end{aligned} \tag{3.53}$$

Therefore, $\sqrt{|\det[T_{kl}(x)]|}$ transforms as an oriented or pseudo-scalar field of weight 1. It is also called a pseudo-scalar density field. Let $\det[T_{kl}(x)] \neq 0$. The inverse matrix elements $\Theta^{ij}(x)$ satisfy

$$\Theta^{ik}(x)T_{kl}(x) = T_{lk}(x)\Theta^{ki}(x) = \delta^{i}{}_{l}.$$

Differentiating the equation above, we obtain that

$$\begin{aligned}\left[\frac{\partial}{\partial x^j}\Theta^{ik}(x)\right] T_{kl}(x) &= -\,\Theta^{ik}(x)\left[\frac{\partial}{\partial x^j}T_{kl}(x)\right],\\ \frac{\partial}{\partial x^j}T_{ml}(x) &= -\,T_{mi}(x)T_{kl}(x)\left[\frac{\partial\Theta^{ik}(x)}{\partial x^j}\right].\end{aligned}$$

The partial derivatives $\frac{\partial T_{ml}(x)}{\partial x^j}$ do *not* obey tensorial transformations. □

We shall now write the rules (3.31), (3.32), (3.33), and (3.34) for the coordinate components of a tensor field. These rules are furnished by the following equations:

$$\begin{aligned}{}^{r}{}_{s}\mathbf{T}(x) + {}^{r}{}_{s}\mathbf{W}(x) &= \left[T^{i_1\ldots i_r}{}_{j_1\ldots j_s}(x) + W^{i_1\ldots i_r}{}_{j_1\ldots j_s}(x)\right]\frac{\partial}{\partial x^{i_1}} \otimes \cdots\\ &\quad \otimes \frac{\partial}{\partial x^{i_r}} \otimes dx^{j_1} \otimes \cdots \otimes dx^{j_s},\end{aligned} \tag{3.54}$$

$$\begin{aligned}\lambda(x)\left[{}^{r}{}_{s}\mathbf{T}(x)\right] &= \left[\lambda(x)T^{i_1\ldots i_r}{}_{j_1\ldots j_s}(x)\right]\frac{\partial}{\partial x^{i_1}} \otimes \cdots\\ &\quad \otimes \frac{\partial}{\partial x^{i_r}} \otimes dx^{j_1} \otimes \cdots \otimes dx^{j_s},\end{aligned} \tag{3.55}$$

$$\begin{aligned}{}^{r}{}_{s}\mathbf{T}(x) \otimes {}^{t}{}_{w}\mathbf{B}(x) &= \left[T^{i_1\ldots i_r}{}_{j_1\ldots j_s}(x)\cdot B^{k_1\ldots k_t}{}_{l_1\ldots l_w}(x)\right]\frac{\partial}{\partial x^{i_1}} \otimes \cdots\\ &\quad \otimes \frac{\partial}{\partial x^{i_r}} \otimes \frac{\partial}{\partial x^{k_1}} \otimes \cdots \otimes \frac{\partial}{\partial x^{k_t}} \otimes dx^{j_1} \otimes \cdots\\ &\quad \otimes dx^{j_s} \otimes dx^{l_1} \otimes \cdots \otimes dx^{l_w},\end{aligned} \tag{3.56}$$

$$\begin{aligned}{}^{h}{}_{k}C\left[{}^{r}{}_{s}\mathbf{T}(x)\right] = {}& T^{i_1\ldots i_{h-1}c\,i_{h+1}\ldots i_r}{}_{j_1\ldots j_{k-1}c\,j_{k+1}\ldots j_s}(x)\frac{\partial}{\partial x^{i_1}} \otimes \cdots \otimes \frac{\partial}{\partial x^{i_{h-1}}}\\ &\otimes \frac{\partial}{\partial x^{i_{h+1}}} \otimes \cdots \otimes \frac{\partial}{\partial x^{i_r}} \otimes dx^{j_1} \otimes \cdots \otimes dx^{j_{k-1}}\\ &\otimes dx^{j_{k+1}} \otimes \cdots \otimes dx^{j_s}.\end{aligned} \tag{3.57}$$

Example 3.3.12 Let us choose $N = 3$ and choose D to be a proper subset of $\mathbb{R}^3$. A tangent vector field and a cotangent vector field are furnished respectively by

$$\begin{aligned}\vec{\mathbf{V}}(x) &= V^i(x)\frac{\partial}{\partial x^i},\\ \tilde{\mathbf{F}}(x) &= F_j(x)dx^j.\end{aligned}$$

Therefore, by (3.56) and (3.57), we obtain

$$\vec{\mathbf{V}}(x) \otimes \tilde{\mathbf{F}}(x) = \left[V^i(x)F_j(x)\right]\frac{\partial}{\partial x^i} \otimes dx^j,$$
$${}^1{}_1\mathcal{C}\left[\vec{\mathbf{V}}(x) \otimes \tilde{\mathbf{F}}(x)\right] = V^i(x)F_i(x).$$

Let us restrict the tangent and cotangent fields on a twice-differentiable curve $\mathcal{X}$ into D. Then, we obtain

$${}^1{}_1\mathcal{C}\left[\vec{\mathbf{V}}(\mathcal{X}(t)) \otimes \tilde{\mathbf{F}}(\mathcal{X}(t))\right] = V^j(\mathcal{X}(t))\, F_j(\mathcal{X}(t)).$$

If we now choose $\vec{\mathbf{V}}(\mathcal{X}(t)) = \vec{\mathcal{X}}'(t)$, the tangent vector along the image of the curve, then we get

$${}^1{}_1\mathcal{C}\left[\vec{\mathcal{X}}'(t) \otimes \tilde{\mathbf{F}}(\tilde{\mathcal{X}}(t))\right] = F_j(\mathcal{X}(t))\,\frac{d\mathcal{X}^j(t)}{dt}.$$

If we regard the image of the curve as a particle trajectory and $\tilde{\mathbf{F}}(x)$ as a covariant vector field of an external force, then ${}^1{}_1\mathcal{C}[\vec{\mathcal{X}}'(t) \otimes \tilde{\mathbf{F}}(\mathcal{X}(t))]$ is the temporal rate of the work performed by the external force (according to Newtonian physics). □

Exercises 3.3

1. Let $T^{i_1 \ldots i_r}{}_{j_1 \ldots j_s}(x)$ be the differentiable components of an $(r+s)$th-order tensor field in D of $\mathbb{R}^N$. Prove that the partial derivatives $\frac{\partial}{\partial x^k}T^{i_1 \ldots i_r}{}_{j_1 \ldots j_s}(x)$ do *not* transform as tensor components under a twice-differentiable coordinate transformation.

2. Let $\tilde{\mathbf{A}}(x) = A_i(x)dx^i$ be a 1-form field such that partial derivatives $\frac{\partial A_i(x)}{\partial x^j}$ exist and are continuous in $D_s \subset D \subset \mathbb{R}^N$. Moreover, let $\widehat{x}^k = \widehat{X}^k(x)$ be a general coordinate transformation such that each of $\widehat{X}^k \in C^2(D_s \subset \mathbb{R}^N; \mathbb{R})$. Prove that under such a transformation, $\left[\frac{\partial A_i(x)}{\partial x^j} - \frac{\partial A_j(x)}{\partial x^i}\right] dx^i \otimes dx^j$ is a $(0+2)$th-order tensor field in D_s of $\mathbb{R}^N$.

3. Let $T_{jk}(x)$ be the components of a $(0+2)$th-order tensor field in D of $\mathbb{R}^N$. Show that the cofactors $C^{jk}(x)$ of the entries $T_{jk}(x)$ in the $\det[T_{jk}(x)]$

transform as components of a relative, second-order contravariant tensor field of weight 2.

4. Consider the numerical identity tensor field in $\mathbb{R}^N$

$$\mathbf{I}^{\cdot}{}_{\cdot}(x) := \delta^i_j \frac{\partial}{\partial x^i} \otimes dx^j.$$

Prove that

(i) ${}^1{}_2\mathcal{C}\,[\mathbf{I}^{\cdot}{}_{\cdot}(x) \otimes \mathbf{I}^{\cdot}{}_{\cdot}(x)] = \mathbf{I}^{\cdot}_{\cdot}(x)$.

(ii) $\{{}^1{}_1\mathcal{C}\,[\mathbf{I}^{\cdot}{}_{\cdot}(x) \otimes \mathbf{I}^{\cdot}{}_{\cdot}(x)]\}\,(\tilde{\mathbf{w}}(x), \vec{\mathbf{v}}(x)) = N\left[w_j(x)v^j(x)\right]$ for every pair of global fields $\tilde{\mathbf{w}}(x)$ and $\vec{\mathbf{v}}(x)$.

3.4 Differential Forms and Exterior Derivatives

A 0-form is defined to be a scalar (or $(0+0)$th-order tensor) field $f(x) \equiv f(x^1, \ldots, x^N)$. A 1-form is defined to be a $(0+1)$th-order tensor (or covariant vector) field $\tilde{\mathbf{T}}(x) = T_j(x)dx^j$, as in (3.16), (3.19), and (3.20).

Now, let us go back to section 2.3, which deals with the totally antisymmetric tensors, alternating operations, wedge products, and Grassman algebra.

(See the references [13] and [22].)

Let ${}_p\mathbf{W}(x)$ be a totally antisymmetric $(0+p)$th-order tensor field in $D \subset \mathbb{R}^N$. (See (2.53).) Therefore, ${}_p\mathbf{W}(x)$ belongs to the space $\Lambda^p(\tilde{T}_x(\mathbb{R}^N))$. Let another antisymmetric field ${}_q\mathbf{U}(x)$ belong to $\Lambda^q(\tilde{T}_x(\mathbb{R}^N))$. The wedge product between ${}_p\mathbf{W}(x)$ and ${}_q\mathbf{U}(x)$ (according to (2.63)) is

$$\begin{aligned}&[{}_p\mathbf{W}(x) \wedge {}_q\mathbf{U}(x)]\,(\vec{\mathbf{a}}_1(x), \ldots, \vec{\mathbf{a}}_p(x), \vec{\mathbf{a}}_{p+1}(x), \ldots, \vec{\mathbf{a}}_{p+q}(x))\\ &:= \left[\frac{1}{p!q!}\right] \sum_{\sigma \in S_{p+q}} [\mathrm{sgn}(\sigma)]\,[{}_p\mathbf{W}(x) \otimes {}_q\mathbf{U}(x)]\,\left(\mathbf{a}_{\sigma(1)}(x), \ldots, \mathbf{a}_{\sigma(p+q)}(x)\right)\end{aligned} \tag{3.58}$$

for all vector fields $\vec{\mathbf{a}}_1(x), \ldots, \vec{\mathbf{a}}_{p+q}(x)$ in $T_x(\mathbb{R}^N)$.

We shall denote the indices of the components of a p-form relative to the basis $\{dx^i\}_1^N$ by i, j, k, l, m, n, etc. According to (2.84), we can express ${}_p\mathbf{W}(x)$ as

$$\begin{aligned}{}_p\mathbf{W}(x) &= W_{i_1 \ldots i_p}(x)dx^{i_1} \otimes \cdots \otimes dx^{i_p}\\ &= (1/p!)\, W_{i_1 \ldots i_p}(x)dx^{i_1} \wedge \cdots \wedge dx^{i_p}\\ &= \sum_{i_1 < \cdots < i_p}^{N} W_{i_1 < \cdots < i_p}(x)dx^{i_1} \wedge \cdots \wedge dx^{i_p}\\ &= W_{i_1 < \cdots < i_p}(x)dx^{i_1} \wedge \cdots \wedge dx^{i_p}.\end{aligned} \tag{3.59}$$

The totally antisymmetric tensor field ${}_p\mathbf{W}(x)$ is called a **pth-order differential form**, or in short a **p-form**. The wedge product (3.58) in terms of the components is expressible as

$$\begin{aligned}&[{}_p\mathbf{W}(x) \wedge {}_q\mathbf{U}(x)]_{i_1\ldots i_p i_{p+1}\ldots i_{p+q}}\, dx^{i_1} \wedge \cdots \wedge dx^{i_p} \wedge dx^{i_{p+1}} \wedge \cdots \wedge dx^{i_{p+q}} \\ &= W_{i_1\ldots i_p}(x) U_{i_{p+1}\ldots i_{p+q}}(x) dx^{i_1} \wedge \cdots \wedge dx^{i_p} \wedge dx^{i_{p+1}} \wedge \cdots \wedge dx^{i_{p+q}}.\end{aligned} \tag{3.60}$$

Analogous to theorem 2.3.7, the wedge product of differential forms is *distributive* and *associative.* Moreover, it has a strange rule for interchange that is worth repeating. It is given by

$${}_q\mathbf{U}(x) \wedge {}_p\mathbf{W}(x) = (-1)^{pq} \left[{}_p\mathbf{W}(x) \wedge {}_q\mathbf{U}(x)\right]. \tag{3.61}$$

Example 3.4.1 Choosing $p = q = 1$, we obtain from (3.61)

$$dx^j \wedge dx^i = -(dx^i \wedge dx^j), \tag{3.62}$$

$$dx^1 \wedge dx^1 = dx^2 \wedge dx^2 = \cdots = dx^N \wedge dx^N = \mathbf{O}_{..}(x). \tag{3.63}$$

(The components of $\mathbf{O}_{..}(x)$ are all identically zero.) □

Example 3.4.2 Let us choose $N = 3$ and $p = 1$. Consider two 1-forms $\tilde{\mathbf{A}}(x)$ and $\tilde{\mathbf{B}}(x)$. Therefore,

$$\begin{aligned}\tilde{\mathbf{A}}(x) \wedge \tilde{\mathbf{B}}(x) &= [A_i(x)B_j(x)]\, dx^i \wedge dx^j \\ &= [A_1(x)B_2(x) - A_2(x)B_1(x)]\, dx^1 \wedge dx^2 \\ &\quad + [A_2(x)B_3(x) - A_3(x)B_2(x)]\, dx^2 \wedge dx^3 \\ &\quad + [A_3(x)B_1(x) - A_1(x)B_3(x)]\, dx^3 \wedge dx^1.\end{aligned}$$

This expression is analogous to the *cross product* between vector fields in the vector calculus. □

Example 3.4.3 Let us choose $N = 2n$, a positive even integer. Consider a 2 form

$${}_2\phi(q;p) \equiv {}_2\phi(q^1, \ldots, q^n; p_1, \ldots, p_n) := \sum_{j=1}^{n} (dp_j \wedge dq^j).$$

Now, by (3.62) and (3.63), we have

$$\begin{aligned}&\left(dp_1 \wedge dq^1\right) \wedge \left(dp_1 \wedge dq^1\right) = -dp_1 \wedge dp_1 \wedge dq^1 \wedge dq^1 = -\mathbf{O}_{....}(q,p), \\ &\left(dp_1 \wedge dq^1\right) \wedge \left(dp_2 \wedge dq^2\right) = (-1)^{(2\cdot 2)} \left(dp_2 \wedge dq^2\right) \wedge \left(dp_1 \wedge dq^1\right).\end{aligned}$$

Thus, two 2 forms $(dp_i \wedge dq^i)$ and $(dp_j \wedge dq^j)$ (summation *suspended*) commute with each other with respect to the wedge product.

Thus, we can derive that

$$\underbrace{{}_2\phi(q;p) \wedge \cdots \wedge {}_2\phi(q;p)}_{n} = (n!) dp_1 \wedge dq^1 \wedge dp_2 \wedge dq^2 \wedge \cdots \wedge dp_n \wedge dq^n.$$

This example is relevant in statistical mechanics. □

Now we shall introduce the concept of an **exterior derivative**. In the case where f is a (totally) differentiable function from $D \subset \mathbb{R}^N$ into $\mathbb{R}$, the exterior derivative is defined by the rule

$$\begin{aligned} df(x) &:= \frac{\partial f(x)}{\partial x^j} dx^j, \\ df(x)[\vec{\mathbf{V}}(x)] &= V^j(x) \frac{\partial f(x)}{\partial x^j}. \end{aligned} \tag{3.64}$$

Therefore, $df(x)$ is a 1-form field. In the case where $\tilde{\mathbf{T}}(x) = T_j(x) dx^j$ is a differentiable 1-form, its exterior derivative is defined as

$$\begin{aligned} d\tilde{\mathbf{T}}(x) &:= [dT_i(x)] \wedge dx^i = \left[\frac{\partial T_i(x)}{\partial x^j}\right] dx^j \wedge dx^i \\ &= (1/2) \left[\frac{\partial T_i(x)}{\partial x^j} - \frac{\partial T_j(x)}{\partial x^i}\right] dx^j \wedge dx^i. \end{aligned} \tag{3.65}$$

In general, the exterior derivative d is a function that maps a differentiable p-form ${}_p\mathbf{W}(x)$ in $\Lambda^p(\tilde{T}_x(\mathbb{R}^N))$ into a $(p+1)$-form $d[{}_p\mathbf{W}(x)]$ in $\Lambda^{p+1}(\tilde{T}_x(\mathbb{R}^N))$. The precise definition is furnished by

$$\begin{aligned} d[{}_p\mathbf{W}(x)] &:= (1/p!) \left[dW_{j_1 \ldots j_p}(x)\right] \wedge dx^{j_1} \wedge \cdots \wedge dx^{j_p} \\ &:= (1/p!) \left[\frac{\partial W_{j_1 \ldots j_p}(x)}{\partial x^k} dx^k\right] \wedge dx^{j_1} \wedge \cdots \wedge dx^{j_p} \\ &= \sum_{j_1 < \cdots < j_p}^{N} \sum_{k=1}^{N} \left[\frac{\partial W_{j_1 \ldots j_p}(x)}{\partial x^k}\right] dx^k \wedge dx^{j_1} \wedge \cdots \wedge dx^{j_p}. \end{aligned} \tag{3.66}$$

In the special case of $p = 0$ and $p = 1$, (3.64) and (3.65) respectively can be recovered from (3.66).

Example 3.4.4 Consider a differentiable 2-form

$${}_2\mathbf{F}(x) = (1/2) F_{ij}(x) dx^i \wedge dx^j = F_{i<j}(x) dx^i \wedge dx^j. \tag{3.67}$$

Then the exterior derivative of ${}_2\mathbf{F}(x)$, according to (3.66), is

$$\begin{aligned} d[{}_2\mathbf{F}(x)] &= (1/2) \left[\frac{\partial F_{ij}(x)}{\partial x^k}\right] dx^k \wedge dx^i \wedge dx^j \\ &= (1/6) \left[\frac{\partial F_{ij}(x)}{\partial x^k} + \frac{\partial F_{jk}(x)}{\partial x^i} + \frac{\partial F_{ki}(x)}{\partial x^j}\right] dx^i \wedge dx^j \wedge dx^k. \end{aligned} \tag{3.68}$$

This example is relevant in relativistic electromagnetic field theory. □

The salient features of the exterior derivative will be summarized below.

Theorem 3.4.5 *The exterior derivatives on differential forms must satisfy the following equations:*

(i)
$$d\,[{}_p\mathbf{W}(x) + {}_p\mathbf{A}(x)] = d\,[{}_p\mathbf{W}(x)] + d\,[\mathbf{A}_p(x)]\,. \tag{3.69}$$

(ii)
$$\begin{aligned} &d\,[{}_p\mathbf{W}(x) \wedge {}_q\mathbf{A}(x)] \\ &\quad = \{d\,[{}_p\mathbf{W}(x)] \wedge {}_q\mathbf{A}(x)\} + (-1)^p\,\{{}_p\mathbf{W}(x) \wedge d\,[{}_q\mathbf{A}(x)]\}\,. \end{aligned} \tag{3.70}$$

(iii)
$${}_{p+1}\widehat{\mathbf{X}}'\,\{d\,\mathbf{W}_p(x)\} = d\left\{{}_p\widehat{\mathbf{X}}'\,[{}_p\mathbf{W}(x)]\right\}. \tag{3.71}$$

Proof: Proofs of parts (i) and (ii) are left to the reader. The proof of part (iii) is the following. For two differentiable 0-forms f and g, the Leibnitz rule

$$d[f(x)g(x)] = f(x)dg(x) + g(x)df(x) \tag{3.72}$$

holds. Now, two special p-form and q-form are chosen to be

$$\begin{aligned} {}_p\mathbf{W}(x) &= f(x)dx^{i_1} \wedge \cdots \wedge dx^{i_p}, \\ {}_q\mathbf{A}(x) &= g(x)dx^{j_1} \wedge \cdots \wedge dx^{j_q}. \end{aligned}$$

For such a case,

$$\begin{aligned} &d[{}_p\mathbf{W}(x) \wedge {}_q\mathbf{A}(x)] \\ &= d[f(x)g(x)] \wedge dx^{i_1} \cdots \wedge dx^{i_p} \wedge dx^{j_1} \wedge \cdots \wedge dx^{j_q} \\ &= [g(x)df(x) + f(x)dg(x)] \wedge dx^{i_1} \wedge \cdots \wedge dx^{i_p} \wedge dx^{j_1} \wedge \cdots \wedge dx^{j_q} \\ &= df(x)dx^{i_1} \wedge \cdots \wedge dx^{i_p} \wedge [g(x)dx^{j_1} \wedge \cdots \wedge dx^{j_q}] \\ &\quad +(-1)^p[f(x)dx^{i_1} \wedge \ldots dx^{i_p}] \wedge [dg(x) \wedge dx^{j_1} \wedge \cdots \wedge dx^{j_q}] \\ &= \{d[{}_p\mathbf{W}(x)] \wedge [{}_q\mathbf{A}(x)]\} + (-1)^p\{{}_p\mathbf{W}(x) \wedge d[{}_q\mathbf{A}(x)]\}. \end{aligned}$$

By the result above and part (i), the proof of part (ii) for general ${}_p\mathbf{W}(x)$ and ${}_q\mathbf{A}(x)$ follows. ■

Example 3.4.6 Let $\tilde{\mathbf{W}}(x) = W_i(x)dx^i$ and $\tilde{\mathbf{A}}(x) = A_j(x)dx^j$ be two differentiable 1-forms. Then, by (3.70) and (3.61),

$$\begin{aligned} d[\tilde{\mathbf{W}}(x) \wedge \tilde{\mathbf{A}}(x)] &= \{d[\tilde{\mathbf{W}}(x)] \wedge \tilde{\mathbf{A}}(x)\} - \{\tilde{\mathbf{W}}(x) \wedge d[\tilde{\mathbf{A}}(x)]\}, \\ d[\tilde{\mathbf{W}}(x) \wedge \tilde{\mathbf{W}}(x)] &= \mathbf{O}_{\ldots}(x). \end{aligned}$$

□

Now, we shall state and prove an important property of the exterior derivative in the following lemma of Poincaré.

Lemma 3.4.7 *Let ${}_p\mathbf{W}(x)$ be a continuously twice-differentiable p-form in $D \subset \mathbb{R}^N$. Then,*

$$d^2\{{}_p\mathbf{W}(x)\} := d\{d[{}_p\mathbf{W}(x)]\} \equiv {}_{p+2}\mathbf{O}(x). \tag{3.73}$$

Proof. By (3.66),

$$\begin{aligned} d^2[{}_p\mathbf{W}(x)] &= \sum_{1\le j_1<\cdots<j_p}^{N} \sum_{l=1}^{N}\sum_{k=1}^{N} \left[\frac{\partial^2 W_{j_1\ldots j_p}(x)}{\partial x^l \partial x^k} dx^l\right] \wedge dx^k \wedge dx^{j_1} \wedge \cdots \wedge dx^{j_p} \\ &= (1/2) \sum_{1\le j_1<\cdots<j_p}^{N} \sum_{l=1}^{N}\sum_{k=1}^{N} \left[\frac{\partial^2 W_{j_1\ldots j_p}(x)}{\partial x^l \partial x^k} - \frac{\partial^2 W_{j_1\ldots j_p}(x)}{\partial x^k \partial x^l}\right] \\ &\qquad dx^l \wedge dx^k \wedge dx^{j_1} \wedge \cdots \wedge dx^{j_p} \\ &\equiv {}_{p+2}\mathbf{O}(x). \end{aligned}$$

■

Remark: The proof above hinges on the cancellation of the mixed second partial derivatives of functions in the class C^2. Note that in problem 2 of exercises 3.1, we have cited a function for which mixed second partials *exist* but *do not commute.*

Example 3.4.8 Consider $N = 3$ and a function $f \in C^2(D \subset \mathbb{R}^3; \mathbb{R})$. In that case,

$$\begin{aligned} df(x) &= \frac{\partial f(x)}{\partial x^j} dx^j, \\ d^2[f(x)] &= \frac{\partial^2 f(x)}{\partial x^k \partial x^j} dx^k \wedge dx^j = \frac{1}{2}\left[\frac{\partial^2 f(x)}{\partial x^k \partial x^j} - \frac{\partial^2 f(x)}{\partial x^j \partial x^k}\right] dx^k \wedge dx^j \\ &\equiv \mathbf{O}_{..}(x). \end{aligned}$$

The statement above is equivalent to the familiar vector field identity

$$\text{curl}\,[\text{grad}\, f(x)] \equiv \boldsymbol{\nabla} \times [\boldsymbol{\nabla} f(x)] \equiv \vec{\mathbf{O}}(x).$$

□

Example 3.4.9 Consider $N = 3$ and a continuously twice-differentiable 1-form $\tilde{\mathbf{V}}(x) = V_j(x) dx^j$. By (3.65), we obtain

$$d\tilde{\mathbf{V}}(x) = \frac{1}{2}\left[\frac{\partial V_j(x)}{\partial x^i} - \frac{\partial V_i(x)}{\partial x^j}\right] dx^i \wedge dx^j.$$

Using (3.66) and (3.69), we get

$$\begin{aligned} d^2[\tilde{\mathbf{V}}(x)] &= \frac{1}{6}\left\{\frac{\partial}{\partial x^k}\left[\frac{\partial V_j(x)}{\partial x^i} - \frac{\partial V_i(x)}{\partial x^j}\right] + \frac{\partial}{\partial x^i}\left[\frac{\partial V_k(x)}{\partial x^j} - \frac{\partial V_j(x)}{\partial x^k}\right]\right. \\ &\quad \left. + \frac{\partial}{\partial x^j}\left[\frac{\partial V_i(x)}{\partial x^k} - \frac{\partial V_k(x)}{\partial x^i}\right]\right\} dx^i \wedge dx^j \wedge dx^k \\ &\equiv \mathbf{O}_{\dots}(x). \end{aligned}$$

The identity above is equivalent to the vector field identity

$$\text{div}\left[\text{curl}\,\vec{\mathbf{V}}(x)\right] \equiv \boldsymbol{\nabla}\cdot\left[\boldsymbol{\nabla}\times\vec{\mathbf{V}}(x)\right] \equiv 0. \qquad \square$$

A differentiable p-form ${}_p\mathbf{W}(x)$ is called **closed** if $d[{}_p\mathbf{W}(x)] \equiv {}_{p+1}\mathbf{O}(x)$. A p-form ${}_p\mathbf{A}(x)$ is called **exact** provided there exists a differentiable $(p-1)$-form ${}_{p-1}\mathbf{B}(x)$ such that ${}_p\mathbf{A}(x) = d[{}_{p-1}\mathbf{B}(x)]$. **Poincaré's lemma** shows that every exact form is closed. We can naturally ask whether or not every closed form is exact. The answer is affirmative provided that the domain $D^* \subset \mathbb{R}^N$ under consideration is star-shaped with respect to an interior point $x_0 \equiv (x_0^1, \dots, x_0{}^N)$. (See fig. 3.10.)

We shall now state the converse of Poincaré's lemma.

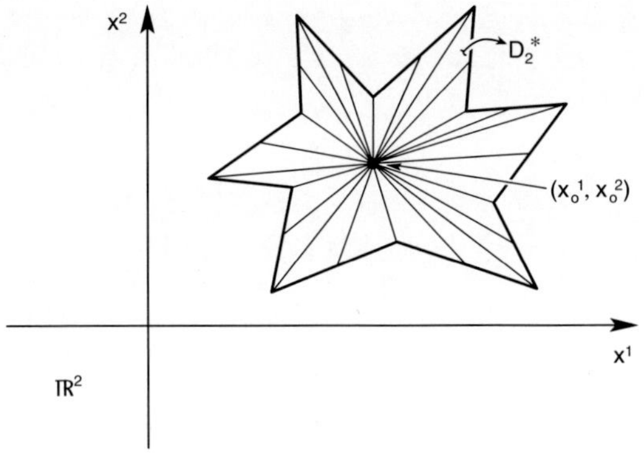

Figure 3.10: A star-shaped domain in $\mathbb{R}^2$.

Theorem 3.4.10 *Let D^* be an open domain of $\mathbb{R}^N$ such that it is star-shaped with respect to an interior point $x_0 \in D^*$. Moreover, let ${}_p\mathbf{W}(x)$ be a differentiable p-form satisfying $d[{}_p\mathbf{W}(x)] \equiv {}_{p+1}\mathbf{O}(x)$ in D^*. Then, there exists a continuously twice-differentiable $(p-1)$-form ${}_{p-1}\mathbf{A}(x)$ in D^* such that*

$$ {}_p\mathbf{W}(x) = d[{}_{p-1}\mathbf{A}(x)]. \tag{3.74}$$

(For the proof, see Flanders's book [13].)

Remark: The $(p-1)$th form ${}_{p-1}\mathbf{A}(x)$ in (3.74) is *not* unique. We can replace it by the $(p-1)$th form

$$ {}_{p}\widehat{\mathbf{A}}(x) = {}_{p-1}\mathbf{A}(x) - d\left[{}_{p-2}\boldsymbol{\lambda}(x)\right]. \tag{3.75}$$

Here, the $(p-2)$th form ${}_{p-2}\boldsymbol{\lambda}(x)$ is an arbitrary one of the differentiability class C^3. By substituting (3.75) in the right-hand side of (3.74), the p-form ${}_{p}\mathbf{W}(x)$ remains unaltered by the Poincaré lemma. The equations (3.74) and (3.75) are applicable to gauge field theory.

Example 3.4.11 Let us choose $N = 4$ and consider special relativistic electromagnetic fields. (See (3.67), (3.68), and (3.70).) The electromagnetic field is identified with a differentiable 2-form

$$ {}_{2}\mathbf{F}(x) = \left(\frac{1}{2}\right) F_{ij}(x) dx^i \wedge dx^j .$$

Four out of eight Maxwell equations can be cast into the tensor field equations:

$$\begin{aligned} d\left[{}_{2}\mathbf{F}(x)\right] &= \mathbf{O}_{...}(x), \\ \frac{\partial F_{ij}(x)}{\partial x^k} + \frac{\partial F_{jk}(x)}{\partial x^i} + \frac{\partial F_{ki}(x)}{\partial x^j} &= 0. \end{aligned} \tag{3.76}$$

In a star-shaped domain D^* of $\mathbb{R}^4$ representing space-time, the equations above imply by theorem 3.4.10 the existence of a continuously twice-differentiable 1-form $\tilde{\mathbf{A}}(x)$ such that

$$\begin{aligned} {}_{2}\mathbf{F}(x) &= d[\tilde{\mathbf{A}}(x)], \\ F_{ij}(x) &= \frac{\partial A_j(x)}{\partial x^i} - \frac{\partial A_i(x)}{\partial x^j}. \end{aligned} \tag{3.77}$$

The 1-form $\tilde{\mathbf{A}}(x)$ is called the **four-potential field**. By (3.75), we can make a transformation

$$\begin{aligned} \widehat{\tilde{\mathbf{A}}}(x) &= \tilde{\mathbf{A}}(x) - d\lambda(x), \\ \widehat{A}_i(x) &= A_i(x) - \frac{\partial \lambda(x)}{\partial x^i} \end{aligned} \tag{3.78}$$

such that the electromagnetic field components $F_{ij}(x)$ remain unchanged. Equation (3.78) is called a **gauge transformation**. □

Now, we shall briefly discuss the integration of a p-form in an open, connected domain D of $\mathbb{R}^N$. Suppose that f is a Riemann- or Lebesgue-integrable function over $D \cup \partial D$. Then, we define

$$\int_D f(x) dx^1 \wedge \cdots \wedge dx^N := \int_D f(x^1, \ldots, x^N) dx^1 \ldots dx^N . \tag{3.79}$$

The right-hand side of (3.79) indicates a multiple integral. Therefore, for a special N-form ${}_N\mathbf{W}(x) := f(x)dx^1 \wedge \cdots \wedge dx^N$, we can write

$$\int_D {}_N\mathbf{W}(x) = \int_D f(x^1, \ldots, x^N) dx^1 \ldots dx^N. \tag{3.80}$$

For a general N-form ${}_N\mathbf{A}(x)$, we define

$$\begin{aligned} \int_D {}_N\mathbf{A}(x) &= (1/N!) \int_D A_{i_1 \ldots i_N}(x) dx^{i_1} \ldots dx^{i_N} \\ &= \int_D A_{i_1 < \cdots < i_N}(x) dx^{i_1} \ldots dx^{i_N}. \end{aligned} \tag{3.81}$$

Let us consider now a differentiable $(p+1)$th-dimensional parametrized, non-degenerate (or regular) hypersurface $D_{p+1} \subset D \subset \mathbb{R}^N$. (Here, $p+1 \leq N$.) It is furnished by the equations

$$\begin{aligned} & x = \xi(u), \\ & x^k = [\pi^k \circ \xi](u) =: \xi^k(u^1, \ldots, u^{p+1}), \\ & \text{Rank} \left[\frac{\partial \xi^k(u)}{\partial u^\alpha} \right] = p+1, \\ & u \equiv (u^1, \ldots, u^{p+1}) \in \mathcal{D}_{p+1} \subset \mathbb{R}^{p+1}, \\ & \alpha \in \{1, \ldots, p+1\}. \end{aligned} \tag{3.82}$$

(See fig. 3.11 for the special case of $N = 3$ and $p = 1$.)

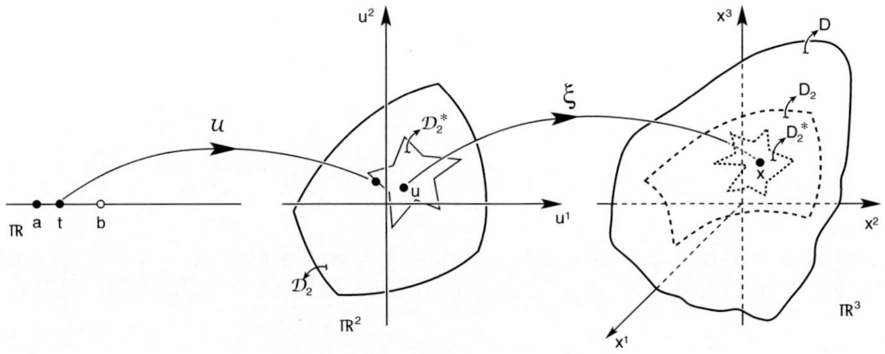

Figure 3.11: A star-shaped domain D_2^* in $\mathbb{R}^3$.

Let us consider a star-shaped domain $\mathcal{D}^*_{p+1} \subset \mathcal{D}_{p+1} \subset \mathbb{R}^{p+1}$. It is mapped by ξ into $D^*_{p+1} \subset D_{p+1} \subset D \subset \mathbb{R}^N$. (The domain D^*_{p+1} of the $(p+1)$th-dimensional hyperspace is called a star-shaped domain in D_{p+1}.) Let the continuous, orientable, piecewise-differentiable boundaries $\partial\mathcal{D}^*_{p+1}$ and ∂D^*_{p+1} be provided by the following parametric equations:

$$\begin{aligned}
&u = \mathcal{U}(t),\\
&u^\alpha = [\pi^\alpha \circ \mathcal{U}](t) =: \mathcal{U}^\alpha(t^1, \ldots, t^p),\\
&\text{Rank}\left[\frac{\partial \mathcal{U}^\alpha(t)}{\partial t^A}\right] = p,\\
&t = (t^1, \ldots, t^p) \in \Delta_p \subset \mathbb{R}^p,\\
&x = [\xi \circ \mathcal{U}]\,(t),\\
&x^k = \xi^k\left[\mathcal{U}^1(t), \ldots, \mathcal{U}^{p+1}(t)\right].
\end{aligned} \tag{3.83}$$

Let ${}_{p+1}\mathbf{A}(x)$ be an integrable $(p+1)$th form over the $(p+1)$th-dimensional hypersurface $D_{p+1} \subset D \subset \mathbb{R}^N$. The integral of ${}_{p+1}\mathbf{A}(x)$ over D^*_{p+1} is given by

$$\begin{aligned}
\int\limits_{D^*_{p+1}} {}_{p+1}\mathbf{A}(x) &= \int\limits_{D^*_{p+1}} A_{i_1<\cdots<i_{p+1}}(x)dx^{i_1} \wedge \cdots \wedge dx^{i_{p+1}}\\
&= \frac{1}{(p+1)!} \int\limits_{\mathcal{D}^*_{p+1}} A_{i_1<\cdots<i_{p+1}}(\xi(u)) \frac{\partial(x^{i_1}, \ldots, x^{i_{p+1}})}{\partial(u^{\alpha_1}, \ldots, u^{\alpha_{p+1}})}\\
&\qquad du^{\alpha_1} \ldots du^{\alpha_{p+1}}.
\end{aligned} \tag{3.84}$$

Now, we are in a position to state the **generalized Stokes' theorem**.

Theorem 3.4.12 *Let D^*_{p+1} be an open, star-shaped, $(p+1)$th-dimensional domain in $\mathbb{R}^N$ $(p+1 \le N)$ with a continuous, piecewise-differentiable, orientable, p-dimensional boundary ∂D^*_{p+1}. Then, for any continuously differentiable p-form ${}_p\mathbf{W}(x)$,*

$$\int\limits_{D^*_{p+1}} d\left[{}_p\mathbf{W}(x)\right] = \int\limits_{\partial D^*_{p+1}} {}_p\mathbf{W}(x). \tag{3.85}$$

(For the proof of this theorem, we refer to the book by Spivak [35].)

Example 3.4.13 Let us choose $N = 1$, $p = 0$, and $D = (a, b) \subset \mathbb{R}$. Let f be a continuously differentiable function over $[a, b]$. The boundary $\partial(a, b) = \{a\} \cup \{b\}$. Then the Stokes' theorem (3.85) yields (in an extended sense)

$$\int\limits_{(a,b)} df(x) = \int_a^b \frac{df(x)}{dx} dx = f(b) - f(a).$$

The equation above is the reiteration of the fundamental theorem of calculus. □

Example 3.4.14 Let us choose $N = 2$ and $p = 1$. Moreover, we choose $D_2^* \subset \mathbb{R}^2$ to be the rectangle $(a, b) \times (c, d)$. The boundary ∂D_2^* of the rectangle is to be traversed in the counterclockwise manner (to maintain the right orientation). For a continuously differentiable 1-form $\tilde{\mathbf{A}}(x) = A_j(x)dx^j$, (3.85) provides

$$\int\limits_{(a,b)\times(c,d)} d\left[\tilde{\mathbf{A}}(x)\right] = \int\limits_{\partial[(a,b)\times(c,d)]} \tilde{\mathbf{A}}(x),$$

$$\int\limits_{(a,b)\times(c,d)} \left[\frac{\partial A_2(x)}{\partial x^1} - \frac{\partial A_1(x)}{\partial x^2}\right] dx^1 dx^2 = \int\limits_{\partial[(a,b)\times(c,d)]} \left[A_1(x)dx^1 + A_2(x)dx^2\right]. \tag{3.86}$$

This is just the usual Stokes' (or Green's) theorem in the two-dimensional vector calculus. □

Example 3.4.15 Consider the case of $N = 3$ and $p = 2$. Let a continuously differentiable 2-form be given by ${}_2\mathbf{F}(x)$ in D of $\mathbb{R}^3$. For a star-shaped domain $D_3^* \subset D$, we have from (3.85) (and (3.76))

$$\int\limits_{D_3^*} d\left[{}_2\mathbf{F}(x)\right] = \int\limits_{\partial D_3^*} {}_2\mathbf{F}(x),$$

$$\frac{1}{6}\int\limits_{D_3^*} \left[\frac{\partial F_{ij}(x)}{\partial x^k} + \frac{\partial F_{jk}(x)}{\partial x^i} + \frac{\partial F_{ki}(x)}{\partial x^j}\right] dx^i dx^j dx^k = (1/2)\int\limits_{\partial D_3^*} F_{ij}(x) dx^i \wedge dx^j \tag{3.87}$$

$$= \int\limits_{\partial \mathcal{D}_3^*} \left[F_{12}(\xi(u)) \frac{\partial(x^1, x^2)}{\partial(u^1, u^2)} + F_{23}(\xi(u)) \frac{\partial(x^2, x^3)}{\partial(u^1, u^2)} + F_{31}(\xi(u)) \frac{\partial(x^3, x^1)}{\partial(u^1, u^2)}\right] du^1 du^2.$$

We can introduce a vector *density* by the up-star operation (see (2.104))

$${}^*F^k(x) = (1/2)\varepsilon^{ijk} F_{ij}(x). \tag{3.88}$$

By (3.87), we can express

$$\int\limits_{D_3^*} \operatorname{div}\left[{}^*\vec{\mathbf{F}}(x)\right] dx^1 dx^2 dx^3 \equiv \int\limits_{D_3^*} \left[\nabla \cdot {}^*\vec{\mathbf{F}}(x)\right] dx^1 dx^2 dx^3$$

$$= \int\limits_{D_3^*} \left[\frac{\partial\, {}^*F^1(x)}{\partial x^1} + \frac{\partial\, {}^*F^2(x)}{\partial x^2} + \frac{\partial\, {}^*F^3(x)}{\partial x^3}\right] dx^1 dx^2 dx^3 \tag{3.89}$$

$$= \int\limits_{\partial D_3^*} \left[F_{12}(x) dx^1 dx^2 + F_{23}(x) dx^2 dx^3 + F_{31}(x) dx^3 dx^1\right].$$

For an application to the static magnetic field $(H^1(x), H^2(x), H^3(x))$ and the vector potential $\tilde{\mathbf{A}}(x) = A_j(x)dx^j$, we put

$$\begin{aligned} H^k(x) &:= {}^*F^k(x), \\ F_{ij}(x) &= \frac{\partial A_j(x)}{\partial x^i} - \frac{\partial A_i(x)}{\partial x^j}. \end{aligned} \tag{3.90}$$

By (3.65), (3.77), and (3.89), we obtain

$$\int_{D_3^*} \left[\frac{\partial H^1(x)}{\partial x^1} + \frac{\partial H^2(x)}{\partial x^2} + \frac{\partial H^3(x)}{\partial x^3}\right] dx^1 dx^2 dx^3 = \int_{\partial D_3^*} d\tilde{\mathbf{A}}(x) = \int_{\partial^2[D_3^*]} \tilde{\mathbf{A}}(x).$$

In algebraic topology, the boundary of a boundary of a subset is always the null subset $(\partial^2 D = \emptyset)$. Therefore, we conclude that

$$\int_{D_3^*} \left[\frac{\partial H^1(x)}{\partial x^1} + \frac{\partial H^2(x)}{\partial x^2} + \frac{\partial H^3(x)}{\partial x^3}\right] dx^1 dx^2 dx^3 = 0. \tag{3.91}$$

The physical meaning of the equation above is that the total magnetic charge inside a star-shaped three-dimensional domain is precisely zero. Thus, according to (3.90), a *magnetic monopole cannot exist* inside a star-shaped three-dimensional domain. □

The definition (3.66) and equation (3.71) show that the exterior derivative produces a $(p+1)$th-order totally antisymmetric tensor out of a pth-order totally antisymmetric tensor. There exists another derivative of tensor fields generating higher-order tensor fields. This is called a **Lie derivative**. We shall discuss it briefly in appendix 2.

We shall conclude this section with the statement of a deep theorem by Darboux.

Theorem 3.4.16 *Let* ${}_2\mathbf{F}(x) \neq \mathbf{O}_{..}(x)$ *be a* 2*-form defined over a convex domain* $D \subset \mathbb{R}^N$. *Then there exists a basis set* $\{\tilde{\mathbf{e}}^1(x), \ldots, \tilde{\mathbf{e}}^N(x)\}$ *such that*

$$\begin{aligned} &{}_2\mathbf{F}(x) = \tilde{\mathbf{e}}^1(x) \wedge \tilde{\mathbf{e}}^2(x) + \tilde{\mathbf{e}}^3(x) \wedge \tilde{\mathbf{e}}^4(x) + \cdots + \tilde{\mathbf{e}}^{2r-1}(x) \wedge \tilde{\mathbf{e}}^{2r}(x); \\ &2 \leq 2r \leq N. \end{aligned} \tag{3.92}$$

(For the proof, consult the book by Sternberg [36].)

Exercises 3.4

1. A 2-form ${}_2\mathbf{W}(x)$ in $\mathbb{R}^4$ is furnished by

$${}_2\mathbf{W}(x) := dx^1 \wedge dx^2 + dx^3 \wedge dx^4.$$

Calculate ${}_2\mathbf{W}(x) \wedge {}_2\mathbf{W}(x) \wedge {}_2\mathbf{W}(x)$.

2. Prove that

$$
\begin{aligned}
{}_p\mathbf{W}(x) \wedge {}_q\mathbf{A}(x) = {} & [1/(p+q)!]\, \delta^{m_1 \ldots m_p n_1 \ldots n_q}{}_{j_1 \cdots j_p j_{p+1} \cdots j_{p+q}} \\
& W_{m_1 \ldots m_p}(x) A_{n_1 \ldots n_q}(x) dx^{j_1} \wedge \cdots \wedge dx^{j_p} \wedge dx^{j_{p+1}} \wedge \cdots \wedge dx^{j_{p+q}}.
\end{aligned}
$$

3. Let n be a non-negative integer such that $2n+1 \leq N$. Show that every $(2n+1)$-form ${}_{2n+1}\mathbf{W}(x)$ satisfies

$$
{}_{2n+1}\mathbf{W}(x) \wedge {}_{2n+1}\mathbf{W}(x) \equiv {}_{4n+2}\mathbf{O}(x).
$$

4. Let U and V be continuously twice-differentiable functions in $D \subset \mathbb{R}^3$. Let a 1-form be given by

$$
\begin{aligned}
&\tilde{\mathbf{A}}(x) := \\
&\left[x^2 - V(x)\frac{\partial U(x)}{\partial x^1}\right] dx^1 - \left[x^1 - V(x)\frac{\partial U(x)}{\partial x^2}\right] dx^2 + \left[1 - V(x)\frac{\partial U(x)}{\partial x^3}\right] dx^3.
\end{aligned}
$$

Prove that $d\tilde{\mathbf{A}}(x) \equiv \mathbf{O}_{..}(x)$ if and only if

$$
\frac{\partial U(x)}{\partial x^3} = \frac{\partial V(x)}{\partial x^3} \equiv 0 \quad \text{and} \quad \frac{\partial(u,v)}{\partial(x^1,x^2)} = 2.
$$

5. Consider a 1-form defined by

$$
\begin{aligned}
&\tilde{\mathbf{W}}(x) := -\left[\frac{x^2}{(x^1)^2 + (x^2)^2}\right] dx^1 + \left[\frac{x^1}{(x^1)^2 + (x^2)^2}\right] dx^2, \\
&(x^1, x^2) \in \mathbb{R}^2 - \{(0,0)\}.
\end{aligned}
$$

(i) Prove that $d\tilde{\mathbf{W}}(x) = \mathbf{O}_{..}(x)$ in $\mathbb{R}^2 - \{(0,0)\}$.

(ii) Show the non-existence of a differentiable function f in $\mathbb{R}^2 - \{(0,0)\}$ such that $\tilde{\mathbf{W}}(x) = df(x)$. Explain.

Chapter 4

Differentiable Manifolds with Connections

4.1 The Affine Connection and Covariant Derivative

We shall introduce a new notation for the **generalized directional derivative.** (See (3.7).)

$$\begin{aligned}
&\partial_j f := \frac{\partial}{\partial x^j} f(x),\\
&\partial_{(q)} f \equiv \partial_q f := \vec{\mathbf{e}}_q(x)[f],\\
&df(x) = (\partial_j f)dx^j = (\partial_q f)\tilde{\mathbf{e}}^q(x),\\
&\vec{\mathbf{V}}(x)[f] = V^j(x)\partial_j f = V^q(x)\partial_q f.
\end{aligned} \tag{4.1}$$

We have used the indices $i, j, k, \ldots$ for coordinate components and the indices $p, q, r, \ldots$, etc., for non-coordinate (or **non-holonomic**) components).

Example 4.1.1 Let us choose $N = 2$ and two basis vectors

$$\vec{\mathbf{e}}_1(x) = \frac{\partial}{\partial x^1} + \frac{\partial}{\partial x^2}, \ \vec{\mathbf{e}}_2(x) = \frac{\partial}{\partial x^1} - \frac{\partial}{\partial x^2}.$$

The corresponding 1-forms are

$$\tilde{\mathbf{e}}^1(x) = (1/2)(dx^1 + dx^2), \ \tilde{\mathbf{e}}^2(x) = (1/2)(dx^1 - dx^2).$$

Consider the polynomial function $f(x) := x^1 x^2$ in $\mathbb{R}^2$.

$$\partial_{(1)} f = \vec{\mathbf{e}}_1(x)[f] = x^1 + x^2, \ \partial_{(2)} f = \vec{\mathbf{e}}_2(x)[f] = x^2 - x^1.$$

Therefore, it follows that

$$(\partial_q f)\tilde{\mathbf{e}}^q(x) = [\partial_{(1)} f]\tilde{\mathbf{e}}^1(x) + [\partial_{(2)} f]\tilde{\mathbf{e}}^2(x) = x^2 dx^1 + x^1 dx^2 = (\partial_j f)dx^j. \quad \square$$

Now we shall introduce a new concept of differentiation of tensor fields. It is called the **affine connection**, **covariant derivative**, or **generalized gradient**. It is denoted by the symbol ∇. The operator ∇ assigns to each tangent vector field $\vec{\mathbf{V}}(x)$ of class $C^r (r \geq 1)$ a $(1+1)$th-order tensor field $\nabla\vec{\mathbf{V}}(x)$ of class C^{r-1}. The tensor field is called the **covariant derivative** of the vector field $\vec{\mathbf{V}}(x)$. The covariant derivative satisfies the following two axioms.

I. $$\nabla\left[\vec{\mathbf{V}}(x) + \vec{\mathbf{W}}(x)\right] = \nabla\left[\vec{\mathbf{V}}(x)\right] + \nabla\left[\vec{\mathbf{W}}(x)\right]. \tag{4.2}$$

II. $$\nabla\left[f(x)\vec{\mathbf{V}}(x)\right] = \vec{\mathbf{V}}(x) \otimes df(x) + f(x)\left[\nabla\vec{\mathbf{V}}(x)\right]. \tag{4.3}$$

Here, f is any differentiable function and $\vec{\mathbf{V}}(x)$, $\vec{\mathbf{W}}(x)$ are arbitrary vector fields of class C^r in D of $\mathbb{R}^N$.

Let $\{\vec{\mathbf{e}}_p(x)\}_1^N := \{\vec{\mathbf{e}}_1(x), \ldots, \vec{\mathbf{e}}_N(x)\}$ and $\{\tilde{\mathbf{e}}^q(x)\}_1^N$ be a differentiable basis field and the corresponding conjugate basis field at $x \in D$ of $\mathbb{R}^N$. By (4.3), the mixed second-order tensor field $\nabla\vec{\mathbf{e}}_p(x)$ belongs to ${}^1{}_1\mathcal{T}_x(\mathbb{R}^N)$ with a basis set $\{\vec{\mathbf{e}}_p(x) \otimes \tilde{\mathbf{e}}^q(x)\}_1^{N^2}$. Therefore, we can express by a linear combination

$$\begin{aligned} \nabla\vec{\mathbf{e}}_p(x) &= \Gamma^u_{qp}(x)\vec{\mathbf{e}}_u(x) \otimes \tilde{\mathbf{e}}^q(x), \\ \Gamma^u_{qp}(x) &:= \left[\nabla\vec{\mathbf{e}}_p(x)\right]\left(\tilde{\mathbf{e}}^u(x), \vec{\mathbf{e}}_q(x)\right). \end{aligned} \tag{4.4}$$

The set of N^3 real-valued functions Γ^u_{qp} are called the **connection coefficients**. (See the references [5], [11] & [19].)

We shall now compute the components of $\nabla\vec{\mathbf{V}}(x)$ explicitly in terms of the connection coefficients. By the rules (4.2) and (4.3) and equation (4.4),

$$\begin{aligned} \nabla\vec{\mathbf{V}}(x) &= \nabla\left[V^p(x)\vec{\mathbf{e}}_p(x)\right] = \left[\vec{\mathbf{e}}_p(x) \otimes dV^p(x)\right] + V^p(x)\left[\nabla\vec{\mathbf{e}}_p(x)\right] \\ &= \vec{\mathbf{e}}_p(x) \otimes \left[dV^p(x) + \Gamma^p_{uq}(x)V^q(x)\tilde{\mathbf{e}}^u(x)\right]. \end{aligned}$$

By (4.1), $dV^p(x) = \{\vec{\mathbf{e}}_u(x)[V^p]\}\,\tilde{\mathbf{e}}^u(x)$. Therefore,

$$\begin{aligned} \nabla\vec{\mathbf{V}}(x) &= \left\{\vec{\mathbf{e}}_u(x)[V^p] + \Gamma^p_{uq}(x)V^q(x)\right\}\vec{\mathbf{e}}_p(x) \otimes \tilde{\mathbf{e}}^u(x) \\ &= \left[\partial_u V^p + \Gamma^p_{uq}(x)V^q(x)\right]\vec{\mathbf{e}}_p(x) \otimes \tilde{\mathbf{e}}^u(x). \end{aligned} \tag{4.5}$$

Therefore, the components $\nabla_u V^p \equiv V^p{}_{;u}$ of the tensor $\nabla\vec{\mathbf{V}}(x)$ are furnished by

$$\begin{aligned}\nabla_u V^p \equiv V^p{}_{;u} &:= \vec{\mathbf{e}}_u(x)[V^p] + \Gamma^p_{uq}(x)V^q(x)\\ &= \partial_u V^p + \Gamma^p_{uq}(x)V^q(x).\end{aligned} \tag{4.6}$$

(In the sequel, we shall use the symbol $\nabla_u V^p$.)

Example 4.1.2 Let us choose the coordinate or the natural basis set. In that case, (4.6) yields

$$\begin{aligned}\vec{\mathbf{V}}(x) &= V^k(x)\frac{\partial}{\partial x^k},\\ \nabla_i V^k &= \vec{\mathbf{e}}_i[V^k] + \Gamma^k_{ij}(x)V^j(x)\\ &= \partial_i V^k + \Gamma^k_{ij}(x)V^j(x),\\ \partial_i &:= \frac{\partial}{\partial x^i}.\end{aligned} \tag{4.7}$$

Therefore, the **covariant divergence** is given by

$$\nabla_k V^k = \partial_k V^k + \Gamma^k_{kj}(x)V^j(x). \qquad \square$$

The N^2 **connection** 1-**forms** $\tilde{\mathbf{w}}^p{}_q(x)$ are defined by

$$\tilde{\mathbf{w}}^p{}_q(x) := \Gamma^p_{uq}(x)\tilde{\mathbf{e}}^u(x). \tag{4.8}$$

(*Remark:* Connection 1-forms are *not tensorial.*)

Therefore, the covariant derivative in (4.5) can be expressed as

$$\nabla\vec{\mathbf{V}}(x) = \vec{\mathbf{e}}_p(x) \otimes \left[dV^p(x) + \tilde{\mathbf{w}}^p_q(x)V^q(x)\right]. \tag{4.9}$$

Example 4.1.3 Consider the coordinate basis. By (4.8),

$$\tilde{\mathbf{w}}^i{}_j(x) = \Gamma^i_{kj}(x)dx^k.$$

In the case where Γ^i_{kj} are differentiable functions, the exterior derivatives of $\tilde{\mathbf{w}}^i{}_j(x)$ (by (3.65)) are given by

$$d\tilde{\mathbf{w}}^i_j(x) = (1/2)\left(\partial_l\Gamma^i_{kj} - \partial_k\Gamma^i_{lj}\right) dx^l \wedge dx^k. \tag{4.10}$$

The equations above will be useful later. $\square$

Now we shall derive the transformation properties of connection coefficients under a general transformation of basis sets

$$\widehat{\vec{\mathbf{e}}}_p(x) = \lambda^q_p(x)\vec{\mathbf{e}}_q(x), \quad \widehat{\tilde{\mathbf{e}}}^p(x) = \mu^p_q(x)\tilde{\mathbf{e}}^q(x). \tag{4.11}$$

(See (2.54).)

Theorem 4.1.4 *Under the general differentiable transformation of basis sets in* (4.11), *the connection coefficients transform as*

$$\widehat{\Gamma}^q{}_{rp}(x) = \mu^q{}_u(x)\lambda^w{}_r(x)\lambda^s{}_p(x)\Gamma^u{}_{ws}(x) + \mu^q{}_u(x)\partial_r\left[\lambda^u{}_p\right]. \tag{4.12}$$

Proof. For the components of a differentiable vector field $\vec{\mathbf{V}}(x)$, we have

$$dV^u(x) = d\left[\lambda^u{}_p(x)\widehat{V}^p(x)\right] = \lambda^u{}_p(x)d\widehat{V}^p(x) + \widehat{V}^p(x)d\lambda^u{}_p(x).$$

By (4.5) and (4.11), we have

$$\begin{aligned}
&\nabla\vec{\mathbf{V}}(x) = \vec{\mathbf{e}}_u(x)\otimes\left[dV^u(x) + \Gamma^u{}_{ws}(x)V^s(x)\tilde{\mathbf{e}}^w(x)\right]\\
&= \left[\mu^q{}_u(x)\widehat{\vec{\mathbf{e}}}_q(x)\right]\otimes\\
&\qquad\left[\lambda^u{}_p(x)d\widehat{V}^p(x) + \widehat{V}^p(x)d\lambda^u{}_p(x) + \lambda^s{}_p(x)\lambda^w{}_r(x)\Gamma^u{}_{ws}(x)\widehat{V}^p(x)\widehat{\tilde{\mathbf{e}}}^r(x)\right]\\
&= \widehat{\vec{\mathbf{e}}}_q(x)\otimes\Big\{d\widehat{V}^q(x)+\\
&\qquad\left[\mu^q{}_u(x)\partial_r\lambda^u{}_p(x) + \mu^q{}_u(x)\lambda^s{}_p(x)\lambda^w{}_r(x)\Gamma^u{}_{ws}(x)\right]\widehat{V}^p(x)\widehat{\tilde{\mathbf{e}}}^r(x)\Big\}.
\end{aligned}$$

Identifying the expression inside the square bracket above with $\widehat{\Gamma}^q{}_{rp}(x)$, equation (4.12) follows. ■

(*Remark:* Equation (4.12) shows that the connection coefficients $\Gamma^u_{ws}(x)$ *do not* transform as components of a tensor field *unless* $d\lambda^u_p(x) \equiv \tilde{\mathbf{O}}(x)$.)

Example 4.1.5 Let us consider the coordinate basis set. In that case, (4.12) yields

$$\widehat{\Gamma}^l{}_{mn}(\widehat{x}) = \frac{\partial\widehat{X}^l(x)}{\partial x^i}\frac{\partial X^j(\widehat{x})}{\partial\widehat{x}^m}\frac{\partial X^k(\widehat{x})}{\partial\widehat{x}^n}\Gamma^i{}_{jk}(x) + \frac{\partial\widehat{X}^l(x)}{\partial x^i}\frac{\partial^2 X^i(\widehat{x})}{\partial\widehat{x}^m\partial\widehat{x}^n}. \tag{4.13}$$

The coefficients $\Gamma^i_{jk}(x)$ *do not* transform as tensor field components (compare with (3.48)) *unless*

$$\frac{\partial^2 X^l(\widehat{x})}{\partial\widehat{x}^m\partial\widehat{x}^n} = 0.$$

The *general solution* of the system of partial differential equations above yields the **general non-homogeneous linear** (or **affine**) transformation

$$x^i = X^i(\widehat{x}) = k^i{}_j\widehat{x}^j + c^i.$$

Here, c^i and $k^i{}_j$ (with $\det[k^i{}_j] \neq 0$) are arbitrary constants of integration. □

Example 4.1.6 Consider the dimension $N = 2$. Let us choose the coordinate transformation from the Cartesian to the polar coordinates furnished by

$$\begin{aligned}
x^1 &= X^1(\widehat{x}) := \widehat{x}^1 \cos \widehat{x}^2, \\
x^2 &= X^2(\widehat{x}) := \widehat{x}^1 \sin \widehat{x}^2, \\
\widehat{D}_s &:= \left\{ (\widehat{x}^1, \widehat{x}^2) \in \mathbb{R}^2 : \widehat{x}^1 > 0, -\pi < \widehat{x}^2 < \pi \right\}.
\end{aligned}$$

The Jacobian matrix is given by

$$\begin{bmatrix} \dfrac{\partial X^1(\widehat{x})}{\partial \widehat{x}^1}, & \dfrac{\partial X^1(\widehat{x})}{\partial \widehat{x}^2} \\ \dfrac{\partial X^2(\widehat{x})}{\partial \widehat{x}^1}, & \dfrac{\partial X^2(\widehat{x})}{\partial \widehat{x}^2} \end{bmatrix} = \begin{bmatrix} \cos \widehat{x}^2, & -\widehat{x}^1 \sin \widehat{x}^2 \\ \sin \widehat{x}^2, & \widehat{x}^1 \cos \widehat{x}^2 \end{bmatrix}.$$

Let us compute with (4.13) the eight functions $\widehat{\Gamma}^i{}_{jk}(\widehat{x})$ by *assigning* $\Gamma^l{}_{mn}(x) \equiv 0$ for the Cartesian coordinates:

$$\begin{aligned}
\widehat{\Gamma}^1{}_{11}(\widehat{x}) &= 0 + \frac{\partial \widehat{X}^1(x)}{\partial x^1} \frac{\partial^2 X^1(\widehat{x})}{(\partial \widehat{x}^1)^2} + \frac{\partial \widehat{X}^1(x)}{\partial x^2} \frac{\partial^2 X^2(\widehat{x})}{(\partial \widehat{x}^1)^2} \equiv 0, \\
\widehat{\Gamma}^1{}_{22}(\widehat{x}) &= 0 + \frac{\partial \widehat{X}^1(x)}{\partial x^1} \frac{\partial^2 X^1(\widehat{x})}{(\partial \widehat{x}^2)^2} + \frac{\partial \widehat{X}^1(x)}{\partial x^2} \frac{\partial^2 X^2(\widehat{x})}{(\partial \widehat{x}^2)^2} = -\widehat{x}^1, \\
\widehat{\Gamma}^2{}_{12}(\widehat{x}) &= (\widehat{x}^1)^{-1} = \widehat{\Gamma}^2{}_{21}(\widehat{x}), \\
\widehat{\Gamma}^1{}_{12}(\widehat{x}) &= \widehat{\Gamma}^1{}_{21}(\widehat{x}) = \widehat{\Gamma}^2{}_{22}(\widehat{x}) = \widehat{\Gamma}^2{}_{11}(\widehat{x}) \equiv 0.
\end{aligned}$$

□

Now, we shall define the **directional covariant derivative** of a vector field $\vec{\mathbf{V}}(x)$ along the direction of another vector field $\vec{\mathbf{U}}(x)$. It is denoted by the symbol $\nabla_{\vec{\mathbf{U}}}[\vec{\mathbf{V}}(x)]$. The operator $\nabla_{\vec{\mathbf{U}}}$ maps a vector field $\vec{\mathbf{V}}(x)$ of class $C^r (r \geq 1)$ into another vector field of class C^{r-1}. The definition is furnished by

$$\nabla_{\vec{\mathbf{U}}}[\vec{\mathbf{V}}(x)](\cdot) := \left[\nabla \vec{\mathbf{V}}(x)\right] \left(\cdot, \vec{\mathbf{U}}(x)\right), \tag{4.14}$$

$$\nabla_{\vec{\mathbf{U}}}[\vec{\mathbf{V}}(x)](\tilde{\mathbf{W}}(x)) = \left[\nabla \vec{\mathbf{V}}(x)\right] \left(\tilde{\mathbf{W}}(x), \vec{\mathbf{U}}(x)\right) \tag{4.15}$$

for all $\tilde{\mathbf{W}}(x)$ in $\tilde{T}_x(\mathbb{R}^N)$.

$$\begin{aligned}
\nabla_{\vec{\mathbf{U}}}[\vec{\mathbf{V}}(x)](\tilde{\mathbf{e}}^q(x)) &= \{\vec{\mathbf{e}}_w(x)[V^u] + \Gamma^u{}_{wp}(x) V^p(x)\} \left[\vec{\mathbf{e}}_u(x) \otimes \tilde{\mathbf{e}}^w(x)\right] \left(\tilde{\mathbf{e}}^q(x), \vec{\mathbf{U}}(x)\right) \\
&= \{\vec{\mathbf{e}}_w(x)[V^q] + \Gamma^q{}_{wp}(x) V^p(x)\} \, U^w(x) \\
&= \vec{\mathbf{U}}(x)[V^q] + \Gamma^q{}_{wp} U^w(x) V^p(x).
\end{aligned}$$

Thus, we have

$$\begin{aligned}
\nabla_{\vec{\mathbf{U}}}[\vec{\mathbf{V}}(x)] &= \left\{ \vec{\mathbf{U}}(x)[V^q] + \Gamma^q{}_{wp}(x) U^w(x) V^p(x) \right\} \vec{\mathbf{e}}_q(x), \\
\left[\nabla_{\vec{\mathbf{U}}} \vec{\mathbf{V}}\right]^q &= U^p \nabla_p V^q.
\end{aligned} \tag{4.16}$$

As special cases of (4.16), we can derive from (4.6)

$$\nabla_{\vec{e}_w}[\vec{\mathbf{V}}(x)] = (\nabla_w V^q)\vec{\mathbf{e}}_q(x), \tag{4.17}$$

$$\nabla_{\vec{e}_w}[\vec{\mathbf{e}}_p(x)] = \Gamma^q{}_{wp}(x)\vec{\mathbf{e}}_q(x). \tag{4.18}$$

The main properties of the operator $\nabla_{\vec{\mathbf{V}}}$ will be summarized below.

Theorem 4.1.7 *The covariant derivatives along directions of vector fields under suitable differentiability conditions must satisfy:*

(i) $$\nabla_{\vec{\mathbf{V}}}[\vec{\mathbf{Y}}(x) + \vec{\mathbf{Z}}(x)] = \nabla_{\vec{\mathbf{V}}}[\vec{\mathbf{Y}}(x)] + \nabla_{\vec{\mathbf{V}}}[\vec{\mathbf{Z}}(x)]\,; \tag{4.19}$$

(ii) $$\nabla_{\vec{\mathbf{Y}}+\vec{\mathbf{Z}}}[\vec{\mathbf{V}}(x)] = \nabla_{\vec{\mathbf{Y}}}[\vec{\mathbf{V}}(x)] + \nabla_{\vec{\mathbf{Z}}}[\vec{\mathbf{V}}(x)]\,; \tag{4.20}$$

(iii) $$\nabla_{f\vec{\mathbf{V}}}[\vec{\mathbf{Y}}(x)] = f(x)\nabla_{\vec{\mathbf{V}}}[\vec{\mathbf{Y}}(x)]\,; \tag{4.21}$$

(iv) $$\nabla_{\vec{\mathbf{V}}}[g(x)\vec{\mathbf{Y}}(x)] = \{\vec{\mathbf{V}}(x)[g]\}\vec{\mathbf{Y}}(x) + g(x)\{\nabla_{\vec{\mathbf{V}}}[\vec{\mathbf{Y}}(x)]\}\,. \tag{4.22}$$

Proof. We shall prove part (iv). (The proofs for parts (i), (ii), and (iii) are left to the reader.) By (4.15), the left-hand side of (4.22) yields

$$\nabla_{\vec{\mathbf{V}}}[g(x)\vec{\mathbf{Y}}(x)](\tilde{\mathbf{W}}(x)) = \nabla[g(x)\vec{\mathbf{Y}}(x)](\tilde{\mathbf{W}}(x), \vec{\mathbf{V}}(x)).$$

Using (4.3) and linearity, we obtain

$$\begin{aligned}
&\nabla_{\vec{\mathbf{V}}}[g(x)\vec{\mathbf{Y}}(x)](\tilde{\mathbf{W}}(x)) \\
&\quad = [\vec{\mathbf{Y}}(x) \otimes dg(x)](\tilde{\mathbf{W}}(x), \vec{\mathbf{V}}(x)) + [g(x)\nabla\vec{\mathbf{Y}}(x)](\tilde{\mathbf{W}}(x), \vec{\mathbf{V}}(x)) \\
&\quad = [\vec{\mathbf{Y}}(x)(\tilde{\mathbf{W}}(x))]\{dg(x)[\vec{\mathbf{V}}]\} + \{g(x)\nabla_{\vec{\mathbf{V}}}[\vec{\mathbf{Y}}(x)]\}(\tilde{\mathbf{W}}(x)),
\end{aligned}$$

or

$$\{\nabla_{\vec{\mathbf{V}}}[g(x)\vec{\mathbf{Y}}(x)] - [dg(x)[\vec{\mathbf{V}}]]\vec{\mathbf{Y}}(x) - g(x)\nabla_{\vec{\mathbf{V}}}[\vec{\mathbf{Y}}(x)]\}(\tilde{\mathbf{W}}(x)) \equiv 0.$$

Since the only tangent vector that maps every $\tilde{\mathbf{W}}(x)$ into zero is $\vec{\mathbf{O}}(x)$, equation (4.22) follows. ∎

Example 4.1.8 In the coordinate basis, (4.17) yields

$$\nabla_{\partial/\partial x^i}[\vec{\mathbf{V}}(x)] = (\nabla_i V^j)\frac{\partial}{\partial x^j},$$

$$\{\nabla_{\partial/\partial x^i}[\vec{\mathbf{V}}(x)]\}(dx^j) = \nabla_i V^j. \qquad \square$$

The definition of the covariant derivative in (4.14) and (4.15) can be extended to a tensor field ${}^{r}{}_{s}\mathbf{T}(x)$ of class C^k $(k \geq 1)$. The corresponding axioms are the following:

I. When ${}^{r}{}_{s}\mathbf{T}(x)$ is of class C^k, the $\nabla_{\vec{\mathbf{V}}}[{}^{r}{}_{s}\mathbf{T}(x)]$ is of the same order and class C^{k-1} $(k \geq 1)$.

II.
$$\nabla_{\vec{\mathbf{V}}}[f(x)] = \vec{\mathbf{V}}(x)[f]. \tag{4.23}$$

III.
$$\nabla_{\vec{\mathbf{V}}}[{}^{r}{}_{s}\mathbf{T}(x) + {}^{r}{}_{s}\mathbf{B}(x)] = \nabla_{\vec{\mathbf{V}}}[{}^{r}{}_{s}\mathbf{T}(x)] + \nabla_{\vec{\mathbf{V}}}[{}^{r}{}_{s}\mathbf{B}(x)]. \tag{4.24}$$

IV.
$$\begin{aligned} \nabla_{\vec{\mathbf{V}}}[{}^{r}{}_{s}\mathbf{T}(x) \otimes {}^{m}{}_{n}\mathbf{G}(x)] = \nabla_{\vec{\mathbf{V}}}[{}^{r}{}_{s}\mathbf{T}(x)] \otimes {}^{m}{}_{n}\mathbf{G}(x) \\ + {}^{r}{}_{s}\mathbf{T}(x) \otimes \nabla_{\vec{\mathbf{V}}}[{}^{m}{}_{n}\mathbf{G}(x)]. \end{aligned} \tag{4.25}$$

V.
$$\nabla_{\vec{\mathbf{V}}}\left\{{}^{i}{}_{j}\mathcal{C}[{}^{r}{}_{s}\mathbf{T}(x)]\right\} = {}^{i}{}_{j}\mathcal{C}\left\{\nabla_{\vec{\mathbf{V}}}[{}^{r}{}_{s}\mathbf{T}(x)]\right\}. \tag{4.26}$$

Equations (4.23), (4.24), and (4.25) imply equations (4.2) and (4.3). (Some caution is required to apply (4.26).)

We shall now express the covariant derivative of a 1-form in terms of the connection coefficients.

Lemma 4.1.9 *The covariant derivative of a differentiable 1-form $\tilde{\mathbf{W}}(x)$ is given by*

$$\nabla_{\vec{\mathbf{V}}}[\tilde{\mathbf{W}}(x)] = V^p(x)\left\{\vec{\mathbf{e}}_p(x)[W_q] - \Gamma^u{}_{pq}(x)W_u(x)\right\}\tilde{\mathbf{e}}^q(x), \tag{4.27}$$

$$\nabla_{\vec{\mathbf{V}}}[\tilde{\mathbf{W}}(x)]\vec{\mathbf{e}}_q(x) = [\nabla_p W_q]V^p(x) := [\partial_p W_q - \Gamma^u{}_{pq}(x)W_u(x)]V^p(x), \tag{4.28}$$

$$\nabla_{\vec{\mathbf{e}}_p}[\tilde{\mathbf{W}}(x)](\vec{\mathbf{e}}_q(x)) = \nabla_p W_q. \tag{4.29}$$

Proof. Recall that

$${}^{1}{}_{1}\mathcal{C}[\vec{\mathbf{V}}(x) \otimes \tilde{\mathbf{W}}(x)] = W^p(x)U_p(x) = [\tilde{\mathbf{W}}(x)](\vec{\mathbf{V}}(x)).$$

Therefore, by (4.23), (4.26), (4.25), and (4.16), we have

$$\begin{aligned} &\vec{\mathbf{V}}(x)[W_p] = \nabla_{\vec{\mathbf{V}}}[\tilde{\mathbf{W}}(x)(\vec{\mathbf{e}}_p(x))] = \nabla_{\vec{\mathbf{V}}}\left\{{}^{1}{}_{1}\mathcal{C}[\vec{\mathbf{e}}_p(x) \otimes \tilde{\mathbf{W}}(x)]\right\} \\ &= {}^{1}{}_{1}\mathcal{C}\left\{\nabla_{\vec{\mathbf{V}}}[\vec{\mathbf{e}}_p(x) \otimes \tilde{\mathbf{W}}(x)]\right\} \\ &= {}^{1}{}_{1}\mathcal{C}\left\{[\nabla_{\vec{\mathbf{V}}}\vec{\mathbf{e}}_p(x)] \otimes \tilde{\mathbf{W}}(x) + \vec{\mathbf{e}}_p(x) \otimes \nabla_{\vec{\mathbf{V}}}\tilde{\mathbf{W}}(x)\right\} \\ &= \tilde{\mathbf{W}}(x)[\nabla_{\vec{\mathbf{V}}}(\vec{\mathbf{e}}_p(x))] + [\nabla_{\vec{\mathbf{V}}}\tilde{\mathbf{W}}(x)](\vec{\mathbf{e}}_p(x)) \\ &= \Gamma^q{}_{rp}(x)V^r(x)W_q(x) + [\nabla_{\vec{\mathbf{V}}}\tilde{\mathbf{W}}(x)](\vec{\mathbf{e}}_p(x)), \end{aligned}$$

or

$$[\nabla_{\vec{\mathbf{V}}}\tilde{\mathbf{W}}(x)](\vec{\mathbf{e}}_q(x)) = \left\{\vec{\mathbf{e}}_p(x)[W_q] - \Gamma^u{}_{pq}(x)W_u(x)\right\}V^p(x).$$

Thus, (4.27) and (4.28) follow. ∎

Example 4.1.10 Let $\tilde{\mathbf{A}}(x)$ be a differentiable 1-form. Using (4.28), we have

$$\nabla_j A_k = \frac{\partial}{\partial x^j} A_k(x) - \Gamma^i{}_{jk}(x) A_i(x).$$

Therefore, we obtain the 2-form

$$\left(\frac{1}{2}\right)(\nabla_j A_k - \nabla_k A_j)\, dx^j \wedge dx^k = d\tilde{A}(x) + \left(\frac{1}{2}\right)[\Gamma^i_{jk}(x) - \Gamma^i_{kj}(x)] A_i(x) dx^j \wedge dx^k.$$

The equation above (and also (4.13)) show that the components $\Gamma^i{}_{jk}(x) - \Gamma^i{}_{kj}(x)$ transform as tensor field components. (However, in a non-coordinate or a **non-holonomic basis**, $\Gamma^u{}_{pq}(x) - \Gamma^u{}_{qp}(x)$ *do not* transform as tensor field components!) □

Now, we shall obtain the covariant derivative of a differentiable tensor field in terms of the connection coefficients.

Theorem 4.1.11 *Let ${}^r{}_s\mathbf{T}(x)$ be a differentiable tensor field of class C^1 in the domain $D \subset \mathbb{R}^N$. Then, the covariant derivative of ${}^r{}_s\mathbf{T}(x)$ along the vector field $\vec{\mathbf{V}}(x)$ is furnished by*

$$\begin{aligned}
\nabla_{\vec{\mathbf{V}}}[{}^r{}_s\mathbf{T}(x)] &= V^u(x)\, \{\vec{\mathbf{e}}_u(x)[T^{p_1\ldots p_r}{}_{q_1\ldots q_s}] + \Gamma^{p_1}{}_{uw} T^{wp_2\ldots p_r}{}_{q_1\ldots q_s} \\
&\quad + \Gamma^{p_2}{}_{uw} T^{p_1 w\ldots p_r}{}_{q_1\ldots q_s} + \cdots + \Gamma^{p_r}{}_{uw} T^{p_1 p_2 \ldots w}{}_{q_1\ldots q_s} \\
&\quad - \Gamma^w{}_{uq_1} T^{p_1\ldots p_r}{}_{wq_2\ldots q_s} - \Gamma^w{}_{uq_2} T^{p_1\ldots p_r}{}_{q_1 w\ldots q_s} - \cdots \\
&\quad - \Gamma^w{}_{uq_s} T^{p_1\ldots p_r}{}_{q_1 q_2\ldots w}\}\, \vec{\mathbf{e}}_{p_1}(x) \otimes \cdots \otimes \vec{\mathbf{e}}_{p_r}(x) \otimes \tilde{\mathbf{e}}^{q_1}(x) \otimes \ldots \\
&\quad \otimes \tilde{\mathbf{e}}^{q_s}(x) \\
&=: V^u(x)(\nabla_u T^{p_1\ldots p_r}{}_{q_1\ldots q_s}) \vec{\mathbf{e}}_{p_1}(x) \otimes \cdots \otimes \vec{\mathbf{e}}_{p_r} \otimes \tilde{\mathbf{e}}^{q_1}(x) \otimes \ldots \\
&\quad \otimes \tilde{\mathbf{e}}^{q_s},
\end{aligned} \tag{4.30}$$

$$\begin{aligned}
&\{\nabla_{\vec{\mathbf{V}}}[{}^r{}_s\mathbf{T}(x)]\}\, (\tilde{\mathbf{e}}^{p_1}(x), \ldots, \tilde{\mathbf{e}}^{p_r}(x), \vec{\mathbf{e}}_{q_1}(x), \ldots, \vec{\mathbf{e}}_{q_s}(x)) \\
&= \{\partial_u T^{p_1\ldots p_r}{}_{q_1\ldots q_s} + \Gamma^{p_1}{}_{uw} T^{wp_2\ldots p_r} + \cdots - \Gamma^w{}_{uq_1} T^{p_1\ldots p_r}{}_{wq_2\ldots q_s} - \ldots\}\, V^u(x) \\
&= (\nabla_u T^{p_1\ldots p_r}{}_{q_1\ldots q_s})\, V^u(x),
\end{aligned} \tag{4.31}$$

$$\begin{aligned}
&\{\nabla_{\vec{e}_u}[{}^r{}_s\mathbf{T}(x)]\}\, (\tilde{\mathbf{e}}^{p_1}(x), \ldots, \tilde{\mathbf{e}}^{p_r}(x), \vec{\mathbf{e}}_{q_1}(x), \ldots, \vec{\mathbf{e}}_{q_s}(x)) \\
&\quad = \nabla_u T^{p_1\ldots p_r}{}_{q_1\ldots q_s}.
\end{aligned} \tag{4.32}$$

The proof is left as an exercise.

We can generalize the definition of the covariant derivative ∇ for any differentiable tensor field ${}^{r}{}_{s}\mathbf{T}(x)$. It is given by

$$\begin{aligned}&\{\nabla[{}^{r}{}_{s}\mathbf{T}(x)]\}\left(\tilde{\mathbf{W}}^{1}(x),\ldots,\tilde{\mathbf{W}}^{r}(x),\vec{\mathbf{U}}_{1}(x),\ldots,\vec{\mathbf{U}}_{s}(x),\vec{\mathbf{V}}(x)\right)\\&:=\{\nabla_{\vec{\mathbf{V}}}[{}^{r}{}_{s}\mathbf{T}(x)]\}\left(\tilde{\mathbf{W}}^{1}(x),\ldots,\tilde{\mathbf{W}}^{r}(x),\vec{\mathbf{U}}_{1}(x),\ldots,\vec{\mathbf{U}}_{s}(x)\right)\end{aligned} \tag{4.33}$$

for all $\tilde{\mathbf{W}}^{1},\ldots\tilde{\mathbf{W}}^{r},\vec{\mathbf{U}}_{1},\ldots\vec{\mathbf{U}}_{s},\vec{\mathbf{V}}$ fields. Therefore, we can deduce that

$$\begin{aligned}&\{\nabla[{}^{r}{}_{s}\mathbf{T}(x)]\}\left(\tilde{\mathbf{e}}^{p_1}(x),\ldots,\tilde{\mathbf{e}}^{p_r}(x),\vec{\mathbf{e}}_{q_1}(x),\ldots,\vec{\mathbf{e}}_{q_s}(x),\vec{\mathbf{e}}_{u}\right)\\&=\nabla_{u}T^{p_1\ldots p_r}{}_{q_1\ldots q_s}.\end{aligned} \tag{4.34}$$

Example 4.1.12 Consider the **identity tensor** $\mathbf{I}^{\cdot}{}_{\cdot}(x):=\delta^{p}{}_{q}\vec{\mathbf{e}}_{p}(x)\otimes\tilde{\mathbf{e}}^{q}(x)$. By the equation (4.30) we have

$$\begin{aligned}\nabla_{\vec{\mathbf{V}}}[\mathbf{I}^{\cdot}{}_{\cdot}(x)] &= V^{u}(x)\left\{\vec{\mathbf{e}}_{u}(x)[\delta^{p}{}_{q}]+\Gamma^{p}{}_{uw}(x)\delta^{w}{}_{q}-\Gamma^{w}{}_{uq}(x)\delta^{p}{}_{w}\right\}\vec{\mathbf{e}}_{p}(x)\otimes\tilde{\mathbf{e}}^{p}(x)\\&= V^{u}(x)\left\{0+\Gamma^{p}{}_{uq}(x)-\Gamma^{p}_{uq}(x)\right\}\vec{\mathbf{e}}_{p}(x)\otimes\tilde{\mathbf{e}}^{p}(x)\\&= \mathbf{O}^{\cdot}{}_{\cdot}(x),\\\nabla_{u}(\delta^{p}{}_{q}) &\equiv 0.\end{aligned}$$

□

Example 4.1.13 In the coordinate basis, the components of the covariant derivative of a tensor field from the equation (4.30) can be written as:

$$\begin{aligned}\nabla_{k}T^{i_1\ldots i_r}{}_{j_1\ldots j_s} &= \partial_{k}T^{i_1\ldots i_r}{}_{j_1\ldots j_s}+\Gamma^{i_1}{}_{kl}T^{li_2\ldots i_r}{}_{j_1\ldots j_s}\\&\quad+\Gamma^{i_2}{}_{kl}T^{i_1l\ldots i_r}{}_{j_1\ldots j_s}+\cdots+\Gamma^{i_r}{}_{kl}T^{i_1i_2\ldots l}{}_{j_1\ldots j_s}\\&\quad-\Gamma^{l}{}_{kj_1}T^{i_1\ldots i_r}{}_{lj_2\ldots j_s}-\Gamma^{l}{}_{kj_2}T^{i_1\ldots i_r}{}_{j_1l\ldots j_s}\\&\quad-\cdots-\Gamma^{l}{}_{kj_s}T^{i_1\ldots i_r}{}_{j_1j_2\ldots l}.\end{aligned} \tag{4.35}$$

If we open up the summations in the right hand side, then we find that there are actually $1+N(r+s)$ terms added there! □

Exercises 4.1

1. Let a domain be specified by $D:=\{x\in\mathbb{R}^N : x^1>0, x^2>0,\ldots,x^N>0\}$. Let a vector field $\vec{\mathbf{V}}$ and connection coefficients be prescribed by $\vec{\mathbf{V}}(x):=x^j\frac{\partial}{\partial x^j}$ and $\Gamma^{i}_{jk}(x):=(x^i/x^jx^k)$, respectively. Prove that $(\nabla_jV^i)(\nabla_iV^k)=\delta^{k}{}_{j}+N(N^2+2)(x^k/x^j)$.

2. Show that

$$\nabla[\delta^{12\ldots M}{}_{i_1i_2\ldots i_M}dx^{i_1}\wedge dx^{i_2}\wedge\cdots\wedge dx^{i_M}]={}^{M}{}_{M+1}\mathbf{O}(x)$$

for $2\le M\le N-1$.

3. Let ${}_p\mathbf{A}(x)$ be a differentiable p-form. Prove that

$$\nabla[{}_p\mathbf{A}(x)] = (-1)^p d[{}_p\mathbf{A}(x)].$$

4.2 Covariant Derivatives of Tensors along a Curve

Now, we shall study a tangent vector field $\vec{\mathbf{V}}(x)$ *restricted* to a curve $\mathcal{X}(t)$. Recall from (3.24) that the tangent vector *along* (the image of) the curve $\mathcal{X}$ is given by

$$\begin{aligned} \vec{\mathcal{X}}'(t) &= \frac{d\mathcal{X}^i(t)}{dt}\left(\frac{\partial}{\partial x^i}\right)_{|\mathcal{X}(t)} = \mathcal{X}'^q(t)\vec{\mathbf{e}}_q(\mathcal{X}(t)), \\ t &\in [a,b] \subset \mathbb{R}. \end{aligned} \tag{4.36}$$

The covariant derivative of a vector field $\vec{\mathbf{V}}(x)$, restricted to the curve $\mathcal{X}(t)$ and along the tangential direction $\vec{\mathcal{X}}'(t)$, is defined to be

$$\begin{aligned} \nabla_{\vec{\mathcal{X}}'}[\vec{\mathbf{V}}(\mathcal{X}(t))] &:= \mathcal{X}'^q(t)[\nabla_q V^p]_{|\mathcal{X}(t)}\vec{\mathbf{e}}_p(\mathcal{X}(t)) \\ &= \frac{d\mathcal{X}^i(t)}{dt}\left[(\nabla_i V^j)\frac{\partial}{\partial x^j}\right]_{|\mathcal{X}(t)}. \end{aligned} \tag{4.37}$$

A vector field $\vec{\mathbf{V}}(\mathcal{X}(t))$ is said to be **parallelly propagated** along the curve $\mathcal{X}(t)$ provided

$$\nabla_{\vec{\mathcal{X}}'}[\vec{\mathbf{V}}(\mathcal{X}(t))] = \vec{\mathbf{O}}(\mathcal{X}(t)). \tag{4.38}$$

(See fig. 4.1.)

Remark: The parallel propagation of a vector is strictly a *local* concept.

Example 4.2.1 Let us choose the coordinate basis. By (4.37) and (4.38), we derive for parallel transport of $\vec{\mathbf{V}}$ along $\mathcal{X}(t)$ the equation

$$\nabla_{\vec{\mathcal{X}}'}[\vec{\mathbf{V}}(\mathcal{X}(t))] = \left[\frac{\partial V^i}{\partial x^j} + \Gamma^i{}_{jk}(x)V^k(x)\right]_{|\mathcal{X}(t)} \frac{d\mathcal{X}^j(t)}{dt}\left[\frac{\partial}{\partial x^i}\right]_{|\mathcal{X}(t)} = \vec{\mathbf{O}}(\mathcal{X}(t)).$$

Using the chain rule, we obtain from above that for parallel propagation

$$\frac{DV^i}{dt}(\mathcal{X}(t)) := \frac{d}{dt}V^i(\mathcal{X}(t)) + \left[\Gamma^i{}_{jk}(x)V^k(x)\right]_{|\mathcal{X}(t)}\frac{d\mathcal{X}^j(t)}{dt} = 0. \tag{4.39}$$

Using (4.39), we can express

$$\frac{DV^i}{dt}(\mathcal{X}(t)) = [\nabla_j V^i]_{|\mathcal{X}(t)}\frac{d\mathcal{X}^j(t)}{dt} = 0.$$

□

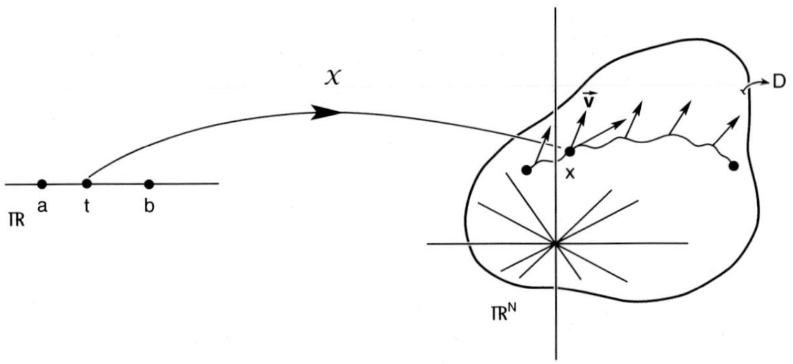

Figure 4.1: Parallel propagation of a vector along a curve.

Remark: For parallel propagation, we derive from (4.39)

$$\frac{dV^i}{dt}(\mathcal{X}(t)) = -\left[\Gamma^i_{jk}(x)V^k(x)\right]_{|\mathcal{X}(t)}\frac{d\mathcal{X}^j(t)}{dt}.$$

Therefore, the rate of change of a parallel vector field $\vec{\mathbf{V}}$ is proportional to *both* $V^k(x)$ and $\frac{d\mathcal{X}^j(t)}{dt}$. Moreover, the components $\frac{d\mathcal{X}^j(t)}{dt}$ depend on the choice of the coordinate chart.

An **affine geodesic** $\mathcal{X}$ is a non-degenerate, twice-differentiable curve into $D \subset \mathbb{R}^N$ such that

$$\nabla_{\vec{\mathcal{X}}'}\left[\vec{\mathcal{X}}'(t)\right] = \lambda(t)\vec{\mathcal{X}}'(t) \tag{4.40}$$

for some integrable function λ over $[a,b] \subset \mathbb{R}$.

Now, recall the reparametrization of a curve provided by (3.26) and (3.27). In an altered notation, a reparametrization is furnished by

$$\begin{aligned} \tau &= \mathcal{T}(t), \quad \frac{d\mathcal{T}(t)}{dt} \neq 0, \\ x &= \mathcal{X}(t) = \mathcal{X}^{\#}(\tau), \\ \frac{df(t)}{dt} &= \frac{d\mathcal{T}(t)}{dt} \cdot \frac{df^{\#}(\tau)}{d\tau}. \end{aligned} \tag{4.41}$$

Theorem 4.2.2 *There exists a reparametrization* (4.41) *for an affine geodesic in* (4.40) *such that*

$$\nabla_{\vec{\mathcal{X}}^{\#\prime}} \left[\vec{\mathcal{X}}^{\#\prime}(t) \right] = \vec{\mathbf{O}}(\mathcal{X}^{\#}(t)), \tag{4.42}$$

$$\frac{d^2\mathcal{X}^{\#i}(\tau)}{d\tau^2} + \Gamma^i{}_{jk}\left(\mathcal{X}^{\#}(\tau)\right) \frac{d\mathcal{X}^{\#j}(\tau)}{d\tau}\frac{d\mathcal{X}^{\#k}(\tau)}{d\tau} = 0. \tag{4.43}$$

Proof. By (4.37), (4.39), and (4.40), the geodesic equation can be written as

$$\frac{d^2\mathcal{X}^i(t)}{dt^2} + \Gamma^i{}_{jk}\left(\mathcal{X}(t)\right) \frac{d\mathcal{X}^j(t)}{dt}\frac{d\mathcal{X}^k(t)}{dt} = \lambda(t)\frac{d\mathcal{X}^i(t)}{dt}. \tag{4.44}$$

Using (4.41) and the chain rule, the last equation yields

$$\frac{d}{dt}\left\{ \frac{d\mathcal{T}(t)}{dt}\frac{d\mathcal{X}^{\#i}(\tau)}{d\tau} \right\} + \Gamma^i{}_{jk}\left(\mathcal{X}^{\#}(\tau)\right) \left[\frac{d\mathcal{T}(t)}{dt}\right]^2 \frac{d\mathcal{X}^{\#j}(\tau)}{d\tau}\frac{d\mathcal{X}^{\#k}(\tau)}{d\tau}$$
$$= \lambda(t)\frac{d\mathcal{T}(t)}{dt}\frac{d\mathcal{X}^{\#i}(\tau)}{d\tau},$$

or

$$\frac{d^2\mathcal{X}^{\#i}(\tau)}{d\tau^2} + \Gamma^i{}_{jk}\left(\mathcal{X}^{\#}(\tau)\right) \frac{d\mathcal{X}^{\#j}(\tau)}{d\tau}\frac{d\mathcal{X}^{\#k}(\tau)}{d\tau}$$
$$= \left[\frac{d\mathcal{T}(t)}{dt}\right]^{-2} \left[\lambda(t)\frac{d\mathcal{T}(t)}{dt} - \frac{d^2\mathcal{T}(t)}{dt^2}\right] \frac{d\mathcal{X}^{\#i}(\tau)}{d\tau}.$$

Let the function $\mathcal{T}$ satisfy the linear, second-order ordinary differential equation

$$\frac{d^2\mathcal{T}(t)}{dt^2} - \lambda(t)\frac{d\mathcal{T}(t)}{dt} = 0.$$

The general solution of the ordinary differential equation above is given by

$$\tau = \mathcal{T}(t) = k\left\{ \int \exp\left[\int \lambda(t)dt\right] dt \right\} + c. \tag{4.45}$$

Here, $k \neq 0$ and c are two arbitrary constants of integration. Any solution $\mathcal{T}$ in (4.45) will render (4.42) and (4.43) valid. ■

Remark: A parameter τ in (4.45) is called an **affine parameter** along the geodesic. A non-homogeneous linear transformation

$$\widehat{\tau} = c_1\tau + c_2, \quad c_1 \neq 0, \tag{4.46}$$

produces another affine parameter.

Example 4.2.3 Let us choose the dimension $N = 2$ and the Euclidean plane. Moreover, let $D := \{(x^1, x^2) \in \mathbb{R}^2 : x^1 > 0, -\pi < x^2 < \pi\}$ for the polar coordinate system. We *assign* the eight connection coefficients as

$$\Gamma^1{}_{11}(x) = \Gamma^1{}_{12}(x) = \Gamma^1{}_{21}(x) = \Gamma^2{}_{11}(x) \equiv 0,$$
$$\Gamma^1{}_{22}(x) = -x^1, \;\; \Gamma^2{}_{12}(x) = \Gamma^2{}_{21}(x) = (x^1)^{-1}.$$

(We have borrowed these connection coefficients from example 4.1.6.)

The affine geodesic equations (4.43), *dropping* the symbol # and changing τ to t, yield

$$\frac{d^2\mathcal{X}^1(t)}{dt^2} - \mathcal{X}^1(t)\left[\frac{d\mathcal{X}^2(t)}{dt}\right]^2 = 0,$$

$$[\mathcal{X}^1(t)]^{-2}\frac{d}{dt}\left\{[\mathcal{X}^1(t)]^2\frac{d\mathcal{X}^2(t)}{dt}\right\} = 0.$$

These are two coupled, second-order, semi-linear ordinary differential equations. The second equation admits the first integral

$$[\mathcal{X}^1(t)]^2\frac{d\mathcal{X}^2(t)}{dt} = h,$$

where h is the constant of integration. Two cases arise. If $h = 0$, the general solutions of the geodesic equations are

$$r \equiv x^1 = \mathcal{X}^1(t) = kt + c_1,$$

$$\phi \equiv x^2 = \mathcal{X}^2(t) = c_2.$$

Here, $k \neq 0$, c_1, and c_2 are constants of integration. The geodesic, in polar coordinates of $\mathbb{R}^2$, represents a portion of a *radial straight line.*

In the second case, $h \neq 0$. We can eliminate $\frac{d\mathcal{X}^2(t)}{dt}$ from the first geodesic equation to obtain

$$\frac{d^2\mathcal{X}^1(t)}{dt^2} - (h)^2[\mathcal{X}^1(t)]^{-3} = 0.$$

This equation can be solved directly. But to have the solution in such a way that the geometry is transparent, we solve it by an unorthodox method. We reparametrize the curve *again* by putting

$$t = T(x^2),$$
$$r \equiv x^1 = \mathcal{X}^1(t) = \mathcal{R}(x^2) \equiv \mathcal{R}(\phi),$$
$$\frac{d\mathcal{X}^1(t)}{dt} = \frac{d\mathcal{X}^2(t)}{dt}\frac{d\mathcal{R}(x^2)}{dx^2} = \frac{h}{[\mathcal{R}(x^2)]^2}\frac{d\mathcal{R}(x^2)}{dx^2},$$
$$\frac{d^2\mathcal{X}^1(t)}{dt^2} = \frac{h}{[\mathcal{R}(x^2)]^2}\frac{d}{dx^2}\left\{\frac{h}{[\mathcal{R}(x^2)]^2}\frac{d\mathcal{R}(x^2)}{dx^2}\right\}.$$

Substituting the equations above into the first geodesic equation, we get

$$\frac{d}{dx^2}\left\{[\mathcal{R}(x^2)]^{-2}\frac{d\mathcal{R}(x^2)}{dx^2}\right\} - [\mathcal{R}(x^2)]^{-1} = 0.$$

Now we introduce another function $\mathcal{U}$ by putting

$$\mathcal{U}(x^2) := [\mathcal{R}(x^2)]^{-1},$$

$$\frac{d\mathcal{U}(x^2)}{dx^2} = -[\mathcal{R}(x^2)]^{-2}\frac{d\mathcal{R}(x^2)}{dx^2}.$$

Expressing the geodesic equation in terms of $\mathcal{U}$, we finally obtain the linear, second-order ordinary differential equation

$$\frac{d^2\mathcal{U}(x^2)}{(dx^2)^2} + \mathcal{U}(x^2) = 0.$$

The general solution of the equation above is

$$\mathcal{U}(x^2) = A\cos(x^2 - \tilde{\omega}).$$

Here, $A \neq 0$ and $\tilde{\omega}$ are two arbitrary constants of integration. The general solution can also be expressed as

$$(\cos\tilde{\omega})(x^1\cos x^2) + (\sin\tilde{\omega})(x^1\sin x^2) = (A)^{-1}.$$

Introducing the usual Cartesian coordinates by

$$x = x^1\cos x^2, \quad y = x^1\sin x^2,$$

the geodesic curve can be symbolically expressed as

$$(\text{const.})x + (\text{const.})y = (\text{non-zero const.}).$$

The equation above represents a portion of a straight line *not* passing through the origin in $\mathbb{R}^2$. □

We shall now generalize the definition in (4.37) to the covariant derivative of a differentiable tensor field along (the image of) a curve $\mathcal{X}$. It is given by

$$\begin{aligned}&\nabla_{\vec{\mathcal{X}}'}[{}^r{}_s\mathbf{T}(\mathcal{X}(t))]\\ &\quad := \vec{\mathcal{X}}'^q(t)\,[\nabla_q T^{p_1\ldots p_r}{}_{w_1\ldots w_s}]_{|\mathcal{X}(t)}\,\vec{\mathbf{e}}_{p_1}(\mathcal{X}(t))\otimes\cdots\otimes\tilde{\mathbf{e}}^{w_s}(\mathcal{X}(t)).\end{aligned} \tag{4.47}$$

Example 4.2.4 In the coordinate basis

$$\begin{aligned}&\nabla_{\vec{\mathcal{X}}'}[{}^r{}_s\mathbf{T}(\mathcal{X}(t))]\\ &\quad = \frac{d\mathcal{X}^k(t)}{dt}\,[\nabla_k T^{i_1\ldots i_r}{}_{j_1\ldots j_s}]_{|\mathcal{X}(t)}\left(\frac{\partial}{\partial x^{i_1}}\otimes\cdots\otimes dx^{j_s}\right)_{|\mathcal{X}(t)},\end{aligned} \tag{4.48}$$

$$
\begin{aligned}
&\nabla_{\vec{\mathcal{X}}'}[{}^r{}_s\mathbf{T}(\mathcal{X}(t))]\left(dx^{i_1}\otimes\cdots\otimes\frac{\partial}{\partial x^{j_s}}\right)_{|\mathcal{X}(t)} \\
&= \left[\nabla_k T^{i_1\ldots i_r}{}_{j_1\ldots j_s}\right]_{|\mathcal{X}(t)}\frac{d\mathcal{X}^k(t)}{dt} =: \frac{DT^{i_1\ldots i_r}{}_{j_1\ldots j_s}(\mathcal{X}(t))}{dt}.
\end{aligned}
\tag{4.49}
$$

For a special case of $\mathbf{T}^{\cdot}_{\ \cdot}(x)$,

$$
\begin{aligned}
\frac{DT^i_j(\mathcal{X}(t))}{dt} &= \left[\nabla_k T^i_j\right]_{|\mathcal{X}(t)}\frac{d\mathcal{X}^k(t)}{dt} \\
&= \left[\partial_k T^i{}_j + \Gamma^i{}_{kl}(x)T^l{}_j(x) - \Gamma^l{}_{kj}(x)T^i{}_l(x)\right]_{|\mathcal{X}(t)}\frac{d\mathcal{X}^k(t)}{dt} \\
&= \frac{dT^i_j(\mathcal{X}(t))}{dt} + \left[\Gamma^i{}_{kl}(x)T^l{}_j(x) - \Gamma^l{}_{kj}(x)T^i{}_l(x)\right]_{|\mathcal{X}(t)}\frac{d\mathcal{X}^k(t)}{dt}.
\end{aligned}
$$

$\frac{DT^i{}_i(\mathcal{X}(t))}{dt} = \frac{dT^i{}_i(\mathcal{X}(t))}{dt}$. □

Exercises 4.2

1. Consider the upper half-plane $D := \{x \in \mathbb{R}^2 : x^1 \in \mathbb{R}, x^2 > 0\}$. Let the connection coefficients in D be prescribed by

$$
\begin{aligned}
&\Gamma^1{}_{jk}(x) \equiv 0,\ \ \Gamma^2_{12}(x) \equiv \Gamma^2_{21}(x) = \Gamma^2_{11}(x) \equiv 0, \\
&\Gamma^2_{22}(x) := (1/2x^2).
\end{aligned}
$$

Integrate the affine geodesic equations using an affine parameter τ. Prove that If $\frac{d\mathcal{X}^1(\tau)}{d\tau} \neq 0$, the geodesics are furnished by semi-cubical parabolas

$$(x^2)^3 = (kx^1 + c)^2.$$

(Here, $k \neq 0$ and c are constants of integration.)

2. Prove the Leibnitz rule of derivatives along a curve $\mathcal{X}$:

$$
\begin{aligned}
&\frac{D}{dt}\left[A^{i_1\ldots i_r}{}_{j_1\ldots j_s}(\mathcal{X}(t))\cdot B^{k_1\ldots k_p}{}_{l_1\ldots l_q}(\mathcal{X}(t))\right] \\
&\quad = A^{i_1\ldots i_r}{}_{j_1\ldots j_s}(\mathcal{X}(t))\cdot\frac{D}{dt}B^{k_1\ldots k_p}{}_{l_1\ldots l_q}(\mathcal{X}(t)) \\
&\quad + \left[\frac{D}{dt}A^{i_1\ldots i_r}{}_{j_1\ldots j_s}(\mathcal{X}(t))\right]\cdot B^{k_1\ldots k_p}{}_{l_1\ldots l_q}(\mathcal{X}(t)).
\end{aligned}
$$

4.3 Lie Bracket, Torsion, and Curvature Tensor

The **commutator** or **Lie bracket** between two vector fields is defined by

$$\left[\vec{\mathbf{U}}, \vec{\mathbf{V}}\right] := \vec{\mathbf{U}}(x)\vec{\mathbf{V}}(x) - \vec{\mathbf{V}}(x)\vec{\mathbf{U}}(x), \tag{4.50}$$

$$\left[\vec{\mathbf{U}}, \vec{\mathbf{V}}\right][f] := \vec{\mathbf{U}}(x)\left\{\vec{\mathbf{V}}(x)[f]\right\} - \vec{\mathbf{V}}(x)\left\{\vec{\mathbf{U}}(x)[f]\right\}, \tag{4.51}$$

for every f belonging to $C^2(D \subset \mathbb{R}^N; \mathbb{R})$. Thus, $[\vec{\mathbf{U}}, \vec{\mathbf{V}}]$ belongs to $T_x(\mathbb{R}^N)$.

Remarks: (i) The Lie bracket is relevant in Lie derivatives. (See appendix 2.) (ii) Commutators are familiar in quantum mechanics.

The salient features of the bracket operation are listed below.

Theorem 4.3.1 *Let $\vec{\mathbf{U}}, \vec{\mathbf{V}}, \vec{\mathbf{W}}$ be three vector fields of class C^3 in D of $\mathbb{R}^N$. Moreover, let f, g be real-valued functions of class C^3 in D. Then the following equations hold:*

(1)
$$\left[\vec{\mathbf{U}}, \vec{\mathbf{V}}\right] = -\left[\vec{\mathbf{V}}, \vec{\mathbf{U}}\right]. \tag{4.52}$$

(2) The Jacobi identity:

$$\left[\vec{\mathbf{U}}, \left[\vec{\mathbf{V}}, \vec{\mathbf{W}}\right]\right] + \left[\vec{\mathbf{V}}, \left[\vec{\mathbf{W}}, \vec{\mathbf{U}}\right]\right] + \left[\vec{\mathbf{W}}, \left[\vec{\mathbf{U}}, \vec{\mathbf{V}}\right]\right] \equiv \vec{\mathbf{O}}(x). \tag{4.53}$$

(3)
$$\begin{aligned}\left[f\vec{\mathbf{V}}, g\vec{\mathbf{W}}\right] = \left\{f(x)\vec{\mathbf{V}}(x)[g]\right\}\vec{\mathbf{W}}(x) - \left\{g(x)\vec{\mathbf{W}}(x)[f]\right\}\vec{\mathbf{V}}(x) \\ + f(x)g(x)[\vec{\mathbf{V}}, \vec{\mathbf{W}}].\end{aligned} \tag{4.54}$$

The proof is left to the reader.

Let basis fields $\vec{\mathbf{e}}_p(x) = E^i{}_p(x)\frac{\partial}{\partial x^i}$ and $\vec{\mathbf{e}}_q(x) = E^j{}_q(x)\frac{\partial}{\partial x^j}$ be of class C^2 in D of $\mathbb{R}^N$. Furthermore, let f be an arbitrary function of class C^2 in D. Then, by (4.51), we have

$$\begin{aligned}[\vec{\mathbf{e}}_p, \vec{\mathbf{e}}_q][f] &= \left[E^i{}_p(x)\frac{\partial E^j{}_q(x)}{\partial x^i} - E^i{}_q(x)\frac{\partial E^j{}_p(x)}{\partial x^i}\right]\frac{\partial f(x)}{\partial x^j} \\ &=: \mathcal{X}^j{}_{pq}(x)\frac{\partial f(x)}{\partial x^j}, \\ [\vec{\mathbf{e}}_p, \vec{\mathbf{e}}_q] &= \mathcal{X}^j{}_{pq}(x)\frac{\partial}{\partial x^j}.\end{aligned} \tag{4.55}$$

In the case where the conjugate 1-forms are given by $\tilde{\mathbf{e}}^u(x) = \tilde{E}^u{}_k(x)dx^k$, we obtain from (4.55) that

$$[\vec{\mathbf{e}}_p, \vec{\mathbf{e}}_q][\tilde{\mathbf{e}}^u] = \mathcal{X}^j{}_{pq}\tilde{E}^u{}_j(x) =: \mathcal{X}^u{}_{pq}(x) \equiv -\mathcal{X}^u{}_{qp}(x). \tag{4.56}$$

$$[\vec{\mathbf{e}}_p, \vec{\mathbf{e}}_q] = \mathcal{X}^u{}_{pq}(x)\vec{\mathbf{e}}_u(x). \tag{4.57}$$

The functions $\mathcal{X}^u{}_{pq}(x)$ are called the **structure coefficients** of the basis field $\{\vec{\mathbf{e}}_p(x)\}_1^N$. *The structure coefficients are not components of a tensor field.*

Example 4.3.2 In a coordinate basis,

$$\begin{aligned} \left[\frac{\partial}{\partial x^i}, \frac{\partial}{\partial x^j}\right] &\equiv \vec{\mathbf{O}}(x), \\ \mathcal{X}^k{}_{ij}(x) &\equiv 0, \end{aligned} \tag{4.58}$$

or, in other words, the basis vectors $\frac{\partial}{\partial x^i}$ and $\frac{\partial}{\partial x^j}$ *commute.* □

Remarks: (i) The basis set $\{\frac{\partial}{\partial x^i}\}_1^N$ is called a **coordinate or natural basis set**. The basis set $\{\vec{e}_q(x)\}_1^N$ for which $\mathcal{X}^u{}_{pq}(x) \equiv 0$ is called a **holonomic basis**. If $\mathcal{X}^u{}_{pq}(x) \not\equiv 0$, the corresponding basis set is called a **non-holonomic basis**.

(ii) The complex operators $(-i)\frac{\partial}{\partial x^j}$ represent **momentum operators** in quantum mechanics.

Example 4.3.3 Consider $N = 3$ and three vector fields furnished by

$$\begin{aligned} &\vec{\mathbf{L}}_1(x) := x^2\frac{\partial}{\partial x^3} - x^3\frac{\partial}{\partial x^2}, \quad \vec{\mathbf{L}}_2(x) := x^3\frac{\partial}{\partial x^1} - x^1\frac{\partial}{\partial x^3}, \\ &\vec{\mathbf{L}}_3(x) := x^1\frac{\partial}{\partial x^2} - x^2\frac{\partial}{\partial x^3}. \end{aligned}$$

The commutators can be worked out to obtain

$$\begin{aligned} &[\vec{\mathbf{L}}_1, \vec{\mathbf{L}}_2] = [\vec{\mathbf{L}}_2, \vec{\mathbf{L}}_3] = \vec{\mathbf{L}}_1, \quad [\vec{\mathbf{L}}_3, \vec{\mathbf{L}}_1] = \vec{\mathbf{L}}_2, \quad [\vec{\mathbf{L}}_i, \vec{\mathbf{L}}_j] = \mathcal{X}^k{}_{ij}(x)\,\vec{\mathbf{L}}_k(x), \\ &\mathcal{X}^k{}_{ij}(x) = \varepsilon_{ijl}\delta^{lk}. \end{aligned}$$

Thus, the structure coefficients are *constant-valued.* □

Remarks: (i) In the case where $\mathcal{X}^k{}_{ij}(x) = C^k{}_{ij}$ are constant-valued, the commutation relation (4.57) indicates that the vectors $\vec{\mathbf{e}}_p(x)$ are **generators of a Lie algebra**.

(ii) In example 4.3.3, the vectors $\vec{\mathbf{L}}_j(x)$ generate the Lie algebra corresponding to the **rotation group** $SO(3, \mathbb{R})$. The complex operators $(-i)\vec{L}_j(x)$ represent the **angular momentum operators** in quantum mechanics.

Now, we shall introduce the **torsion operator T**, which maps a pair of vectors into another vector by the following rule:

$$\mathbf{T}\left(\vec{\mathbf{U}}(x), \vec{\mathbf{V}}(x)\right) := \nabla_{\vec{\mathbf{U}}}\left[\vec{\mathbf{V}}(x)\right] - \nabla_{\vec{\mathbf{V}}}\left[\vec{\mathbf{U}}(x)\right] - \left[\vec{\mathbf{U}}(x), \vec{\mathbf{V}}(x)\right]. \tag{4.59}$$

Note that

$$\mathbf{T}\left(\vec{\mathbf{V}}(x), \vec{\mathbf{U}}(x)\right) \equiv -\mathbf{T}\left(\vec{\mathbf{U}}(x), \vec{\mathbf{V}}(x)\right), \tag{4.60}$$

$$\mathbf{T}\left(f(x)\vec{\mathbf{U}}(x), g(x)\vec{\mathbf{V}}(x)\right) = f(x)g(x)\mathbf{T}\left(\vec{\mathbf{U}}(x), \vec{\mathbf{V}}(x)\right). \tag{4.61}$$

The torsion tensor field $\mathbf{T}^{\cdot}{}_{..}(x)$ is defined by

$$\left[\mathbf{T}^{\cdot}{}_{..}(x)\right]\left(\tilde{\mathbf{W}}(x), \vec{\mathbf{U}}(x), \vec{\mathbf{V}}(x)\right) := \tilde{\mathbf{W}}(x)\left[\mathbf{T}\left(\vec{\mathbf{U}}(x), \vec{\mathbf{V}}(x)\right)\right], \tag{4.62}$$

for all $\tilde{\mathbf{W}}, \vec{\mathbf{U}}, \vec{\mathbf{V}}$ fields. By (4.60) and (4.62), we have the antisymmetry

$$\left[\mathbf{T}^{\cdot}{}_{..}(x)\right]\left(\tilde{\mathbf{W}}(x), \vec{\mathbf{V}}(x), \vec{\mathbf{U}}(x)\right) \equiv -\left[\mathbf{T}^{\cdot}{}_{..}(x)\right]\left(\tilde{\mathbf{W}}(x), \vec{\mathbf{U}}(x), \vec{\mathbf{V}}(x)\right). \tag{4.63}$$

Let us work out the components of $\mathbf{T}^{\cdot}{}_{..}(x)$ relative to a general basis set $\{\vec{\mathbf{e}}_p(x)\}_1^N$ and the corresponding $\{\tilde{\mathbf{e}}^q(x)\}_1^N$. By (4.62) and (4.63), we have

$$\begin{aligned} T^p{}_{qu}(x) &= \left[\mathbf{T}^{\cdot}{}_{..}(x)\right]\left(\tilde{\mathbf{e}}^p(x), \vec{\mathbf{e}}_q(x), \vec{\mathbf{e}}_u(x)\right) \\ &= \tilde{\mathbf{e}}^p(x)\left[\nabla_{\vec{\mathbf{e}}_q}\left[\vec{\mathbf{e}}_u\right]\right] - \tilde{\mathbf{e}}^p(x)\left[\nabla_{\vec{\mathbf{e}}_u}\left[\vec{\mathbf{e}}_q\right]\right] - \tilde{\mathbf{e}}^p(x)\left[\vec{\mathbf{e}}_q, \vec{\mathbf{e}}_u\right]. \end{aligned}$$

Using (4.18) and (4.57),

$$T^p{}_{qu}(x) = \Gamma^p{}_{qu}(x) - \Gamma^p{}_{uq}(x) - \mathcal{X}^p{}_{qu}(x) \equiv -T^p{}_{uq}(x). \tag{4.64}$$

By (4.58), the components of the torsion tensor relative to the coordinate basis are

$$T^i{}_{jk}(x) = \Gamma^i{}_{jk}(x) - \Gamma^i{}_{kj}(x) \equiv -T^i{}_{kj}(x). \tag{4.65}$$

Introducing the symmetric part of $\Gamma^i{}_{jk}(x)$ as

$$S^i{}_{jk}(x) := (1/2)\left[\Gamma^i{}_{jk}(x) + \Gamma^i{}_{kj}(x)\right] \equiv S^i{}_{kj}(x), \tag{4.66}$$

we can express

$$\Gamma^i{}_{jk}(x) \equiv S^i{}_{jk}(x) + (1/2)T^i{}_{jk}(x). \tag{4.67}$$

The equation above shows that $\Gamma^i{}_{jk}(x)$ is symmetric if and only if $T^i{}_{jk}(x) \equiv 0$ or the torsion is identically zero.

Example 4.3.4 Let us define the connection coefficients in the coordinate basis as

$$\Gamma^i{}_{jk}(x) = x^i (x^j)^2 (x^k)^3$$

for every $x \in \mathbb{R}^N$. Using (4.65), the torsion components are given by

$$T^i{}_{jk}(x) = \Gamma^i{}_{jk}(x) - \Gamma^i{}_{kj}(x) = x^i(x^j x^k)^2(x^k - x^j) \equiv -T^i{}_{kj}(x).$$

The torsion components vanish at each of the $(N-1)$-dimensional **hyperplanes** given by

$$x^i = 0, \quad x^j = x^k,$$
$$i, j, k \in \{1, 2, \ldots, N\}.$$

Using (4.66), the symmetric part of the connection coefficient is provided by

$$\begin{aligned} S^i{}_{kj}(x) = (1/2)\left[\Gamma^i{}_{jk}(x) + \Gamma^i{}_{kj}(x)\right] &= (1/2)x^i(x^j x^k)^2(x^k + x^j) \\ &\equiv S^i{}_{kj}(x). \end{aligned}$$

$S^i{}_{jk}(1, 1, \ldots, 1) \equiv 1.$ □

Now, we shall derive the commutators of covariant derivatives. By (4.1), (4.31), and (4.55), we get

$$\begin{aligned} \nabla_q \nabla_p f - \nabla_p \nabla_q f &= [\vec{\mathbf{e}}_q, \vec{\mathbf{e}}_p]\, f - [\Gamma^u{}_{qp}(x) - \Gamma^u{}_{pq}(x)]\, \nabla_u f \\ &= -T^u{}_{qp}(x) \nabla_u f \equiv T^u{}_{pq}(x) \partial_u f, \end{aligned} \tag{4.68}$$

$$\nabla_i \nabla_j f - \nabla_j \nabla_i f = -T^k{}_{ij}(x) \frac{\partial f(x)}{\partial x^k} = \left[\Gamma^k{}_{ji}(x) - \Gamma^k{}_{ij}(x)\right] \partial_k f. \tag{4.69}$$

Now, we shall define the **curvature operator R**, which maps a pair of vectors into a *linear operator* over $T_x(\mathbb{R}^N)$. The exact definition is provided by

$$\begin{aligned} \left[\mathbf{R}\left(\vec{\mathbf{U}}(x), \vec{\mathbf{V}}(x)\right)\right]\left[\vec{\mathbf{Y}}(x)\right] &:= \nabla_{\vec{\mathbf{U}}}\left[\nabla_{\vec{\mathbf{V}}} \vec{\mathbf{Y}}(x)\right] - \nabla_{\vec{\mathbf{V}}}\left[\nabla_{\vec{\mathbf{U}}} \vec{\mathbf{Y}}(x)\right] \\ &\quad - \nabla_{[\vec{\mathbf{U}}, \vec{\mathbf{V}}]}\left[\vec{\mathbf{Y}}(x)\right] \\ &\equiv -\left[\mathbf{R}\left(\vec{\mathbf{V}}, \vec{\mathbf{U}}\right)\right]\left[\vec{\mathbf{Y}}(x)\right] \end{aligned} \tag{4.70}$$

for all $\vec{\mathbf{U}}, \vec{\mathbf{V}}, \vec{\mathbf{Y}}$ in $T_x(\mathbb{R}^N)$. By (4.19), (4.20), (4.21), and (4.22), it follows that

$$\left[\mathbf{R}\left(f(x)\vec{\mathbf{U}}, \vec{\mathbf{V}}\right)\right]\left[\vec{\mathbf{Y}}\right] = f(x)\left[\mathbf{R}\left(\vec{\mathbf{U}}, \vec{\mathbf{V}}\right)\right]\left[\vec{\mathbf{Y}}\right], \tag{4.71}$$

$$\begin{aligned} &\left[\mathbf{R}\left(\vec{\mathbf{U}}, \vec{\mathbf{V}}\right)\right]\left[f(x)\vec{\mathbf{Y}}(x) + g(x)\vec{\mathbf{Z}}(x)\right] \\ &\quad = f(x)\left[\mathbf{R}\left(\vec{\mathbf{U}}(x), \vec{\mathbf{V}}(x)\right)\right]\left[\vec{\mathbf{Y}}(x)\right] + g(x)\left[\mathbf{R}\left(\vec{\mathbf{U}}(x), \vec{\mathbf{V}}(x)\right)\right]\left[\vec{\mathbf{Z}}(x)\right]. \end{aligned} \tag{4.72}$$

The $(1+3)$th-order **curvature tensor** field $\mathbf{R}^{\cdot}{}_{\ldots}(x)$ is defined by

$$\begin{aligned} &[\mathbf{R}^{\cdot}{}_{\ldots}(x)]\left(\tilde{\mathbf{W}}(x), \vec{\mathbf{Y}}(x), \vec{\mathbf{U}}(x), \vec{\mathbf{V}}(x)\right) \\ &\quad := \tilde{\mathbf{W}}(x)\left[\mathbf{R}\left(\vec{\mathbf{U}}(x), \vec{\mathbf{V}}(x)\right)\left[\vec{\mathbf{Y}}(x)\right]\right] \end{aligned} \tag{4.73}$$

for all $\tilde{\mathbf{W}}, \vec{\mathbf{Y}}, \vec{\mathbf{U}}$, and $\vec{\mathbf{V}}$ fields.

Theorem 4.3.5 *The components of the curvature tensor field* $\mathbf{R}^{\cdot}...(x)$ *relative to a general basis field* $\{\vec{\mathbf{e}}_p(x) \otimes \tilde{\mathbf{e}}^q(x) \otimes \tilde{\mathbf{e}}^r(x) \otimes \tilde{\mathbf{e}}^s(x)\}_1^N$ *are furnished by*

$$\begin{aligned} &R^p{}_{qrs}(x) \\ &:= [\mathbf{R}^{\cdot}...(x)]\,(\tilde{\mathbf{e}}^p(x) \otimes \vec{\mathbf{e}}_q(x) \otimes \vec{\mathbf{e}}_r(x) \otimes \vec{\mathbf{e}}_s(x)) \\ &= \partial_r \Gamma^p_{sq} - \partial_s \Gamma^p_{rq} + \Gamma^p_{ru}(x)\Gamma^u_{sq}(x) - \Gamma^p_{su}(x)\Gamma^u_{rq}(x) - \chi^u_{rs}(x)\Gamma^p_{uq}(x) \\ &\equiv -R^p_{qsr}(x). \end{aligned} \tag{4.74}$$

Proof. By (4.73) and (4.70), we have

$$\begin{aligned} R^p{}_{qrs}(x) &= \tilde{\mathbf{e}}^p(x)\,[\mathbf{R}\,(\vec{\mathbf{e}}_r(x), \vec{\mathbf{e}}_s(x))\,[\vec{\mathbf{e}}_q(x)]] \\ &= \tilde{\mathbf{e}}^p(x)\left\{\nabla_{\vec{\mathbf{e}}_r}\left[\nabla_{\vec{\mathbf{e}}_s}(\vec{\mathbf{e}}_q(x))\right] - \nabla_{\vec{\mathbf{e}}_s}\left[\nabla_{\vec{\mathbf{e}}_r}(\vec{\mathbf{e}}_q(x))\right] - \nabla_{[\vec{\mathbf{e}}_r, \vec{\mathbf{e}}_s]}\left[\vec{\mathbf{e}}_q(x)\right]\right\}. \end{aligned}$$

Using (4.18), (4.22), (4.57), (4.20), and (4.21), we obtain

$$\begin{aligned} R^p{}_{qrs}(x) &= \tilde{\mathbf{e}}^p(x)\left\{\nabla_{\vec{\mathbf{e}}_r}\left[\Gamma^u{}_{sq}\vec{\mathbf{e}}_u\right] - \nabla_{\vec{\mathbf{e}}_s}\left[\Gamma^u{}_{rq}\vec{\mathbf{e}}_u\right] - \nabla_{\chi^u{}_{rs}\vec{\mathbf{e}}_u}\left[\vec{\mathbf{e}}_q\right]\right\} \\ &= \tilde{\mathbf{e}}^p(x)\left\{\left[\vec{\mathbf{e}}_r\left[\Gamma^u{}_{sq}\right]\right]\vec{\mathbf{e}}_u + \Gamma^u{}_{sq}\nabla_{\vec{\mathbf{e}}_r}(\vec{\mathbf{e}}_u) - \left[\vec{\mathbf{e}}_s\left[\Gamma^u{}_{rq}\right]\right]\vec{\mathbf{e}}_u \right. \\ &\qquad \left. - \Gamma^u{}_{rq}\nabla_{\vec{\mathbf{e}}_s}(\vec{\mathbf{e}}_u) - \chi^u{}_{rs}\nabla_{\vec{\mathbf{e}}_u}(\vec{\mathbf{e}}_q)\right\} \\ &= \vec{\mathbf{e}}_r\left[\Gamma^p{}_{sq}\right] - \vec{\mathbf{e}}_s\left[\Gamma^p{}_{rq}\right] + \Gamma^p{}_{ru}\Gamma^u{}_{sq} - \Gamma^p{}_{su}\Gamma^u{}_{rq} - \chi^u{}_{rs}\Gamma^p{}_{uq}. \qquad \blacksquare \end{aligned}$$

Corollary 4.3.6 *The components of* $\mathbf{R}^{\cdot}...(x)$ *in the coordinate basis are provided by*

$$\begin{aligned} R^i{}_{jmn}(x) &= \frac{\partial}{\partial x^m}\Gamma^i{}_{nj}(x) - \frac{\partial}{\partial x^n}\Gamma^i{}_{mj}(x) + \Gamma^i{}_{mh}(x)\Gamma^h{}_{nj}(x) - \Gamma^i{}_{nh}(x)\Gamma^h{}_{mj}(x) \\ &\equiv -R^i{}_{jnm}. \end{aligned} \tag{4.75}$$

(The proof is left to the reader.)

Example 4.3.7 $\Gamma^i{}_{jk}(x) \equiv 0 \Longrightarrow R^i{}_{jmn}(x) \equiv 0$. (However, $R^i{}_{jmn}(x) \equiv 0$ does *not* imply that $\Gamma^i{}_{jk}(x) \equiv 0$.) $\square$

Example 4.3.8 Let us borrow the connection coefficients in polar coordinates for $\mathbb{R}^2$ from example 4.2.3. The non-zero components are given by

$$\Gamma^1{}_{22}(x) = -x^1, \;\; \Gamma^2{}_{12}(x) = \Gamma^2{}_{21}(x) = (x^1)^{-1}.$$

Therefore, we have

$$\frac{\partial}{\partial x^2}\left[\Gamma^i{}_{jk}(x)\right] \equiv 0.$$

Let us calculate from (4.75) the component

$$\begin{aligned} R^1{}_{212}(x) &= \partial_1 \Gamma^1{}_{22} - \partial_2 \Gamma^1{}_{12} + \Gamma^1{}_{1h}\Gamma^h{}_{22} - \Gamma^1{}_{2h}\Gamma^h{}_{12} \\ &= -1 - (-x^1)(x^1)^{-1} \equiv 0. \end{aligned}$$

Similarly, all other components turn out to be identically zero. Thus, in this example $R^i{}_{jmn}(x) \equiv 0$, although $\Gamma^i{}_{jk}(x) \not\equiv 0$. □

In the preceding two examples, we have $\mathbf{R}^{\cdot}{}_{\dots}(x) \equiv \mathbf{O}^{\cdot}{}_{\dots}(x)$ in domains of $\mathbb{R}^N$ and $\mathbb{R}^2$, respectively. The domain in the manifold M corresponding to the domain where $\mathbf{R}^{\cdot}{}_{\dots}(x) \equiv \mathbf{O}^{\cdot}{}_{\dots}(x)$ is called **flat**. If $\mathbf{R}^{\cdot}{}_{\dots}(x) \not\equiv \mathbf{O}^{\cdot}{}_{\dots}(x)$, the corresponding domain in the manifold M is **curved** in some sense.

Now, we shall work out the commutators of covariant derivatives. By (4.16) and (4.17), we have

$$\nabla_u T^p = \vec{\mathbf{e}}_u(x)[T^p] + \Gamma^p{}_{uq}(x) T^q(x).$$

Since $\nabla_u T^p$ are components of a $(1+1)$th-order tensor, (4.31) yields (with Leibnitz rules)

$$\begin{aligned} \nabla_v \nabla_u T^p &= \vec{\mathbf{e}}_v \left\{ \vec{\mathbf{e}}_u[T^p] + \Gamma^p{}_{uq} T^q \right\} + \Gamma^p{}_{vw} \left\{ \vec{\mathbf{e}}_u[T^w] + \Gamma^w{}_{uq} T^q \right\} \\ &\quad - \Gamma^w{}_{vu} \left\{ \vec{\mathbf{e}}_w[T^p] + \Gamma^p{}_{wq} T^q \right\} \\ &= \vec{\mathbf{e}}_v \vec{\mathbf{e}}_u[T^p] + \left\{ \vec{\mathbf{e}}_v[\Gamma^p{}_{uq}] + \Gamma^p{}_{vw}\Gamma^w{}_{uq} - \Gamma^w{}_{vu}\Gamma^p{}_{wq} \right\} T^q \\ &\quad + \Gamma^p{}_{uq}\vec{\mathbf{e}}_v[T^q] + \Gamma^p{}_{vw}\vec{\mathbf{e}}_u[T^w] - \Gamma^w{}_{vu}\vec{\mathbf{e}}_w[T^p]. \end{aligned}$$

By the equation above, (4.72), and some cancellations of terms, we compute the commutator

$$\begin{aligned} &(\nabla_v \nabla_u - \nabla_u \nabla_v) T^p \\ &= [\vec{\mathbf{e}}_v, \vec{\mathbf{e}}_u][T^p] \\ &\quad + \left\{ \vec{\mathbf{e}}_v[\Gamma^p{}_{uq}] - \vec{\mathbf{e}}_u[\Gamma^p{}_{vq}] + \Gamma^p{}_{vw}\Gamma^w{}_{uq} - \Gamma^p{}_{uw}\Gamma^w{}_{vq} + (\Gamma^w{}_{uv} - \Gamma^w{}_{vu})\Gamma^p{}_{wq} \right\} T^q \\ &\quad + (\Gamma^w{}_{uv} - \Gamma^w{}_{vu})\vec{\mathbf{e}}_w[T^p] \\ &= \left\{ \vec{\mathbf{e}}_v[\Gamma^p{}_{uq}] - \vec{\mathbf{e}}_u[\Gamma^p{}_{vq}] + \Gamma^p{}_{vw}\Gamma^w{}_{uq} - \Gamma^p{}_{uw}\Gamma^w{}_{vq} \right\} T^q \\ &\quad - T^w{}_{vu}\vec{\mathbf{e}}_w[T^q] + (\Gamma^w{}_{uv} - \Gamma^w{}_{vu})\Gamma^p{}_{wq} T^q \\ &= \left\{ \vec{\mathbf{e}}_v[\Gamma^p{}_{uq}] - \vec{\mathbf{e}}_u[\Gamma^p{}_{vq}] + \Gamma^p{}_{vw}\Gamma^w{}_{uq} - \Gamma^p{}_{uw}\Gamma^w{}_{vq} - \chi^w{}_{uv}\Gamma^p{}_{wq} \right\} T^q \\ &\quad - T^w{}_{vu}\nabla_w T^p \\ &= R^p{}_{qvu}(x) T^q(x) - T^q{}_{vu}(x)\nabla_q T^p. \end{aligned} \tag{4.76}$$

Thus, in the coordinate basis,

$$(\nabla_k \nabla_l - \nabla_l \nabla_k) T^j = R^j{}_{ikl}(x) T^i(x) - T^i{}_{kl}(x) \nabla_i T^j. \tag{4.77}$$

Now, we shall deal *briefly* with **Cartan's approach** to connections and curvatures. We have already defined N^2 connection forms $\tilde{\mathbf{w}}^p{}_q(x) = \Gamma^p{}_{uq}(x)\tilde{\mathbf{e}}^u(x)$ in (4.8). Similarly, we can define N **torsion 2-forms**

$$\mathbf{T}^p{}_{..}(x) := (1/2)T^p{}_{qu}(x)\tilde{\mathbf{e}}^q(x) \wedge \tilde{\mathbf{e}}^u(x). \tag{4.78}$$

Moreover, we can introduce N^2 **curvature 2-forms** by

$$\boldsymbol{\Omega}^p{}_{q..}(x) := (1/2)R^p{}_{quv}(x)\tilde{\mathbf{e}}^u(x) \wedge \tilde{\mathbf{e}}^v(x). \tag{4.79}$$

Cartan's structural equations involve the 2-forms $\mathbf{T}^p{}_{..}(x)$ and $\boldsymbol{\Omega}^p{}_{q..}(x)$. These equations are furnished in the following theorem.

Theorem 4.3.9 *Let the N torsion 2-forms $\mathbf{T}^p{}_{..}(x)$ and the N^2 curvature 2-forms $\boldsymbol{\Omega}^p{}_{q..}(x)$ exist and be differentiable in a domain $D \subset \mathbb{R}^N$. Then, Cartan's structural equations*

(i)
$$\mathbf{T}^p{}_{..}(x) = d\tilde{\mathbf{e}}^p(x) + \tilde{\mathbf{w}}^p{}_q(x) \wedge \tilde{\mathbf{e}}^q(x), \tag{4.80}$$

(ii)
$$\boldsymbol{\Omega}^p{}_{q..}(x) = d\tilde{\mathbf{w}}^p{}_q(x) + \tilde{\mathbf{w}}^p{}_u(x) \wedge \tilde{\mathbf{w}}^u{}_q(x), \tag{4.81}$$

hold in that domain.

Proof. (i) Let $\tilde{\mathbf{A}}(x)$ be an *arbitrary* differentiable 1-form so that $d\tilde{\mathbf{A}}(x)$ is a 2-form. By (4.50), (3.9), and (3.16), we obtain

$$\begin{aligned} &2d\tilde{\mathbf{A}}(x)\left(\vec{\mathbf{U}}(x), \vec{\mathbf{V}}(x)\right) \\ &= \vec{\mathbf{U}}(x)\left(\tilde{\mathbf{A}}(x)[\vec{\mathbf{V}}]\right) - \vec{\mathbf{V}}(x)\left(\tilde{\mathbf{A}}(x)[\vec{\mathbf{U}}]\right) - \tilde{\mathbf{A}}(x)[\vec{\mathbf{U}}, \vec{\mathbf{V}}]. \end{aligned} \tag{4.82}$$

By (4.82), (4.8), and (4.64), the right-hand side of (4.80) yields

$$\begin{aligned} &d\tilde{\mathbf{e}}^p\left(\vec{\mathbf{e}}_u, \vec{\mathbf{e}}_v\right) + \tilde{\mathbf{w}}^p{}_q \wedge \tilde{\mathbf{e}}^q\left(\vec{\mathbf{e}}_u, \vec{\mathbf{e}}_v\right) \\ &= (1/2)\left\{\vec{\mathbf{e}}_u[\tilde{\mathbf{e}}^p(\vec{\mathbf{e}}_v)] - \vec{\mathbf{e}}_v[\tilde{\mathbf{e}}^p(\vec{\mathbf{e}}_u)] - \tilde{\mathbf{e}}^p[\vec{\mathbf{e}}_u, \vec{\mathbf{e}}_v]\right\} \\ &\quad + \frac{1}{2}\left\{\Gamma^p{}_{rq}(\tilde{\mathbf{e}}^r \otimes \tilde{\mathbf{e}}^q - \tilde{\mathbf{e}}^q \otimes \tilde{\mathbf{e}}^r)(\vec{\mathbf{e}}_u, \vec{\mathbf{e}}_v)\right\} \\ &= \frac{1}{2}\left\{[0 - 0 - \chi^p{}_{uv}] + (\Gamma^p{}_{uv} - \Gamma^p{}_{vu})\right\} = \left(\frac{1}{2}\right)T^p{}_{uv}(x) = \mathbf{T}^p{}_{..}(x)(\vec{\mathbf{e}}_u, \vec{\mathbf{e}}_v). \end{aligned}$$

(ii) By (4.82), (4.8), and (4.74), the right-hand side of (4.81) yields

$$
\begin{aligned}
& d\tilde{\mathbf{w}}^p{}_q(x)\,(\vec{\mathbf{e}}_u, \vec{\mathbf{e}}_v) + \left[\tilde{\mathbf{w}}^p{}_u(x) \wedge \tilde{\mathbf{w}}^u_q(x)\right](\vec{\mathbf{e}}_u, \vec{\mathbf{e}}_v) \\
&= (1/2)\,\{\vec{\mathbf{e}}_u[\tilde{\mathbf{w}}^p{}_q(\vec{\mathbf{e}}_p)] - \vec{\mathbf{e}}_v[\tilde{\mathbf{w}}^p{}_q(\vec{\mathbf{e}}_u)] - \tilde{\mathbf{w}}^p{}_q[\vec{\mathbf{e}}_u, \vec{\mathbf{e}}_v]\} \\
&\qquad + \{\Gamma^p{}_{rw}\Gamma^w{}_{sq}[\tilde{\mathbf{e}}^r \wedge \tilde{\mathbf{e}}^s](\vec{\mathbf{e}}_u, \vec{\mathbf{e}}_v)\} \\
&= \frac{1}{2}\,\{\vec{\mathbf{e}}_u[\Gamma^p{}_{vq}] - \vec{\mathbf{e}}_v[\Gamma^p{}_{uq}] - \chi^r{}_{uv}\Gamma^p{}_{rq} \\
&\qquad + \Gamma^p{}_{rw}\Gamma^w{}_{sq}(\delta^r{}_u\delta^s{}_v - \delta^r{}_v\delta^s{}_u)\} \\
&= \left(\frac{1}{2}\right) R^p{}_{quv}(x) = \mathbf{\Omega}^p{}_{q..}(x)(\vec{\mathbf{e}}_u, \vec{\mathbf{e}}_v). \qquad \blacksquare
\end{aligned}
$$

Now we shall prove the **Bianchi's differential identities.**

Theorem 4.3.10 *Let N torsion forms $\mathbf{T}^p{}_{..}(x)$ and N^2 curvature forms $\mathbf{\Omega}^p{}_{q..}(x)$ be differentiable in a domain D of $\mathbb{R}^N$. Then the differential identities*

(i)
$$d\mathbf{T}^p{}_{..}(x) \equiv \mathbf{\Omega}^p{}_{q..}(x) \wedge \tilde{\mathbf{e}}^q(x) - \tilde{\mathbf{w}}^p{}_q(x) \wedge \mathbf{T}^q{}_{..}(x), \tag{4.83}$$

(ii)
$$d\mathbf{\Omega}^p{}_{q..}(x) \equiv \mathbf{\Omega}^p{}_{u..}(x) \wedge \tilde{\mathbf{w}}^u{}_q(x) - \tilde{\mathbf{w}}^p{}_u(x) \wedge \mathbf{\Omega}^u{}_{q..}(x), \tag{4.84}$$

hold in D.

Proof. (i) By (4.80), (3.73), and (3.70), we have

$$
\begin{aligned}
d\mathbf{T}^p{}_{..}(x) &= d^2\tilde{\mathbf{e}}^p(x) + d[\tilde{\mathbf{w}}^p{}_q(x) \wedge \tilde{\mathbf{e}}^q(x)] \\
&\equiv \mathbf{O}_{...}(x) + [d\tilde{\mathbf{w}}^p{}_q(x) \wedge \tilde{\mathbf{e}}^q(x) - \tilde{\mathbf{w}}^p{}_q(x) \wedge d\tilde{\mathbf{e}}^q(x)] \\
&= [\mathbf{\Omega}^p{}_{q..}(x) - \tilde{\mathbf{w}}^p{}_u(x) \wedge \tilde{\mathbf{w}}^u{}_q(x)] \wedge \tilde{\mathbf{e}}^q(x) \\
&\quad - \tilde{\mathbf{w}}^p{}_q(x) \wedge [\mathbf{T}^q{}_{..}(x) - \tilde{\mathbf{w}}^q{}_u(x) \wedge \tilde{\mathbf{e}}^u(x)] \\
&= \mathbf{\Omega}^p{}_{q..}(x) \wedge \tilde{\mathbf{e}}^q(x) - \tilde{\mathbf{w}}^p{}_q(x) \wedge \mathbf{T}^q{}_{..}(x).
\end{aligned}
$$

(ii)

$$
\begin{aligned}
d\mathbf{\Omega}^p{}_{q..}(x) &\equiv \mathbf{O}_{...}(x) + [d\tilde{\mathbf{w}}^p{}_u(x) \wedge \tilde{\mathbf{w}}^u{}_q(x) - \tilde{\mathbf{w}}^p{}_u(x) \wedge d\tilde{\mathbf{w}}^u{}_q(x)] \\
&= [\mathbf{\Omega}^p{}_{u..}(x) - \tilde{\mathbf{w}}^p{}_v(x) \wedge \tilde{\mathbf{w}}^v{}_u(x)] \wedge \tilde{\mathbf{w}}^u{}_q(x) \\
&\quad - \tilde{\mathbf{w}}^p{}_u(x) \wedge [\mathbf{\Omega}^u{}_{q..}(x) - \tilde{\mathbf{w}}^u{}_v(x) \wedge \tilde{\mathbf{w}}^v{}_q(x)] \\
&= \mathbf{\Omega}^p{}_{u..}(x) \wedge \tilde{\mathbf{w}}^u{}_q(x) - \tilde{\mathbf{w}}^p{}_u(x) \wedge \mathbf{\Omega}^u{}_{q..}(x). \qquad \blacksquare
\end{aligned}
$$

Corollary 4.3.11 *In a coordinate basis, the identities* (4.83) *and* (4.84) *imply, respectively,*

(*i*)

$$\begin{aligned} & R^k{}_{ijl}(x) + R^k{}_{jli}(x) + R^k{}_{lij}(x) \\ & \equiv [\nabla_j T^k{}_{li} + T^m{}_{ij}(x) T^k{}_{ml}(x)] \\ & \quad +[\nabla_l T^k{}_{ij} + T^m{}_{jl}(x) T^k{}_{mi}(x)] \\ & \quad [\nabla_i T^k{}_{jl} + T^m{}_{li}(x) T^k{}_{mj}(x)], \end{aligned} \tag{4.85}$$

(*ii*)

$$\begin{aligned} & \nabla_j R^m{}_{ikl} + \nabla_k R^m{}_{ilj} + \nabla_l R^m{}_{ijk} + T^h{}_{jk}(x) R^m{}_{ihl}(x) \\ & \quad +T^h{}_{kl}(x) R^m{}_{ihj}(x) + T^h{}_{lj}(x) R^m{}_{ihk}(x) \\ & \equiv 0. \end{aligned} \tag{4.86}$$

(The proof is left to the reader.)

Example 4.3.12 Consider the surface S^2 of a unit sphere. (See fig. 3.3.) We use the usual spherical polar coordinates $x^1 \equiv \theta$, $x^2 \equiv \phi$. The domain of validity of polar coordinates is given by $D := \{(x^1, x^2) \in \mathbb{R}^2 : \ 0 < x^1 < \pi, -\pi < x^2 < \pi\}$. The coordinate basis and corresponding 1-forms are provided by

$$\begin{aligned} & (\vec{\mathbf{e}}_1(x), \vec{\mathbf{e}}_2(x)) = \left(\frac{\partial}{\partial x^1}, \frac{\partial}{\partial x^2} \right) \equiv \left(\frac{\partial}{\partial \theta}, \frac{\partial}{\partial \phi} \right), \\ & (\tilde{\mathbf{e}}^1(x), \tilde{\mathbf{e}}^2(x)) = (dx^1, dx^2) \equiv (d\theta, d\phi), \\ & d\tilde{\mathbf{e}}^p(x) \equiv \mathbf{O}_{..}(x). \end{aligned}$$

We *prescribe* the eight connection coefficients in the coordinate basis by

$$\begin{aligned} & \Gamma^1{}_{11}(x) = \Gamma^1{}_{12}(x) = \Gamma^1{}_{21}(x) = \Gamma^1{}_{12}(x) = \Gamma^2{}_{11}(x) \equiv 0, \\ & \Gamma^1{}_{22}(x) = -(\sin x^1)(\cos x^1), \ \ \Gamma^2{}_{12}(x) = \Gamma^2{}_{21}(x) = \cot x^1. \end{aligned} \tag{4.87}$$

We note that

$$\Gamma^i{}_{kj}(x) \equiv \Gamma^i{}_{jk}(x), \ \ \chi^i{}_{jk}(x) \equiv 0, \ \ T^i{}_{jk} \equiv 0.$$

By (4.8), we compute the four connection 1-forms to be

$$\begin{aligned} \tilde{\mathbf{w}}^1{}_1(x) &= \Gamma^1{}_{j1} dx^j \equiv \tilde{\mathbf{O}}(x), \\ \tilde{\mathbf{w}}^1{}_2(x) &= \Gamma^1{}_{j2}(x) dx^j = \Gamma^1{}_{22}(x) dx^2 = -(\sin x^1)(\cos x^1) dx^2, \\ \tilde{\mathbf{w}}^2{}_1(x) &= \Gamma^2{}_{21}(x) dx^2 = (\cot x^1) dx^2, \\ \tilde{\mathbf{w}}^2{}_2(x) &= (\cot x^1) dx^1, \\ d\tilde{\mathbf{w}}^1{}_1(x) &\equiv \mathbf{O}_{..}(x), \\ d\tilde{\mathbf{w}}^1{}_2(x) &= -(\cos 2x^1) dx^1 \wedge dx^2, \\ d\tilde{\mathbf{w}}^2{}_1(x) &= -(\text{cosec}\, x^1)^2 dx^1 \wedge dx^2, \\ d\tilde{\mathbf{w}}^2{}_2(x) &\equiv \mathbf{O}_{..}(x). \end{aligned} \tag{4.88}$$

In this example, the torsion tensor components are identically zero, so that for the two torsion 2-forms

$$\begin{aligned}\mathbf{T}^{i}{}_{..}(x) &:= (1/2)\,T^{i}{}_{jk}(x)dx^{j} \wedge dx^{k}\\ &\equiv \mathbf{O}_{..}(x).\end{aligned}$$

We shall now compute the components of the curvature tensor by (4.75).

$$\begin{aligned}R^{1}{}_{1kl}(x) &\equiv -R^{1}{}_{1lk}(x) = \partial_{k}\Gamma^{1}{}_{l1} - \partial_{l}\Gamma^{1}{}_{k1} + \Gamma^{1}{}_{k2}(x)\Gamma^{2}{}_{l1}(x)\\ &\qquad -\Gamma^{1}{}_{l2}(x)\Gamma^{2}{}_{k1}(x)\\ &\equiv 0,\\ R^{2}{}_{2kl}(x) &\equiv -R^{2}{}_{2lk}(x) \equiv 0,\\ R^{1}{}_{212}(x) &\equiv -R^{1}{}_{221}(x) = \partial_{1}\Gamma^{1}{}_{22} - 0 + 0 - \Gamma^{1}{}_{22}(x)\Gamma^{2}{}_{12}(x)\\ &= (\sin x^{1})^{2},\\ R^{2}{}_{112}(x) &\equiv -R^{2}{}_{121}(x) \equiv -1.\end{aligned} \tag{4.89}$$

The four curvature 2-forms are furnished by (4.79) as

$$\begin{aligned}\Omega^{1}{}_{1..}(x) &= \Omega^{2}{}_{2..}(x) \equiv \mathbf{O}_{..}(x),\\ \Omega^{1}{}_{2..}(x) &= R^{1}{}_{212}(x)dx^{1} \wedge dx^{2} = (\sin x^{1})^{2}dx^{1} \wedge dx^{2},\\ \Omega^{2}{}_{1..}(x) &= -dx^{1} \wedge dx^{2}.\end{aligned} \tag{4.90}$$

In this example, the structural equation (4.80) reduce to

$$\tilde{\mathbf{w}}^{i}{}_{1}(x) \wedge dx^{1} + \tilde{\mathbf{w}}^{i}{}_{2}(x) \wedge dx^{2} = \mathbf{O}_{..}(x).$$

We can explicitly verify these equation by (4.88).

Now, let us try to check the structural equations (4.81) by (4.88). One of these equations, reduce to

$$\begin{aligned}d\tilde{\mathbf{w}}^{1}{}_{2}(x) + \tilde{\mathbf{w}}^{1}{}_{j}(x) \wedge \tilde{\mathbf{w}}^{j}{}_{2}(x) &= (\sin^{2} x^{1} - \cos^{2} x^{1})dx^{1} \wedge dx^{2}\\ &\qquad -(\sin x^{1})(\cos x^{1})(\cot x^{1})dx^{2} \wedge dx^{1}\\ &= (\sin^{2} x^{1})dx^{1} \wedge dx^{2} = \Omega^{1}{}_{2..}(x).\end{aligned}$$

Similarly, other structural equations can be verified. □

Now, we shall discuss the geometrical significance of the curvature tensor $\mathbf{R}^{\cdot}{}_{...}(x)$. Suppose that a differentiable vector field $\vec{\mathbf{V}}(x)$ is defined in the neighborhood $N_{\delta}(x_0)$ of $x_0 \in D \subset \mathbb{R}^{N}$. Moreover, let the vector field $\vec{\mathbf{V}}(x)$ be propagated parallelly along every differentiable, non-degenerate curve passing through x_0. In that case, we must have by (4.39)

$$\nabla_k V^i{}_{|x_0} = \left[\partial_k V^i + \Gamma^i{}_{kj}(x)V^j(x)\right]_{|x_0} = 0. \tag{4.91}$$

Let (4.91) hold for every $x \in D \subset \mathbb{R}^N$, so that

$$\nabla_k V^i = \partial_k V^i + \Gamma^i{}_{kj}(x)V^j(x) = 0 \tag{4.92}$$

for all $x \in D$.

Suppose that there is a differentiable curve $\mathcal{X}$ with an image Γ inside D. By (4.92), we obtain the difference

$$\begin{aligned}\left[\Delta V^i\right] := V^i(\mathcal{X}(b)) - V^i(\mathcal{X}(a)) &= \int_a^b \frac{dV^i}{dt}(\mathcal{X}(t))dt \\ = \int_\Gamma [\partial_k V^i] dx^k &= -\int_\Gamma \Gamma^i{}_{kj}(x)V^j(x)dx^k.\end{aligned} \tag{4.93}$$

In the case where Γ is a simple closed curve (**Jordan curve**) and is the boundary of a star-shaped surface $D^*{}_2$ interior to D, we can apply the Stokes' theorem summarized in (3.85). (See fig. 3.8, fig. 3.10, fig. 3.11, and fig. 4.2.)

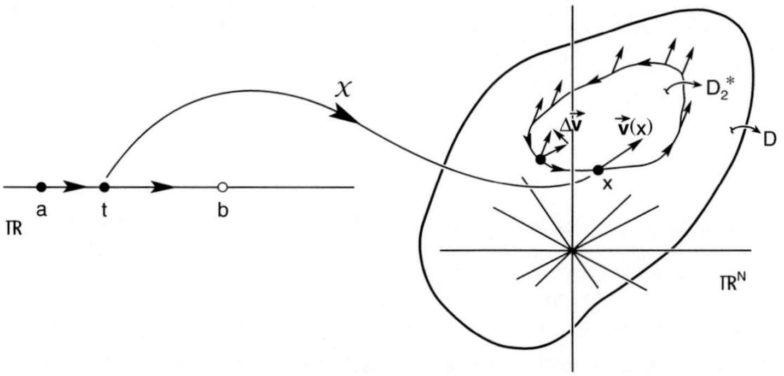

Figure 4.2: Parallel transport along a closed curve.

Applying the Stokes' theorem and equations (4.92) and (4.75) into (4.93), we get the jump discontinuity

$$\begin{aligned}
[\Delta V^i] &:= -\oint_{\partial D_2^*} \Gamma^i{}_{kj}(x) V^j(x) dx^k \\
&= (1/2) \int_{D_2^*} \left[\partial_k(\Gamma^i{}_{lj} V^j) - \partial_l(\Gamma^i{}_{kj} V^j)\right] dx^l \wedge dx^k \\
&= (1/2) \int_{D_2^*} \left[\left(\partial_k \Gamma^i{}_{lj} - \partial_l \Gamma^i{}_{kj}\right) V^j + \left(\Gamma^i{}_{kj}\Gamma^j{}_{lh} - \Gamma^i{}_{lj}\Gamma^j{}_{kh}\right) V^h\right] dx^l \wedge dx^k \\
&= (1/2) \int_{D_2^*} \left[R^i{}_{jkl}(x) V^j(x)\right] dx^l \wedge dx^k. \qquad (4.94)
\end{aligned}$$

In the case where the components $R^i{}_{jkl}$ are continuous, by the assumed differentiability of the V^j's, the integrand $R^i{}_{jkl}(x)V^j(x)$ is continuous. By the mean-value theorem of integration, there exists a point $x_m \in D_2^* \subset D$ such that

$$[\Delta V^i] = (1/2) R^i{}_{jkl}(x_m) V^j(x_m) \int_{D_2^*} dx^l \wedge dx^k. \qquad (4.95)$$

Therefore, the change of the vector components, after being parallelly transported along a closed curve ∂D_2^*, is proportional to the curvature tensor components, the vector components, and the oriented area of the surface D_2^*. If $R^i{}_{jkl}(x) \equiv 0$ in D (or the domain is flat), the changes $[\Delta V^i] \equiv 0$. In such a case, the system of the first-order partial differential equations (4.92) is **integrable**.

Example 4.3.13 Consider the surface of the unit sphere S^2. (See example 4.3.12.) The polar coordinates are $x^1 \equiv \theta$, $x^2 \equiv \phi$. The *non-zero* connection coefficients are provided by

$$\begin{aligned}
\Gamma^1{}_{22}(x) &= -(\sin x^1)(\cos x^1), \\
\Gamma^2{}_{12}(x) &= \Gamma^2{}_{21}(x) = \cot x^1.
\end{aligned}$$

The geodesic equations (4.43) yield

$$\begin{aligned}
&\frac{d^2\chi^1(\tau)}{d\tau^2} - [\sin \mathcal{X}^1(\tau)][\cos \mathcal{X}^1(\tau)] \left[\frac{d\mathcal{X}^2(\tau)}{d\tau}\right]^2 = 0, \\
&\frac{d^2\chi^2(\tau)}{d\tau^2} + 2[\cot \mathcal{X}^1(\tau)] \frac{d\mathcal{X}^1(\tau)}{d\tau}\frac{d\mathcal{X}^2(\tau)}{d\tau} = 0.
\end{aligned}$$

A *special class* of solution of the equations above is furnished by

$$\begin{aligned}
\theta \equiv x^1 &= \mathcal{X}^1(\tau) = (\pi/2), \\
\phi \equiv x^2 &= \mathcal{X}^2(\tau) = k\tau + c.
\end{aligned}$$

Here, $k \neq 0$ and c are arbitrary constants. The equations above yield a portion of the great circle that is the "equator" on S^2. (By the symmetry of the sphere, we can intuitively conclude that great circles are geodesics.) Moreover, the coordinate $\phi \equiv x^2$ is an affine parameter along the equator.

Consider now circles (not necessarily great circles) given by

$$\begin{aligned} &\theta \equiv x^1 = \mathcal{X}^1(\tau) := \theta_0, \\ &\phi \equiv x^2 = \mathcal{X}^2(\tau) = \tau, \\ &\tau \in [-\pi, \pi) \subset \mathbb{R}. \end{aligned} \tag{4.96}$$

Here, $\theta_0 \in (0, \pi)$ is a prescribed constant.

Let a differentiable vector field $\vec{\mathbf{V}}(\mathcal{X}(\phi))$ be parallelly propagated around the circle given by (4.96). By (4.39), the components $V^i(\mathcal{X}(\phi))$ must satisfy the equations

$$\begin{aligned} &\frac{\partial V^1}{\partial \phi}(\theta_0, \phi) - (\sin\theta_0)(\cos\theta_0)V^2(\theta_0, \phi) = 0, \\ &\frac{\partial V^2}{\partial \phi}(\theta_0, \phi) + (\cot\theta_0)V^1(\theta_0, \phi) = 0. \end{aligned} \tag{4.97}$$

The general solution of the system of equations (4.97) is furnished by

$$\begin{aligned} V^1(\theta_0, \phi) &= B(\theta_0)(\sin\theta_0)\sin[(\cos\theta_0)\phi - \alpha(\theta_0)], \\ V^2(\theta_0, \phi) &= B(\theta_0)\cos[(\cos\theta_0)\phi - \alpha(\theta_0)]. \end{aligned} \tag{4.98}$$

Here, B and α are two arbitrary functions of integration.

The jump discontinuities $\Delta V^i(\theta_0)$ around the circle are provided by (4.93) as

$$\begin{aligned} \Delta V^1(\theta_0) &= \left[\lim_{\phi\to\pi_-} V^1(\theta_0, \phi)\right] - V^1(\theta_0, -\pi) \\ &= 2B(\theta_0)(\sin\theta_0)[\cos\alpha(\theta_0)]\sin(\pi\cos\theta_0) \not\equiv 0, \\ \Delta V^2(\theta_0) &= 2B(\theta_0)[\sin\alpha(\theta_0)]\sin(\pi\cos\theta_0) \not\equiv 0. \end{aligned} \tag{4.99}$$

However, in the special case of the equator, $\Delta V^1(\pi/2) = \Delta V^2(\pi/2) = 0$. □

Exercises 4.3

1. Consider the connection 1-forms $\tilde{\mathbf{w}}^q{}_p(x)$. Construct an $N \times N$ "matrix" $[\tilde{\mathbf{w}}(x)]$ with 1-forms $\tilde{\mathbf{w}}^q{}_p(x)$ as entries. Prove that the transformation rules (4.12) can be expressed as a "matrix" equation,

$$[\widehat{\tilde{\mathbf{w}}}(x)] = [\lambda(x)]^{-1}[\tilde{\mathbf{w}}(x)][\lambda(x)] + [\lambda(x)]^{-1}[d\lambda(x)].$$

2. Suppose that ${}^{r}{}_{s}\mathbf{T}(x)$ is a tensor field of class C^2 in $D \subset \mathbb{R}^N$. Prove the **Ricci identities:**

$$\begin{aligned}[\nabla_u\nabla_v - \nabla_v\nabla_u]\,T^{p_1\ldots p_r}{}_{q_1\ldots q_s} &= \sum_{\alpha=1}^{r}[T^{p_1\ldots p_{\alpha-1}wp_{\alpha+1}\ldots p_r}{}_{q_1\ldots q_s}\cdot R^{p_\alpha}{}_{wuv}] \\ &\quad-\sum_{\beta=1}^{s}[T^{p_1\ldots p_r}{}_{q_1\ldots q_{\beta-1}wq_{\beta+1}\ldots q_s}\cdot R^{w}{}_{q_\beta uv}] \\ &\quad-T^{w}{}_{uv}(x)\cdot\nabla_w T^{p_1\ldots p_r}{}_{q_1\ldots q_s}.\end{aligned}$$

3. Suppose there exist two manifolds (M,∇) and $(M,\widehat{\nabla})$ with identical atlases and distinct connection coefficients.

(i) Show that the difference $D^i{}_{jk}(x) := \widehat{\Gamma}^i{}_{jk}(x) - \Gamma^i{}_{jk}(x)$ in a coordinate basis transforms as components of a $(1+2)$th-order tensor field.

(ii) Curvature tensor components are related by the equations

$$\widehat{R}^i{}_{jkl}(x) = R^i{}_{jkl}(x) + \nabla_k D^i{}_{lj} - \nabla_l D^i{}_{kj} + D^i{}_{kh}D^h{}_{lj} - D^i{}_{lh}D^h{}_{kj} + T^h{}_{kl}D^i{}_{hj}.$$

4. Consider a differentiable, symmetric connection $\Gamma^i{}_{kj}(x) \equiv \Gamma^i{}_{jk}(x)$ relative to a coordinate basis in $D \subset \mathbb{R}^N (N \geq 2)$. A **projective transformation** is defined by

$$\widehat{\Gamma}^i{}_{jk}(x) := \Gamma^i{}_{jk}(x) + \delta^j{}_j\psi_k(x) + \delta^i{}_k\psi_j(x).$$

(Here, $\psi_k(x)dx^k$ is a differentiable 1-form.) Define Weyl's **projective curvature tensor** by

$$\begin{aligned}W^i{}_{jkl}(x) := R^i{}_{jkl}(x) &- [1/(N+1)]\delta^i{}_j R^m{}_{mkl}(x) \\ &+[1/(N-1)][\delta^i{}_k R^m{}_{jlm}(x) - \delta^i{}_l R^m{}_{jkm}(x)] \\ &+[1/(N^2-1)][\delta^i{}_k R^m{}_{mjl}(x) - \delta^i{}_l R^m{}_{mjk}(x)].\end{aligned}$$

Prove that under a projective transformation

$$\widehat{W}^i{}_{jkl}(x) \equiv W^i{}_{jkl}(x).$$

Chapter 5

Riemannian and Pseudo-Riemannian Manifolds

5.1 Metric Tensor, Christoffel Symbols, and Ricci Rotation Coefficients

Let an N-dimensional differentiable manifold M be endowed with a metric field $\mathbf{g}_{..}(p) \in {}^0{}_2\mathcal{T}_p(M)$. (See section 2.4.) Let p belong to the open subset $U \subset M$ and (χ, U) be a coordinate chart yielding $x = \chi(p)$. (See fig. 3.1.) The metric field has the isomorphic image $\mathbf{g}_{..}(x) \in {}^0{}_2\mathcal{T}_x(\mathbb{R}^N)$. (For the sake of simplicity, the same symbol $\mathbf{g}_{..}$ has been used!) The axioms (2.113), (2.114), and (2.115) can be virtually repeated as

I1. $$\mathbf{g}_{..}(x)(\vec{\mathbf{A}}(x), \vec{\mathbf{B}}(x)) \in \mathbb{R} \text{ for all } \vec{\mathbf{A}}(x), \vec{\mathbf{B}}(x) \in T_x(\mathbb{R}^N).$$

I2. $$\begin{aligned} &\mathbf{g}_{..}(x)(\vec{\mathbf{B}}(x),\ \vec{\mathbf{A}}(x)) \\ &= \mathbf{g}_{..}(x)(\vec{\mathbf{A}}(x), \vec{\mathbf{B}}(x)) \text{ for all } \vec{\mathbf{A}}(x), \vec{\mathbf{B}}(x) \in T_x(\mathbb{R}^N). \end{aligned} \tag{5.1}$$

I3. $$\begin{aligned} &\mathbf{g}_{..}(x)(\lambda(x)\vec{\mathbf{A}}(x) + \mu(x)\vec{\mathbf{B}}(x), \vec{\mathbf{C}}(x)) \\ &= \lambda(x)\mathbf{g}_{..}(x)(\vec{\mathbf{A}}(x), \vec{\mathbf{C}}(x)) + \mu(x)\mathbf{g}_{..}(x)(\vec{\mathbf{B}}(x), \vec{\mathbf{C}}(x)). \end{aligned} \tag{5.2}$$

I4. $$\begin{aligned} &\mathbf{g}_{..}(x)(\vec{\mathbf{A}}(x), \vec{\mathbf{V}}(x)) = 0 \text{ for all } \vec{\mathbf{V}}(x) \in T_x(\mathbb{R}^N) \text{ iff} \\ &\vec{\mathbf{A}}(x) = \vec{\mathbf{O}}(x). \end{aligned} \tag{5.3}$$

Axioms I1, I2, and I3 imply that $\mathbf{g}..(x)$ is a symmetric tensor field of order $(0+2)$. Axiom I4 is the condition of *non-degeneracy.*

The **separation** σ of a vector field $\vec{\mathbf{V}}(x)$ is given by (2.120) as

$$\sigma(\vec{\mathbf{V}}(x)) := +\sqrt{|\mathbf{g}..(x)(\vec{\mathbf{V}}(x), \vec{\mathbf{V}}(x))|}. \tag{5.4}$$

A **"unit vector field"** $\vec{\mathbf{U}}(x)$ must satisfy

$$\sigma(\vec{\mathbf{U}}(x)) \equiv 1. \tag{5.5}$$

Two **"orthogonal vectors"** $\vec{\mathbf{A}}(x)$ and $\vec{\mathbf{B}}(x)$ satisfy the condition

$$\mathbf{g}..(x)(\vec{\mathbf{A}}(x), \vec{\mathbf{B}}(x)) = 0. \tag{5.6}$$

An **"orthonormal basis"** set $\{\vec{\mathbf{e}}_a(x)\}_1^N$ is characterized by

$$\begin{aligned} &\mathbf{g}..(x)(\vec{\mathbf{e}}_a(x), \vec{\mathbf{e}}_b(x)) = d_{ab} =: \operatorname{diag}\left(\underbrace{1,\ldots,1}_{p}, \underbrace{-1,\ldots,-1}_{n}\right), \\ &|\mathbf{g}..(x)(\vec{\mathbf{e}}_a(x), \vec{\mathbf{e}}_b(x))| = \delta_{ab}, \\ &\operatorname{sgn}(\mathbf{g}..(x)) = p - n \geq 0. \end{aligned} \tag{5.7}$$

We shall *drop* " " in the sequel.

In the case of a **positive-definite metric**, axiom I4 has to be replaced by a *stronger* axiom:

I4+.

$$\begin{aligned} &\mathbf{g}..(x)(\vec{\mathbf{V}}(x), \vec{\mathbf{V}}(x)) \geq 0; \\ &\mathbf{g}..(x)(\vec{\mathbf{V}}(x), \vec{\mathbf{V}}(x)) = 0 \iff \vec{\mathbf{V}}(x) = \vec{\mathbf{O}}(x). \end{aligned} \tag{5.8}$$

In the case of a positive-definite metric, the **norm** or **length** associated with a vector field is given by

$$\|\vec{\mathbf{V}}(x)\| := \sigma(\vec{\mathbf{V}}(x)) = +\sqrt{\mathbf{g}..(x)(\vec{\mathbf{V}}(x), \vec{\mathbf{V}}(x))}. \tag{5.9}$$

(See (2.118).)

Moreover, in the case of a positive-definite metric, the angle field between two *non-zero* vector fields is furnished by

$$\begin{aligned} &\cos[\theta(x)] := \frac{[\mathbf{g}..(x)(\vec{\mathbf{U}}(x), \vec{\mathbf{V}}(x))]}{\|\vec{\mathbf{U}}(x)\| \cdot \|\vec{\mathbf{V}}(x)\|}, \\ &-1 \leq \cos[\theta(x)] \leq 1. \end{aligned} \tag{5.10}$$

(See (2.119).)

Example 5.1.1 Let $\mathbb{E}_N$ be the N-dimensional **Euclidean manifold**. Let $(\chi, \mathbb{E}_N)$ be one of the global **Cartesian charts**. In the Cartesian coordinate basis $\{\frac{\partial}{\partial x^i}\}_1^N$, the components of the metric tensor are

$$\begin{aligned} \mathbf{g}_{..}(x)\left(\frac{\partial}{\partial x^i}, \frac{\partial}{\partial x^j}\right) &= g_{ij}(x) = \delta_{ij}, \\ \|\vec{\mathbf{V}}(x)\|^2 &= \delta_{ij} V^i(x) V^j(x) \geq 0, \\ \cos[\theta(x)] &= \left[\frac{\delta_{ij} U^i(x) V^j(x)}{\sqrt{\delta_{kl}\delta_{mn} U^k(x) U^l(x) V^m(x) V^n(x)}}\right]. \end{aligned} \tag{5.11}$$

□

Example 5.1.2 Let M be an N-dimensional flat manifold with Lorentz metric $\mathbf{g}_{..}(x) = \mathbf{d}_{..}(x)$. (See (2.116).) In the global **Minkowskian chart** (χ, M), the metric field is furnished by

$$\begin{aligned} \mathbf{g}_{..}(x) = \mathbf{d}_{..}(x) &= d_{ij} dx^i \otimes dx^j \\ &= dx^1 \otimes dx^1 + \cdots + dx^{N-1} \otimes dx^{N-1} - dx^N \otimes dx^N. \end{aligned} \tag{5.12}$$

This metric is obviously *not* positive-definite. The signature $\mathrm{sgn}\,(\mathbf{d}_{..}(x)) = N - 2$. □

In the sequel, we shall deal with *three types of basis sets*, namely the general basis set $\{\vec{\mathbf{e}}_p(x)\}_1^N$, the coordinate basis set $\left\{\frac{\partial}{\partial x^i}\right\}_1^N$, and the orthonormal basis set $\{\vec{\mathbf{e}}_a(x)\}_1^N$. (Strictly speaking, the general basis set includes the two others.) Our plan to handle such situations is to break up the lower case latin alphabet into three classes: (a, b, c, d, e, f, etc.), (i, j, k, l, m, n, etc.), and (p, q, r, s, u, v, w, etc.). We shall denote the indices of a tensor component relative to a general basis by p, q, r, s, u, v, w, etc. The indices of a tensor component relative to a coordinate basis are indicated by i, j, k, l, m, n, etc. Moreover, the indices for a tensor component relative to an orthonormal basis are denoted by a, b, c, d, e, f, etc.

Therefore, by (5.11), (5.7), (5.1), (5.2), and (5.3), we can write that

$$\begin{aligned} \mathbf{g}_{..}(x) = g_{pq}(x)\tilde{\mathbf{e}}^p(x) \otimes \tilde{\mathbf{e}}^q(x) &= g_{ij}(x) dx^i \otimes dx^j \\ &= g_{ab}(x)\tilde{\mathbf{e}}^a(x) \otimes \tilde{\mathbf{e}}^b(x) \equiv d_{ab}\tilde{\mathbf{e}}^a(x) \otimes \tilde{\mathbf{e}}^b(x), \end{aligned} \tag{5.13}$$

$g_{qp}(x) = g_{pq}(x)$, $g_{ji}(x) = g_{ij}(x)$, $d_{ba} = d_{ab}$, $g := \det[g_{ij}(x)] \neq 0$, $\det[g_{pq}(x)] \neq 0$, $|\det[d_{ab}]| = 1$.

The components of the conjugate (contravariant) metric tensor field $\mathbf{g}^{..}(x)$ are defined to be the entries of the inverse matrix $[g_{pq}(x)]^{-1}$, etc. Therefore,

$$\begin{aligned} &\mathbf{g}^{..}(x) = g^{pq}(x)\vec{\mathbf{e}}_p(x) \otimes \vec{\mathbf{e}}_q(x) = g^{ij}(x)\frac{\partial}{\partial x^i} \otimes \frac{\partial}{\partial x^j} = d^{ab}\vec{\mathbf{e}}_a(x) \otimes \vec{\mathbf{e}}_b(x), \\ &g^{qp}(x) \equiv g^{pq}(x), \quad g^{ji}(x) \equiv g^{ij}(x), \quad d^{ba} = d^{ab}, \\ &g^{pu}(x) g_{uq}(x) \equiv \delta^p{}_q, \quad g^{ik}(x) g_{kj}(x) \equiv \delta^i{}_j, \quad d^{ac} d_{cb} = \delta^a{}_b. \end{aligned} \tag{5.14}$$

(See (2.129).)

Now, let us discuss briefly the transformation from one basis set to another. By (3.29), (3.43), and (3.46), we obtain

$$\begin{aligned}&\widehat{\vec{\mathbf{e}}}_p(x) = \lambda^q{}_p(x)\vec{\mathbf{e}}_q(x),\ \widehat{\tilde{\mathbf{e}}}^p(x) = \mu^p{}_q(x)\tilde{\mathbf{e}}^q(x),\\ &[\mu^q{}_p(x)] := [\lambda^p{}_q(x)]^{-1};\end{aligned} \tag{5.15}$$

$$\begin{aligned}&\widehat{\vec{\mathbf{X}}}'\left(\frac{\partial}{\partial x^i}\right) = \frac{\partial \widehat{X}^k(x)}{\partial x^i}\frac{\partial}{\partial \widehat{x}^k},\ \widehat{\tilde{\mathbf{X}}}'\left(dx^i\right) = \frac{\partial X^i(\widehat{x})}{\partial \widehat{x}^k}d\widehat{x}^k,\\ &\left[\frac{\partial X^i(\widehat{x})}{\partial \widehat{x}^k}\right] = \left[\frac{\partial \widehat{X}^k(x)}{\partial x^i}\right]^{-1};\end{aligned} \tag{5.16}$$

$$\begin{aligned}&\widehat{\vec{\mathbf{e}}}_a(x) = L^b{}_a(x)\vec{\mathbf{e}}_b(x),\ \ \vec{\mathbf{e}}_a(x) = A^b{}_a(x)\widehat{\vec{\mathbf{e}}}_b(x),\\ &\widehat{\tilde{\mathbf{e}}}^a(x) = A^a{}_b(x)\tilde{\mathbf{e}}^b(x),\ \ \tilde{\mathbf{e}}^a(x) = L^a{}_b(x)\widehat{\tilde{\mathbf{e}}}^b(x),\\ &\left[A^b{}_a(x)\right] := \left[L^b{}_a(x)\right]^{-1},\\ &L^a{}_b(x)d_{ac}L^c{}_e(x) = d_{be}.\end{aligned} \tag{5.17}$$

The last equation defines a **generalized Lorentz transformation**. (The set of all generalized Lorentz transformations constitutes a group denoted by $O(p, n; \mathbb{R})(p + n = N)$.)

The transformation from a coordinate basis to an orthonormal basis is furnished by

$$\begin{aligned}&\vec{\mathbf{e}}_a(x) = \lambda^i{}_a(x)\frac{\partial}{\partial x^i},\ \frac{\partial}{\partial x^i} = \mu^a{}_i(x)\vec{\mathbf{e}}_a(x),\\ &\tilde{\mathbf{e}}^a(x) = \mu^a{}_i(x)dx^i,\ \ dx^i = \lambda^i{}_a(x)\tilde{\mathbf{e}}^a(x).\end{aligned} \tag{5.18}$$

It follows from (5.13) and (5.18) that

$$\begin{aligned}&d_{ab} = \mathbf{g}_{..}(x)\left(\vec{\mathbf{e}}_a(x), \vec{\mathbf{e}}_b(x)\right) = g_{ij}(x)\lambda^i{}_a(x)\lambda^j{}_b(x),\\ &g_{ij}(x) = \mathbf{g}_{..}(x)\left(\frac{\partial}{\partial x^i}, \frac{\partial}{\partial x^j}\right) = d_{ab}\mu^a{}_i(x)\mu^b{}_j(x),\\ &d^{ab} = \mathbf{g}^{..}(x)\left(\tilde{\mathbf{e}}^a(x), \tilde{\mathbf{e}}^b(x)\right) = g^{ij}(x)\mu^a{}_i(x)\mu^b{}_j(x),\\ &g^{ij}(x) = \mathbf{g}^{..}(x)\left(dx^i, dx^j\right) = d^{ab}\lambda^i{}_a(x)\lambda^j{}_b(x),\\ &\delta^i{}_j = \mathbf{g}^{\cdot}{}_{\cdot}(x)\left(dx^i, \frac{\partial}{\partial x^j}\right) \equiv \mathbf{I}^{\cdot}{}_{\cdot}(x)\left(dx^i, \frac{\partial}{\partial x^j}\right) = \lambda^i{}_a(x)\mu^a{}_j(x),\\ &\delta^a{}_b = \mathbf{g}^{\cdot}{}_{\cdot}(x)\left(\tilde{\mathbf{e}}^a(x), \vec{\mathbf{e}}_b(x)\right) \equiv \mathbf{I}^{\cdot}{}_{\cdot}(x)\left(\tilde{\mathbf{e}}^a(x), \vec{\mathbf{e}}_b(x)\right) = \mu^a{}_i(x)\lambda^i{}_b(x).\end{aligned} \tag{5.19}$$

Example 5.1.3 Consider the four-dimensional flat manifold of the special theory of relativity. In the spherical polar coordinate chart, the metric tensor components are provided by

$$D := \left\{x \in \mathbb{R}^4 :\ x^1 > 0,\ 0 < x^2 < \pi,\ -\pi < x^3 < \pi,\ -\infty < x^4 < \infty\right\},$$

$$\begin{aligned}\mathbf{g}_{..}(x) &= dx^1 \otimes dx^1 + (x^1)^2[dx^2 \otimes dx^2 + \sin^2 x^2 dx^3 \otimes dx^3] - dx^4 \otimes dx^4 \\ &= \tilde{\mathbf{e}}^1(x) \otimes \tilde{\mathbf{e}}^1(x) + \tilde{\mathbf{e}}^2(x) \otimes \tilde{\mathbf{e}}^2(x) + \tilde{\mathbf{e}}^3(x) \otimes \tilde{\mathbf{e}}^3(x) - \tilde{\mathbf{e}}^4(x) \otimes \tilde{\mathbf{e}}^4(x), \\ \mathbf{g}^{..}(x) &= \frac{\partial}{\partial x^1} \otimes \frac{\partial}{\partial x^1} + (x^1)^{-2}\left[\frac{\partial}{\partial x^2} \otimes \frac{\partial}{\partial x^2} + (\sin x^2)^{-2}\frac{\partial}{\partial x^3} \otimes \frac{\partial}{\partial x^3}\right] \\ &\quad - \frac{\partial}{\partial x^4} \otimes \frac{\partial}{\partial x^4} \\ &= \vec{\mathbf{e}}^1(x) \otimes \vec{\mathbf{e}}^1(x) + \vec{\mathbf{e}}^2(x) \otimes \vec{\mathbf{e}}^2(x) + \vec{\mathbf{e}}^3(x) \otimes \vec{\mathbf{e}}^3(x) - \vec{\mathbf{e}}^4(x) \otimes \vec{\mathbf{e}}^4(x).\end{aligned}$$

Therefore, we have, for d_{ab}, $d_{11} = d^{11} = d_{22} = d^{22} = d_{33} = d^{33} = -d_{44} = -d^{44} = 1$. The eigenvalues of $[g_{ij}(x)]$ are given by

$$\lambda_1(x) \equiv 1, \;\; \lambda_2(x) = (x^1)^2, \;\; \lambda_3(x) = (x^1 \sin x^2)^2, \;\; \lambda_4(x) \equiv -1.$$

$$\operatorname{sgn}(\mathbf{g}_{..}(x)) = \sum_{k=1}^{4} \operatorname{sgn}(\lambda_k(x)) = +2,$$

$$g := \det\left[g_{i_j}(x)\right] = -\left[(x^1)^2 \sin x^2\right]^2 < 0,$$

$$\det[d_{ab}] = -1.$$

The separation of a vector field $\vec{\mathbf{V}}(x)$ is given by

$$\sigma[\vec{\mathbf{V}}(x)] = +\left|[V^1(x)]^2 + (x^1)^2[(V^2(x))^2 + (\sin x^2 V^3(x))^2] - [V^4(x)]^2\right|^{1/2}.$$

An orthonormal basis $\{\vec{\mathbf{e}}_a(x)\}_1^4$, or **tetrad**, is furnished by inspection as

$$\vec{\mathbf{e}}_1(x) = \frac{\partial}{\partial x^1}, \; \vec{\mathbf{e}}_2(x) = (x^1)^{-1}\frac{\partial}{\partial x^2}, \; \vec{\mathbf{e}}_3(x) = (x^1 \sin x^2)^{-1}\frac{\partial}{\partial x^3}, \; \vec{\mathbf{e}}_4(x) = \frac{\partial}{\partial x^4}.$$

The coefficients in (5.18) for this example are provided by

$$\mu^1{}_i(x) = \delta^1{}_i, \mu^2{}_i(x) = (x^1)\delta^2{}_i, \mu^3{}_i(x) = (x^1 \sin x^2)\delta^3{}_i, \mu^4{}_i(x) = \delta^4{}_i,$$

$$\lambda^i{}_1(x) = \delta^i{}_1, \lambda^i{}_2(x) = (x^1)^{-1}\delta^i{}_2, \lambda^i{}_3(x) = (x^1 \sin x^2)^{-1}\delta^i{}_3, \lambda^i{}_4(x) = \delta^i{}_4.$$

Another orthonormal basis or tetrad is obtained by the transformation

$$\begin{aligned}&\widehat{\vec{\mathbf{e}}}_1(x) = (\cosh\alpha(x))\vec{\mathbf{e}}_1(x) + (\sinh\alpha(x))\vec{\mathbf{e}}_4(x), \\ &\widehat{\vec{\mathbf{e}}}_2(x) = \vec{\mathbf{e}}_2(x), \;\; \widehat{\vec{\mathbf{e}}}_3(x) = \vec{\mathbf{e}}_3(x), \\ &\widehat{\vec{\mathbf{e}}}_4(x) = (\sinh\alpha(x))\vec{\mathbf{e}}_1(x) + (\cosh\alpha(x))\vec{\mathbf{e}}_4(x).\end{aligned}$$

This is a *variable* **Lorentz transformation** called the variable **boost** in special relativity. □

A tensor field ${}^r{}_s\mathbf{T}(x)$ can be expressed in terms of various basis sets as

$$\begin{aligned}{}^r{}_s\mathbf{T}(x) &= T^{p_1\ldots p_r}{}_{q_1\ldots q_s}(x)\vec{\mathbf{e}}_{p_1}(x)\otimes\cdots\otimes\vec{\mathbf{e}}_{p_r}(x)\otimes\tilde{\mathbf{e}}^{q_1}(x)\otimes\cdots\otimes\tilde{\mathbf{e}}^{q_s}(x)\\ &= T^{i_1\ldots i_r}{}_{j_1\ldots j_s}(x)\frac{\partial}{\partial x^{i_1}}\otimes\cdots\otimes\frac{\partial}{\partial x^{i_r}}\otimes dx^{j_1}\otimes\cdots\otimes dx^{j_s} \qquad (5.20)\\ &= T^{a_1\ldots a_r}{}_{b_1\ldots b_s}(x)\vec{\mathbf{e}}_{a_1}(x)\otimes\cdots\otimes\vec{\mathbf{e}}_{a_r}(x)\otimes\cdots\otimes\tilde{\mathbf{e}}^{b_1}(x)\otimes\cdots\otimes\tilde{\mathbf{e}}^{b_s}(x).\end{aligned}$$

Caution: We have used the *same* letter, "T", to denote the components relative to a different bases.

By (3.30), (3.48), (5.15), (5.16), and (5.17), we have for the transformation of tensor field components

$$\begin{aligned}&\widehat{T}^{p_1\ldots p_r}{}_{q_1\ldots q_s}(x) =\\ &\mu^{p_1}{}_{u_1}(x)\ldots\mu^{p_r}{}_{u_r}(x)\lambda^{v_1}{}_{q_1}(x)\ldots\lambda^{v_s}{}_{q_s}(x)T^{u_1\ldots u_r}{}_{v_1\ldots v_s}(x),\end{aligned} \qquad (5.21)$$

$$\begin{aligned}&\widehat{T}^{i_1\ldots i_r}{}_{j_1\ldots j_s}(\widehat{x}) =\\ &\frac{\partial\widehat{X}^{i_1}(x)}{\partial x^{k_1}}\cdots\frac{\partial\widehat{X}^{i_r}(x)}{\partial x^{k_r}}\frac{\partial X^{l_1}(\widehat{x})}{\partial\widehat{x}^{j_1}}\cdots\frac{\partial X^{l_s}(\widehat{x})}{\partial\widehat{x}^{j_s}}T^{k_1\ldots k_r}{}_{l_1\ldots l_s}(x),\end{aligned} \qquad (5.22)$$

$$\begin{aligned}&\widehat{T}^{a_1\ldots a_r}{}_{b_1\ldots b_s}(x) =\\ &A^{a_1}{}_{c_1}(x)\ldots A^{a_r}{}_{c_r}(x)L^{d_1}{}_{b_1}(x)\ldots L^{d_s}{}_{b_s}(x)T^{c_1\ldots c_r}{}_{d_1\ldots d_s}(x),\end{aligned} \qquad (5.23)$$

$$\begin{aligned}&T^{a_1\ldots a_r}{}_{b_1\ldots b_s}(x) =\\ &\mu^{a_1}{}_{i_1}(x)\ldots\mu^{a_r}{}_{i_r}(x)\lambda^{j_1}{}_{b_1}(x)\ldots\lambda^{j_s}{}_{b_s}(x)T^{i_1\ldots i_r}{}_{j_1\ldots j_s}(x),\end{aligned} \qquad (5.24)$$

$$\begin{aligned}&T^{i_1\ldots i_r}{}_{j_1\ldots j_s}(x) =\\ &\lambda^{i_1}{}_{a_1}(x)\ldots\lambda^{i_r}{}_{a_r}(x)\mu^{b_1}{}_{j_1}(x)\ldots\mu^{b_s}{}_{j_s}(x)T^{a_1\ldots a_r}{}_{b_1\ldots b_s}(x).\end{aligned} \qquad (5.25)$$

(See references [2], [28], [34], and [37].)

The **raising** and **lowering tensor component indices** in (2.134) can almost be repeated to obtain

$$g^{uq_1}(x)T^{p_1\ldots p_r}{}_{q_1\ldots q_s}(x) = T^{p_1\ldots p_r u}{}_{q_2\ldots q_s}(x), \qquad (5.26)$$

$$g_{up_r}(x)T^{p_1\ldots p_r}{}_{q_1\ldots q_s}(x) = T^{p_1\ldots p_{r-1}}{}_{uq_1\ldots q_s}(x), \qquad (5.27)$$

$$g^{kj_1}(x)T^{i_1\ldots i_r}{}_{j_1\ldots j_s}(x) = T^{i_1\ldots i_r k}{}_{j_2\ldots j_s}(x), \qquad (5.28)$$

$$g_{ki_r}(x)T^{i_1\ldots i_r}{}_{j_1\ldots j_s}(x) = T^{i_1\ldots i_{r-1}}{}_{kj_1\ldots j_s}(x), \qquad (5.29)$$

$$d^{cb_1}T^{a_1\ldots a_r}{}_{b_1\ldots b_s}(x) = T^{a_1\ldots a_r c}{}_{b_2\ldots b_s}(x), \qquad (5.30)$$

$$d_{ca_r}T^{a_1\ldots a_r}{}_{b_1\ldots b_s}(x) = T^{a_1\ldots a_{r-1}}{}_{cb_1\ldots b_s}(x). \qquad (5.31)$$

Now we shall introduce the notion of a **linear Riemannian connection**. (See (4.2), (4.3), (4.4), (4.59), and (4.62).)

Theorem 5.1.4 *In the domain $D \subset \mathbb{R}^N$ corresponding to a domain of a manifold with a metric, there exists a unique connection such that*

(*i*) *the torsion tensor*

$$\mathbf{T}^{\cdot}{}_{..}(x) \equiv \mathbf{O}^{\cdot}{}_{..}(x), \tag{5.32}$$

and

(*ii*) *the covariant derivative of the metric*

$$\nabla \mathbf{g}_{..}(x) \equiv \mathbf{O}_{...}(x). \tag{5.33}$$

Proof. By the equation (4.64), the equation (5.32) yields

$$T^w{}_{qu}(x) = \Gamma^w{}_{qu}(x) - \Gamma^w{}_{uq}(x) - \chi^w{}_{qu}(x) \equiv 0. \tag{5.34}$$

Using (5.33), (4.30), and (4.31), we obtain that

$$\begin{aligned} \nabla_u g_{pq}(x) &= \partial_u g_{pq} - \Gamma^w{}_{up}(x) g_{wq}(x) - \Gamma^w{}_{uq}(x) g_{pw}(x) \\ &\equiv 0. \end{aligned} \tag{5.35}$$

With cyclic permutations of indices u, p, q in (5.35), we can derive that

$$\begin{aligned} \partial_p g_{qu} + \partial_q g_{up} - \partial_u g_{pq} &= (\Gamma^w{}_{pq} + \Gamma^w{}_{qp}) g_{wu} \\ &\quad + (\Gamma^w{}_{pu} - \Gamma^w{}_{up}) g_{wq} + (\Gamma^w{}_{qu} - \Gamma^w{}_{uq}) g_{wp}. \end{aligned}$$

By (5.34), the equation above yields

$$\begin{aligned} \Gamma^w{}_{pq}(x) g_{wu}(x) &= (1/2)(\partial_p g_{qu} + \partial_q g_{up} - \partial_u g_{pq}) \\ &\quad - (1/2)(\chi^w{}_{qp} g_{wu} + \chi^w{}_{pu} g_{wq} + \chi^w{}_{qu} g_{wp}), \\ \Gamma^v{}_{pq}(x) &= (1/2) g^{vu} (\partial_p g_{qu} + \partial_q g_{pu} - \partial_u g_{pq}) \\ &\quad - (1/2)(\chi^v{}_{qp} + g^{vu} g_{qw} \chi^w{}_{pu} + g^{vu} g_{pw} \chi^w{}_{qu}). \end{aligned} \tag{5.36}$$

Thus, for given $g_{pq}(x)$ and $\chi^p{}_{pq}(x)$, we can derive from (5.34) and (5.35) the *unique* expression (5.36) for the connection coefficients $\Gamma^v{}_{pq}(x)$. ■
(See references [5] and [33].)

Although $\Gamma^v{}_{pq}(x)$ and $\chi^v{}_{pq}(x)$ are *not* components of tensor fields, let us *still denote* the lowering of indices by

$$\begin{aligned} \Gamma_{upq}(x) &:= g_{uv}(x) \Gamma^v{}_{pq}(x), \\ \chi_{upq}(x) &:= g_{uv}(x) \chi^v{}_{pq}(x). \end{aligned} \tag{5.37}$$

Using (5.36) and (5.37), we get

$$\begin{aligned} \Gamma_{upq}(x) &= (1/2)(\partial_p g_{qu} + \partial_q g_{pu} - \partial_u g_{pq}) \\ &\quad - (1/2)(\chi_{pqu}(x) + \chi_{qpu}(x) - \chi_{upq}(x)). \end{aligned} \tag{5.38}$$

In the coordinate basis, by (4.58), (4.65), and (5.34), we get

$$\chi^i{}_{jk}(x) \equiv 0. \tag{5.39}$$

Therefore, (5.38) and (5.36) yield

$$[jk,i] := \Gamma_{ijk}(x) = (1/2)(\partial_j g_{ki} + \partial_k g_{ji} - \partial_i g_{jk}) = [kj,i], \tag{5.40}$$

$$\left\{ {l \atop jk} \right\} := \Gamma^l{}_{jk}(x) = (1/2)g^{li}(x)(\partial_j g_{ki} + \partial_k g_{ji} - \partial_i g_{jk}) = \left\{ {l \atop kj} \right\}. \tag{5.41}$$

The symbols $[jk,i]$ and $\left\{ {l \atop jk} \right\}$ are called **Christoffel symbols of the first and the second kind**, respectively. The Christoffel symbols above are *symmetric* with respect to the interchange of indices j and k. Therefore, there exist $N^2(N+1)/2$ linearly independent components of each of these symbols. The transformation of Christoffel symbols under a twice-differentiable coordinate transformation is provided by (4.13), (5.36), (5.37), (5.40), and (5.41) as

$$\widehat{[jk,i]} = \frac{\partial X^m(\widehat{x})}{\partial \widehat{x}^j}\frac{\partial X^n(\widehat{x})}{\partial \widehat{x}^k}\frac{\partial X^l(\widehat{x})}{\partial \widehat{x}^i}[mn,l] + \frac{\partial X^m(\widehat{x})}{\partial \widehat{x}^i}\frac{\partial^2 X^n(\widehat{x})}{\partial \widehat{x}^j \partial \widehat{x}^k} g_{mn}(x), \tag{5.42}$$

$$\widehat{\left\{ {i \atop jk} \right\}} = \frac{\partial \widehat{X}^i(x)}{\partial x^l}\frac{\partial X^m(\widehat{x})}{\partial \widehat{x}^j}\frac{\partial X^n(\widehat{x})}{\partial \widehat{x}^k}\left\{ {l \atop mn} \right\} + \frac{\partial \widehat{X}^i(x)}{\partial x^l}\frac{\partial^2 X^l(\widehat{x})}{\partial \widehat{x}^j \partial \widehat{x}^k}. \tag{5.43}$$

The main properties of Christoffel symbols in (5.40) and (5.41) are summarized below:

$$g_{kl}(x)\left\{ {l \atop ij} \right\} = [ij,k], \tag{5.44}$$

$$\partial_j g_{ik} = [ij,k] + [jk,i], \tag{5.45}$$

$$\partial_k g^{ij} = -\left[g^{il}\left\{ {j \atop kl} \right\} + g^{jl}\left\{ {i \atop kl} \right\}\right]. \tag{5.46}$$

The cofactor of the entry $g_{ij}(x)$ in the $N \times N$ matrix $[g_{ij}(x)]$ is $g\,g^{ij}(x)$. Therefore, we have

$$\begin{aligned}
&\det[g_{ij}]\delta^k{}_l = g_{lm}(g\,g^{km}),\\
&\frac{\partial}{\partial g_{lm}}\det[g_{ij}] = g\,g^{lm}(x),\\
&\partial_k g = (\partial_k g_{ij})(g\,g^{ij}),\\
&\left\{ {j \atop ij} \right\} = (1/2g)\partial_i g = \partial_i \ln\sqrt{|g|}.
\end{aligned} \tag{5.47}$$

Example 5.1.5 Consider the polar coordinate chart for the subset of the Euclidean plane $\mathbb{E}_2$. The metric tensor is furnished by

$$\mathbf{g}_{..}(x) = dx^1 \otimes dx^1 + (x^1)^2 dx^2 \otimes dx^2,$$

$$D := \left\{x \in \mathbb{R}^2 : x^1 > 0, -\pi < x^2 < \pi\right\}.$$

$$[g_{ij}(x)] = \begin{bmatrix} 1 & 0 \\ 0 & (x^1)^2 \end{bmatrix}, \ [g^{ij}(x)] = \begin{bmatrix} 1 & 0 \\ 0 & (x^1)^{-2} \end{bmatrix}, \ g = (x^1)^2 > 0.$$

The Christoffel symbols from (5.40) and (5.41) are given by

$$[11,1] = [11,2] = [12,1] = [22,2] \equiv 0,$$

$$[12,2] = -[22,1] = x^1,$$

$$\left\{ {1 \atop 11} \right\} = \left\{ {2 \atop 11} \right\} = \left\{ {1 \atop 12} \right\} = \left\{ {2 \atop 22} \right\} \equiv 0,$$

$$\left\{ {2 \atop 12} \right\} = (x^1)^{-1}, \ \left\{ {1 \atop 22} \right\} = -x^1,$$

$$\partial_i \ln \sqrt{|g|} = \partial_i \ln x^1.$$

(See example 4.1.6.) □

If we choose an orthonormal basis field $\{\vec{\mathbf{e}}_a(x)\}_1^N$, then the corresponding metric tensor fields satisfy

$$g_{ab}(x) \equiv d_{ab}, \ g^{ab}(x) \equiv d^{ab}, \ \partial_c g_{ab} \equiv 0, \ \partial_c g^{ab} \equiv 0. \tag{5.48}$$

(Recall that we denote orthonormal indices by a, b, c, d, e etc.) Therefore, by (5.36) and (5.38), we get

$$\begin{aligned} \Gamma^c{}_{ab}(x) &= -(1/2)\left[\chi^c{}_{ba} + d^{ce}\left(d_{bf}\chi^f{}_{ae} + d_{af}\chi^f{}_{be}\right)\right], \\ \Gamma^c{}_{ab}(x) - \Gamma^c{}_{ba}(x) &= \chi^c{}_{ab}(x), \\ \Gamma_{cab}(x) &= -(1/2)\left[\chi_{abc}(x) + \chi_{bac}(x) - \chi_{cab}(x)\right] \\ &\equiv -\Gamma_{bac}(x). \end{aligned} \tag{5.49}$$

The **Ricci rotation coefficients** are defined by a minor permutation of indices as

$$\begin{aligned} &\gamma_{abc}(x) := \Gamma_{bca}(x), \\ &\gamma_{bac}(x) = -\gamma_{abc}(x), \\ &d^{ab}\gamma_{abc}(x) \equiv 0, \\ &\gamma^b{}_{ac}(x) = -\gamma_a{}^b{}_c(x) = -\Gamma^b{}_{ca}(x), \\ &\gamma_a{}^b{}_c(x) - \gamma_c{}^b{}_a(x) = \gamma^b{}_{ca}(x) - \gamma^b{}_{ac}(x) = \chi^b{}_{ca}(x). \end{aligned} \tag{5.50}$$

The number of linearly independent components of $\gamma_{abc}(x)$ is $N^2(N-1)/2$.

By (4.57) and (5.50), we obtain

$$\begin{aligned} [\vec{\mathbf{e}}_a, \vec{\mathbf{e}}_b]\,[f] &= \partial_a\partial_b f - \partial_b\partial_a f = \chi^c{}_{ab}(x)\vec{\mathbf{e}}_c(x)[f] \\ &= \left[\gamma^c{}_{ab}(x) - \gamma^c{}_{ba}(x)\right]\partial_c f. \end{aligned} \tag{5.51}$$

Putting $f(x) := x^j$, we can derive from (5.18) and (5.50) the "**metric equation**"

$$\lambda^i{}_a(x)\partial_i\lambda^j{}_b - \lambda^k{}_b(x)\partial_k\lambda^j{}_a = [\gamma^c{}_{ab}(x) - \gamma^c{}_{ba}(x)]\,\lambda^j{}_c(x). \tag{5.52}$$

Now, we shall state and prove a theorem about Ricci rotation coefficients.

Theorem 5.1.6 *The Ricci rotation coefficients for a differentiable orthonormal basis field* $\{\vec{\mathbf{e}}_a(x)\}_1^N$ *in a domain* $D \subset \mathbb{R}^N$ *must satisfy*

$$\gamma_{abc}(x) = g_{jl}(x)\left(\nabla_k\lambda^l{}_a\right)\lambda^j{}_b(x)\lambda^k{}_c(x). \tag{5.53}$$

Proof. By (5.50), (5.37), (5.19) and (4.18), we can show that

$$\begin{aligned}\gamma_{abc}(x) &= d_{bf}\Gamma^f{}_{ca}(x) = \mathbf{g}_{..}(x)\left(\vec{\mathbf{e}}_b, \Gamma^f{}_{ca}\vec{\mathbf{e}}_f\right)\\ &= \mathbf{g}_{..}(x)\left(\vec{\mathbf{e}}_b, \nabla_{\vec{\mathbf{e}}_c}\left(\vec{\mathbf{e}}_a\right)\right).\end{aligned}$$

Employing (5.18), (4.18), (5.1), (5.2), and theorem 4.1.7, we can derive that

$$\begin{aligned}\mathbf{g}_{..}(x)\left(\vec{\mathbf{e}}_b, \nabla_{\vec{\mathbf{e}}_c}\left(\vec{\mathbf{e}}_a\right)\right) &= \mathbf{g}_{..}(x)\left(\lambda^j{}_b\partial_j, \nabla_{\lambda^k{}_c\partial_k}\left(\lambda^l{}_a\partial_l\right)\right)\\ &= \lambda^j{}_b(x)\lambda^k{}_c(x)\mathbf{g}_{..}(x)\left(\partial_j, \left(\nabla_k\lambda^l{}_a\right)\partial_l\right)\\ &= g_{jl}(x)\left(\nabla_k\lambda^l{}_a\right)\lambda^j{}_b(x)\lambda^k{}_c(x).\end{aligned}$$ ■

Corollary 5.1.7

$$g_{jl}(x)\left(\nabla_i\lambda^l{}_a\right) = \gamma_{abc}(x)\mu^b{}_j(x)\mu^c{}_i(x). \tag{5.54}$$

The proof follows from (5.53) and (5.18).

Example 5.1.8 Consider the surface of a unit sphere S^2. In the spherical polar coordinate chart, the domain of validity and the metric tensor field are given by

$$D := \left\{x \in \mathbb{R}^2 \,:\, 0 < x^1 < \pi,\ -\pi < x^2 < \pi\right\},$$

$$\begin{aligned}\mathbf{g}_{..}(x) &= dx^1 \otimes dx^1 + (\sin x^1)^2 dx^2 \otimes dx^2\\ &= \tilde{\mathbf{e}}^1(x) \otimes \tilde{\mathbf{e}}^1(x) + \tilde{\mathbf{e}}^2(x) \otimes \tilde{\mathbf{e}}^2(x)\\ &= d_{ab}\tilde{\mathbf{e}}^a(x) \otimes \tilde{\mathbf{e}}^b(x).\end{aligned}$$

(See example 4.3.12.)

The non-zero components of the Christoffel symbols from (5.40) and (5.41) are

$$[12, 2] = -[22, 1] = (\sin x^1)(\cos x^1),$$

$$\left\{\begin{matrix}2\\12\end{matrix}\right\} = \cot x^1, \quad \left\{\begin{matrix}1\\22\end{matrix}\right\} = -(\sin x^1)(\cos x^1).$$

An orthonormal basis set is furnished by

$$\tilde{\mathbf{e}}^1(x) = dx^1, \;\; \tilde{\mathbf{e}}^2(x) = (\sin x^1)dx^2,$$
$$\vec{\mathbf{e}}_1(x) = \frac{\partial}{\partial x^1}, \;\; \vec{\mathbf{e}}_2(x) = (\sin x^1)^{-1}\frac{\partial}{\partial x^2},$$
$$\lambda^i{}_1(x) = \delta^i{}_1, \;\; \lambda^i_2(x) = (\sin x^1)^{-1}\delta^i{}_2,$$
$$\mu^1{}_i(x) = \delta^1{}_i, \;\; \mu^2{}_i(x) = (\sin x^1)\delta^2{}_i.$$

The Lie bracket operation on an arbitrary twice-differentiable function yields

$$[\vec{\mathbf{e}}_1, \vec{\mathbf{e}}_2][f] = -(\cot x^1)(\operatorname{cosec} x^1)\frac{\partial f(x)}{\partial x^2} = -(\cot x^1)\vec{\mathbf{e}}_2[f].$$

Therefore, by (4.57), the non-zero components of the structure coefficients $\chi^c{}_{ab}(x)$ are

$$\chi^2{}_{12}(x) = -\chi^2{}_{21}(x) = -\cot x^1.$$

By (5.49), the non-zero components of $\Gamma^c{}_{ab}(x)$ and $\Gamma_{cab}(x)$ are

$$\Gamma^1{}_{22}(x) = -\cot x^1, \;\; \Gamma^2{}_{21}(x) = \cot x^1,$$
$$\Gamma_{122}(x) = -\cot x^1, \;\; \Gamma_{221}(x) = \cot x^1.$$

(Note that $\Gamma^2{}_{12}(x) \equiv 0$!) Using (5.50), we obtain the non-zero components of the Ricci rotation coefficients

$$\gamma_{122}(x) = \Gamma_{221}(x) = \cot x^1,$$
$$\gamma_{212}(x) = \Gamma_{122}(x) = -\cot x^1 = -\gamma_{122}(x). \qquad \square$$

Consider the connection 1-forms $\tilde{\mathbf{w}}^a{}_b(x)$ from (4.8) relative to an orthonormal basis. We can write

$$\begin{aligned} \tilde{\mathbf{w}}^a{}_b(x) &= \Gamma^a{}_{cb}(x)\tilde{\mathbf{e}}^c(x), \\ \tilde{\mathbf{w}}_{ab}(x) &:= d_{ac}\tilde{\mathbf{w}}^c{}_b(x) = \Gamma_{acb}(x)\tilde{\mathbf{e}}^c(x). \end{aligned} \tag{5.55}$$

By (5.50), the above yields

$$\tilde{\mathbf{w}}_{ab}(x) = \gamma_{bac}(x)\tilde{\mathbf{e}}^c(x) = -\tilde{\mathbf{w}}_{ba}(x). \tag{5.56}$$

By the first structural equation (4.80) and the condition (5.56), we obtain

$$\begin{aligned} d\tilde{\mathbf{e}}^a(x) &= -\tilde{\mathbf{w}}^a{}_b(x) \wedge \tilde{\mathbf{e}}^b(x) \\ &= d^{ac}\gamma_{bcf}(x)\tilde{\mathbf{e}}^f(x) \wedge \tilde{\mathbf{e}}^b(x). \end{aligned} \tag{5.57}$$

Example 5.1.9 Let us consider the surface S^2 of the unit sphere and borrow the basis 1-forms and Ricci rotation coefficients from example 5.1.8. These are given by

$$\tilde{\mathbf{e}}^1(x) = dx^1, \ \ \tilde{\mathbf{e}}^2(x) = (\sin x^1)dx^2,$$
$$\gamma_{122}(x) \equiv -\gamma_{212}(x) = \cot x^1.$$

The connection 1-forms from (5.56) are provided by 2×2 matrix (of 1 -forms) as

$$[\tilde{\mathbf{w}}_{ab}(x)] = \begin{bmatrix} 0, & -(\cos x^1)dx^2 \\ (\cos x^1)dx^2, & 0 \end{bmatrix} \equiv -[\tilde{\mathbf{w}}_{ba}(x)]\,.$$

(Compare and contrast these 1-forms with those in example 4.3.12.) Furthermore, we can derive that

$$\mathbf{O}_{..}(x) \equiv d^2(x^1) = d\tilde{\mathbf{e}}^1(x) = -\tilde{\mathbf{w}}_{12}(x)\tilde{\mathbf{e}}^2(x) = (\cos x^1)(dx^2 \wedge dx^2),$$
$$d((\sin x^1)dx^2) = d\tilde{\mathbf{e}}^2(x) = \tilde{\mathbf{w}}_{21}(x)\tilde{\mathbf{e}}^1(x) = (\cos x^1)dx^1 \wedge dx^2.$$

Thus, the equations (5.57) are explicitly verified in this example. □

Now, let us discuss the transformation properties of the relative tensor fields and the oriented relative tensor fields. From (3.35), (3.50), (5.22), and (5.23), the transformation of the relative tensor field of weight w is given by

$$\begin{aligned}\widehat{\theta}^{i_1\ldots i_r}{}_{j_1\ldots j_s}(\widehat{x}) = \left\{\det\left[\frac{\partial X^m(\widehat{x})}{\partial \widehat{x}^n}\right]\right\}^w \frac{\partial \widehat{X}^{i_1}(x)}{\partial x^{k_1}}\cdots\frac{\partial \widehat{X}^{i_r}(x)}{\partial x^{k_r}}\cdots\frac{\partial X^{l_1}(\widehat{x})}{\partial \widehat{x}^{j_1}} \\ \cdots\frac{\partial X^{l_s}(\widehat{x})}{\partial \widehat{x}^{j_s}}\cdots\theta^{k_1\ldots k_r}{}_{l_1\ldots l_s}(x),\end{aligned} \tag{5.58}$$

$$\begin{aligned}\widehat{\theta}^{a_1\ldots a_r}{}_{b_1\ldots b_s}(x) = (\pm 1)^w A^{a_1}{}_{c_1}(x)\ldots A^{a_r}{}_{c_r}(x)L^{d_1}{}_{b_1}(x) \\ \ldots L^{d_s}{}_{b_s}(x)\theta^{c_1\ldots c_r}{}_{d_1\ldots d_s}(x).\end{aligned} \tag{5.59}$$

(Here, we have used the fact that $\det[L^a{}_b(x)] = \pm 1$.)

By (3.38), (3.51), (5.58), and (5.59), the transformation properties of the components of an oriented relative tensor are furnished by

$$\begin{aligned}\widehat{\theta}^{i_1\ldots i_r}{}_{j_1\ldots j_s}(\widehat{x}) = \left\{\operatorname{sgn}\left[\det\left[\frac{\partial X^m(\widehat{x})}{\partial \widehat{x}^n}\right]\right]\right\}\left\{\det\left[\frac{\partial X^m(\widehat{x})}{\partial \widehat{x}^n}\right]\right\}^w \frac{\partial \widehat{X}^{i_1}(x)}{\partial x^{k_1}} \\ \cdots\frac{\partial \widehat{X}^{i_r}(x)}{\partial x^{k_r}}\frac{\partial X^{l_1}(\widehat{x})}{\partial \widehat{x}^{j_1}}\cdots\frac{\partial X^{l_s}(\widehat{x})}{\partial \widehat{x}^{j_s}}\theta^{k_1\ldots k_r}{}_{l_1\ldots l_s}(x),\end{aligned} \tag{5.60}$$

$$\begin{aligned}\widehat{\theta}^{a_1\ldots a_r}{}_{b_1\ldots b_s}(x) = (\pm 1)^{1+w} A^{a_1}{}_{c_1}(x)\ldots A^{a_r}{}_{c_r}(x)L^{d_1}{}_{b_1}(x) \\ \ldots L^{d_s}{}_{b_s}(x)\theta^{c_1\ldots c_r}{}_{d_1\ldots d_s}(x).\end{aligned} \tag{5.61}$$

Comparing (5.59) and (5.61), we conclude that the orthonormal components of an oriented relative tensor transform just like a relative tensor of a higher weight. Moreover, under the subgroup of the *proper* generalized Lorentz transformations (with $\det[L^a{}_b(x)] = +1$), there are no distinctions among orthonormal components of an absolute tensor, a relative tensor, and an oriented relative tensor.

Example 5.1.10 By (3.36), (5.58), and (5.59), the transformation of the totally antisymmetric, numerical relative tensor of weight -1 is given by

$$\widehat{\varepsilon}_{i_1 \dots i_N} = \left\{\det\left[\frac{\partial X^k(\widehat{x})}{\partial \widehat{x}^l}\right]\right\}^{-1} \frac{\partial X^{j_1}(\widehat{x})}{\partial \widehat{x}^{i_1}} \cdots \frac{\partial X^{j_N}(\widehat{x})}{\partial \widehat{x}^{i_N}} \varepsilon_{j_1 \dots j_N}, \tag{5.62}$$

$$\widehat{\varepsilon}_{a_1 \dots a_N} = (\pm 1)^{-1} L^{b_1}{}_{a_1}(x) \dots L^{b_N}{}_{a_N}(x) \varepsilon_{b_1 \dots b_N}. \tag{5.63}$$

The contravariant version of (5.62) and (5.63) is furnished by

$$\widehat{\varepsilon}^{i_1 \dots i_N} = \left\{\det\left[\frac{\partial X^k(\widehat{x})}{\partial \widehat{x}^l}\right]\right\} \frac{\partial \widehat{X}^{i_1}(x)}{\partial x^{j_1}} \cdots \frac{\partial \widehat{X}^{i_N}(x)}{\partial x^{j_N}} \varepsilon^{j_1 \dots j_N}, \tag{5.64}$$

$$\widehat{\varepsilon}^{a_1 \dots a_N} = \pm A^{a_1}{}_{b_1}(x) \dots A^{a_N}{}_{b_N}(x) \varepsilon^{b_1 \dots b_N}. \tag{5.65}$$

□

Example 5.1.11 Let us define Levi-Civita's antisymmetric, oriented tensor fields by (2.139), (2.140), (2.141), (2.142), and (5.13) as

$$\eta_{i_1 \dots i_N}(x) := \sqrt{|\det[g_{kl}(x)]|}\, \varepsilon_{i_1 \dots i_N} \equiv \sqrt{|g|} \varepsilon_{i_1 \dots i_N}, \tag{5.66}$$

$$\eta_{a_1 \dots a_N} := \varepsilon_{a_1 \dots a_N}, \tag{5.67}$$

$$\eta^{i_1 \dots i_N}(x) = \left\{[\mathrm{sgn}(g)]/\sqrt{|g|}\right\} \varepsilon^{i_1 \dots i_N}, \tag{5.68}$$

$$\eta^{a_1 \dots a_N} = \{\mathrm{sgn}[\det[d_{cd}]]\} \varepsilon^{a_1 \dots a_N} \equiv (\pm 1) \varepsilon^{a_1 \dots a_N}, \tag{5.69}$$

$$\widehat{\eta}_{i_1 \dots i_N}(\widehat{x}) = \mathrm{sgn}\left\{\det\left[\frac{\partial X^k(\widehat{x})}{\partial \widehat{x}^l}\right]\right\} \frac{\partial X^{j_1}(\widehat{x})}{\partial \widehat{x}^{i_1}} \cdots \frac{\partial X^{j_N}(\widehat{x})}{\partial \widehat{x}^{i_N}} \eta_{j_1 \dots j_N}(x), \tag{5.70}$$

$$\widehat{\eta}_{a_1 \dots a_N} = (\pm 1) L^{b_1}{}_{a_1}(x) \cdots L^{b_N}{}_{a_N}(x) \eta_{b_1 \dots b_N}, \tag{5.71}$$

$$\widehat{\eta}^{i_1 \dots i_N}(\widehat{x}) = \mathrm{sgn}\left\{\det\left[\frac{\partial X^k(\widehat{x})}{\partial \widehat{x}^l}\right]\right\} \frac{\partial \widehat{X}^{i_1}(x)}{\partial x^{j_1}} \cdots \frac{\partial \widehat{X}^{i_N}(x)}{\partial x^{j_N}} \eta^{j_1 \dots j_N}(x), \tag{5.72}$$

$$\widehat{\eta}^{a_1 \dots a_N} = (\pm 1) A^{a_1}{}_{b_1}(x) \cdots A^{a_N}{}_{b_N}(x) \eta^{b_1 \dots b_N}. \tag{5.73}$$

(Note that the components above relative to an orthonormal basis are *constant-valued.*)

□

Let us consider the totally antisymmetric fields ${}_p\mathbf{W}(x)$ and ${}_{N-p}\mathbf{A}(x)$ for $1 < p < N$. The Hodge-star duality operations can be defined from (2.144) and (2.145) as

$$*W_{i_1\ldots i_{N-p}}(x) := (1/p!)\eta^{j_1\ldots j_p}{}_{i_1\ldots i_{N-p}}(x)W_{j_1\ldots j_p}(x), \tag{5.74}$$

$$*W_{a_1\ldots a_{N-p}}(x) := (1/p!)\eta^{b_1\ldots b_p}{}_{a_1\ldots a_{N-p}}W_{b_1\ldots b_p}(x), \tag{5.75}$$

$$*A_{j_1\ldots j_p}(x) = [1/(N-p)!]\,\eta^{i_1\ldots i_{N-p}}{}_{j_1\ldots j_p}(x)A_{i_1\ldots i_{N-p}}(x), \tag{5.76}$$

$$*A_{b_1\ldots b_p}(x) = [1/(N-p)!]\,\eta^{a_1\ldots a_{N-p}}{}_{b_1\ldots b_p}A_{a_1\ldots a_{N-p}}(x). \tag{5.77}$$

Here, $\eta^{j_1\ldots j_p}{}_{i_1\ldots i_{N-p}}(x) := g^{j_1k_1}(x)\ldots g^{j_pk_p}(x)\eta_{k_1\ldots k_p\,i_1\ldots i_{N-p}}(x)$ and so on.

Example 5.1.12 Let us choose $N = 4$, $p = 2$, and $\text{sgn}(\mathbf{g}..(x)) = +2$. The antisymmetric electromagnetic field tensor is furnished by

$$\mathbf{F}..(x) = (1/2)F_{ij}(x)dx^i \wedge dx^j = (1/2)F_{ab}(x)\tilde{\mathbf{e}}^a(x) \wedge \tilde{\mathbf{e}}^b(x). \tag{5.78}$$

(See example 2.4.12.) The Hodge dual tensor is provided by (5.74) and (5.75) as

$$*F^{ij}(x) = (1/2)\eta^{klij}(x)F_{kl}(x), \tag{5.79}$$

$$*F^{ab}(x) = (1/2)\eta^{cdab}(x)F_{cd}(x) = -(1/2)\varepsilon^{cdab}F_{cd}(x). \tag{5.80}$$

The transformation properties are summarized as

$$*\widehat{F}^{ij}(x) = \left\{\text{sgn}\left[\det\left[\frac{\partial X^m(\widehat{x})}{\partial \widehat{x}^n}\right]\right]\right\}\frac{\partial \widehat{X}^i(x)}{\partial x^k}\frac{\partial \widehat{X}^j(x)}{\partial x^l} *F^{kl}(x), \tag{5.81}$$

$$*\widehat{F}^{ab}(x) = \pm A^a{}_c(x)A^b{}_d(x) *F^{cd}(x). \tag{5.82}$$

Thus, the dual field $*\mathbf{F}..(x)$ transforms as an *oriented* antisymmetric tensor. □

Exercises 5.1

1. Obtain the signature of the metric tensor given by

$$\mathbf{g}..(x) := dx^1 \otimes dx^1 + 3dx^1 \otimes dx^2 + 4dx^2 \otimes dx^2 + dx^3 \otimes dx^3.$$

2. Compute the Christoffel symbols of the two kinds, as well as the Ricci rotation coefficients for the two-dimensional Anti-DeSitter metric

$$\mathbf{g}..(x) = (x^1 + x^2)^{-2}dx^1 \otimes dx^2.$$

3. (i) Prove that for a differentiable metric field $\mathbf{g}..(x)$,

$$dg_{pq}(x) = g_{up}(x)\tilde{\mathbf{w}}^u{}_q(x) + g_{uq}(x)\tilde{\mathbf{w}}^u{}_p(x).$$

(ii) Prove that for three arbitrary differentiable vector fields, a metric tensor must satisfy

$$\vec{\mathbf{W}}(x)[\mathbf{g}_{..}(\vec{\mathbf{U}},\vec{\mathbf{V}})] = \mathbf{g}_{..}(x)(\nabla_{\vec{\mathbf{W}}}\vec{\mathbf{U}},\vec{\mathbf{V}}) + \mathbf{g}_{..}(x)(\vec{\mathbf{U}},\nabla_{\vec{\mathbf{W}}}\vec{\mathbf{V}}).$$

4. Show that the "eigenvalues" $\lambda(x)$ of the equation

$$\det\left[A_{pq}(x) - \lambda(x)B_{pq}(x)\right] = 0$$

are *invariant* under a general transformation of the basis sets.

5. Show that under a general Lorentz transformation, the Ricci rotation coefficients undergo the following transformations:

$$\widehat{\gamma}_{abc}(x) = L^f{}_a(x)L^d{}_b(x)L^e{}_c(x)\gamma_{fde}(x) + d_{ef}L^f{}_b(x)\left[\partial_c L^e{}_a(x)\right].$$

5.2 Covariant Derivatives and the Curvature Tensor

Let ${}^r{}_s\mathbf{T}(x)$ be a differentiable tensor field in $D \subset \mathbb{R}^N$. By (4.31), (4.35), (5.20), (5.41), (5.49), and (5.50), we derive the following rules of covariant derivatives:

$$\begin{aligned}\nabla_k T^{i_1\ldots i_r}{}_{j_1\ldots j_s} = \partial_k T^{i_1\ldots i_r}{}_{j_1\ldots j_s} &+ \sum_{\alpha=1}^{r}\left\{\begin{matrix} i_\alpha \\ kl \end{matrix}\right\} T^{i_1\ldots i_{\alpha-1} l i_{\alpha+1}\ldots i_r}{}_{j_1\ldots j_s} \\ &- \sum_{\beta=1}^{s}\left\{\begin{matrix} l \\ kj_\beta \end{matrix}\right\} T^{i_1\ldots i_r}{}_{j_1\ldots j_{\beta-1} l j_{\beta+1}\ldots j_s},\end{aligned} \tag{5.83}$$

$$\begin{aligned}\nabla_c T^{a_1\ldots a_r}{}_{b_1\ldots b_s} &= \partial_c T^{a_1\ldots a_r}{}_{b_1\ldots b_s} + \sum_{\alpha=1}^{r}\Gamma^{a_\alpha}{}_{cd}T^{a_1\ldots a_{\alpha-1} d a_{\alpha+1}\ldots r}{}_{b_1\ldots b_s} \\ &\quad - \sum_{\beta=1}^{s}\Gamma^d{}_{cb_\beta}T^{a_1\ldots a_r}{}_{b_1\ldots b_{\beta-1} d b_{\beta+1}\ldots b_s} \\ &\equiv \partial_c T^{a_1\ldots a_r}{}_{b_1\ldots b_s} - \sum_{\alpha=1}^{r}\gamma^{a_\alpha}{}_{dc}T^{a_1\ldots a_{\alpha-1} d a_{\alpha+1}\ldots a_r}{}_{b_1\ldots b_s} \\ &\quad + \sum_{\beta=1}^{s}\gamma^d{}_{b_\beta c}T^{a_1\ldots a_r}{}_{b_1\ldots b_{\beta-1} d b_{\beta+1}\ldots b_s}.\end{aligned} \tag{5.84}$$

Here, note that $\left\{\begin{smallmatrix} i \\ jk \end{smallmatrix}\right\}$ are Christoffel symbols and $\gamma_{abc} = d_{ah}\gamma^h{}_{bc}$ are the Ricci rotation coefficients.

(*Caution* should be applied in regard to the positions of the indices.)

Example 5.2.1 Let $\phi(x)$ be a differentiable scalar field or a 0-form.

$$\nabla_i \phi = \partial_i \phi \equiv \frac{\partial \phi(x)}{\partial x^i},$$

$$\nabla_a \phi = \lambda^i{}_a(x) \frac{\partial \phi(x)}{\partial x^i},$$

are components of a covariant vector field. The corresponding 1-form is given by

$$d\phi(x) = (\partial_i \phi) dx^i = (\partial_a \phi) \tilde{\mathbf{e}}^a(x). \qquad \square$$

Example 5.2.2 Consider a differentiable vector field $\vec{\mathbf{A}}(x)$. By (5.83) and (5.84), the covariant derivatives are furnished by

$$\begin{aligned}
&\nabla_j A^i = \partial_j A^i + \left\{ \begin{matrix} i \\ jk \end{matrix} \right\} A^k(x), \\
&\nabla_b A^a = \partial_b A^a - \gamma^a{}_{cb}(x) A^c(x), \\
&\nabla_i A^i = \left(1/\sqrt{|g|}\right) \partial_i \left[\sqrt{|g|} A^i\right], \qquad (5.85) \\
&\nabla_a A^a = \partial_a A^a - \gamma^a{}_{ca}(x) A^c(x), \\
&\nabla_j A_i - \nabla_i A_j = \partial_j A_i - \partial_i A_j, \\
&\nabla_b A_a - \nabla_a A_b = \partial_b A_a - \partial_a A_b + (\gamma^c{}_{ab} - \gamma^c{}_{ba}) A_c(x). \qquad \square
\end{aligned}$$

Example 5.2.3 Consider a differentiable metric field

$$\mathbf{g}_{..}(x) = g_{ij}(x) dx^i \otimes dx^j = d_{ab} \tilde{\mathbf{e}}^a(x) \otimes \tilde{\mathbf{e}}^b(x).$$

By (5.35), (5.41), (5.44), (5.46), (5.83), and (5.84), we obtain

$$\begin{aligned}
&\nabla_k g_{ij} = \partial_k g_{ij} - [ki, j] - [kj, i] \equiv 0, \\
&\nabla_k g^{ij} \equiv 0, \ \nabla_k g^i{}_j \equiv \nabla_k \delta^i{}_j \equiv 0, \\
&\nabla_c g_{ab} \equiv \nabla_c d_{ab} = \gamma_{bac}(x) + \gamma_{abc}(x) \equiv 0, \\
&\nabla_c d^{ab} \equiv 0, \ \nabla_c g^a{}_b \equiv \nabla_c \delta^a{}_b \equiv 0.
\end{aligned} \qquad (5.86)$$

Therefore, the metric tensor components are "*covariantly constant* ." □

Example 5.2.4 Let $\mathbf{T}_{..}(x)$ be a differentiable, *symmetric* tensor field. Then, by (5.83) and (5.84), we get

$$\begin{aligned}
\nabla_k T^{ij} &= \partial_k T^{ij} + \left\{ \begin{matrix} i \\ kl \end{matrix} \right\} T^{lj}(x) + \left\{ \begin{matrix} j \\ kl \end{matrix} \right\} T^{il}(x), \\
\nabla_k T^{ab} &= \partial_c T^{ab} - \gamma^a{}_{ec}(x) T^{eb}(x) - \gamma^b{}_{ec} T^{ae}(x), \\
\nabla_j T^{ij} &= \left(1/\sqrt{|g|}\right) \left[\partial_j \left(\sqrt{|g|} T^{ij} \right) + \left\{ \begin{matrix} i \\ jk \end{matrix} \right\} \left(\sqrt{|g|} T^{jk}(x) \right) \right], \\
\nabla_b T^{ab} &= \partial_b T^{ab} - \gamma^a{}_{eb}(x) T^{eb}(x) - \gamma^b{}_{eb}(x) T^{ae}(x).
\end{aligned} \tag{5.87}$$

This example is relevant in Einstein's theory of gravitation. □

Example 5.2.5 Let $\mathbf{F}^{\cdot\cdot}(x)$ be a differentiable, *antisymmetric* tensor field. The covariant derivatives are given by

$$\begin{aligned}
\nabla_j F^{ij} &= \left(1/\sqrt{|g|}\right) \partial_j \left(\sqrt{|g|} F^{ij} \right), \\
\nabla_b F^{ab} &= \partial_b F^{ab} - \gamma^a{}_{cb}(x) F^{cb}(x) - \gamma^b{}_{cb}(x) F^{ac}(x).
\end{aligned} \tag{5.88}$$

This example is applicable in electromagnetic field theory. □

The **Laplacian** is an invariant, second-order, linear differential operator furnished by

$$\Delta := g^{ij}(x) \nabla_i \nabla_j =: \nabla^i \nabla_i \equiv \nabla^a \nabla_a. \tag{5.89}$$

(Other notations for the *same* operator are ∇^2 or Δ^2.) For a scalar field $V(x)$ of class C^2, the Laplacian of $V(x)$ is provided by (5.85) as

$$\Delta V = \nabla_i \nabla^i V = \left(1/\sqrt{|g|}\right) \partial_i \left[\sqrt{|g|} g^{ij} \partial_j V \right], \tag{5.90}$$

$$\Delta V = \nabla_a \nabla^a V = d^{ab} \left[\partial_b \partial_a V + \gamma^c{}_{ab}(x) \partial_c V \right]. \tag{5.91}$$

Example 5.2.6 Consider the Euclidean plane $\mathbb{E}_2$ and a global Cartesian chart such that

$$\mathbf{g}_{\cdot\cdot}(x) = \delta_{ij} dx^i \otimes dx^j = \delta_{ab} \tilde{\mathbf{e}}^a(x) \otimes \tilde{\mathbf{e}}^b(x)$$

$$\tilde{\mathbf{e}}^a(x) = \delta^a{}_i dx^i, \ \vec{\mathbf{e}}_a(x) = \delta^i{}_a \frac{\partial}{\partial x^i}.$$

The Laplacian of a twice-differentiable function $V(x)$ is given by (5.90) and (5.91) as

$$\Delta V = \frac{\partial^2 V}{(\partial x^1)^2} + \frac{\partial^2 V}{(\partial x^2)^2}.$$

The **harmonic equation**, or the two-dimensional **potential equation**,

$$\Delta V = 0$$

is solved by $V(x^1, x^2) = \mathrm{Re}[f(x^1 + ix^2)]$. Here, f is an *arbitrary* **holomorphic function** of the complex variable $x^1 + ix^2$. □

Example 5.2.7 In a polar coordinate chart of $\mathbb{E}_2$, the metric tensor field is provided by example 5.1.5 as

$$\mathbf{g}_{..}(x) = dx^1 \otimes dx^1 + (x^1)^2(dx^2 \otimes dx^2) = \delta_{ab}\tilde{\mathbf{e}}^a(x) \otimes \tilde{\mathbf{e}}^b(x).$$

The Laplacian from (5.90) is given by

$$\begin{aligned}\Delta V &= (1/x^1)[\partial_1(x^1\partial_1 V) + \partial_2((x^1)^{-1}\partial_2 V)] \\ &= \frac{\partial^2 V}{(\partial x^1)^2} + \frac{1}{x^1}\frac{\partial V}{\partial x^1} + \frac{1}{(x^1)^2}\frac{\partial^2 V}{(\partial x^2)^2}.\end{aligned}$$

We compute the Ricci rotation coefficients from (5.53). The non-zero components are furnished by

$$\gamma_{122}(x) \equiv -\gamma_{212}(x) = (x^1)^{-1} > 0.$$

Therefore, by (5.91), we obtain

$$\begin{aligned}\Delta V &= \left[\partial_{(1)}\partial_{(1)}V + \gamma^c{}_{11}(x)\partial_c V\right] + \left[\partial_{(2)}\partial_{(2)}V + \gamma^c{}_{22}(x)\partial_c V\right] \\ &= \left[\frac{\partial^2 V}{(\partial x^1)^2}\right] + \left[\frac{1}{(x^1)^2}\frac{\partial^2 V}{(\partial x^2)^2} + \frac{1}{x^1}\frac{\partial V}{\partial x^1}\right].\end{aligned}$$

Of course, both computations coincide. □

Example 5.2.8 Consider the Euclidean space $\mathbb{E}_3$ and a coordinate chart with

$$\mathbf{g}_{..}(x) = g_{ij}(x)dx^i \otimes dx^j.$$

Suppose that the **wave function** $\psi(x)$ is a complex-valued function over $D \subset \mathbb{R}^3$ and is of class C^2. The **Schröedinger equation** for a stationary state of a physical system in **wave mechanics** is provided by

$$\begin{aligned}&\Delta\psi + \left(2m/(\hbar)^2\right)[E - V(x)]\,\psi(x) \\ &= \left(1/\sqrt{|g|}\right)\frac{\partial}{\partial x^i}\left[\sqrt{|g|}g^{ij}(x)\frac{\partial\psi}{\partial x^j}\right] + \frac{2m}{(\hbar)^2}[E - V(x)]\,\psi(x) \qquad (5.92) \\ &= 0.\end{aligned}$$

Here, $2\pi\hbar$ is Planck's constant; m, E, and $V(x)$ are the mass, total energy, and potential energy of the wave-mechanical particle. □

The **Riemann-Christoffel curvature tensor** components are furnished in a coordinate basis by (4.74), (4.75), and (5.41) as

$$R^i{}_{jkl}(x) = \partial_k\left\{ {i \atop lj} \right\} - \partial_l\left\{ {i \atop kj} \right\} + \left\{ {i \atop kh} \right\}\left\{ {h \atop lj} \right\} - \left\{ {i \atop lh} \right\}\left\{ {h \atop kj} \right\}. \qquad (5.93)$$

$$R_{ijkl}(x) = \partial_k[lj,i] - \partial_l[kj,i] + [il,h]\left\{ \begin{matrix} h \\ jk \end{matrix} \right\} - [ik,h]\left\{ \begin{matrix} h \\ jl \end{matrix} \right\}, \tag{5.94}$$

$$\begin{aligned} R_{ijkl}(x) = \left(\frac{1}{2}\right) & [\partial_j\partial_k g_{il} + \partial_i\partial_l g_{jk} - \partial_j\partial_l g_{ik} - \partial_i\partial_k g_{jl}] \\ & + g^{hm}\{[jk,m][il,h] - [jl,m][ik,h]\}. \end{aligned} \tag{5.95}$$

The curvature tensor components in an orthonormal basis are provided by (4.74), (5.49), and (5.50) as

$$\begin{aligned} R^a{}_{bcd}(x) = & \partial_d\gamma^a{}_{bc} - \partial_c\gamma^a{}_{bd} + \gamma^a{}_{hc}\gamma^h{}_{bd} - \gamma^a{}_{hd}\gamma^h{}_{bc} \\ & + \gamma^a{}_{bh}\left(\gamma^h{}_{cd} - \gamma^h{}_{dc}\right), \end{aligned} \tag{5.96}$$

$$\begin{aligned} R_{abcd}(x) = & \partial_d\gamma_{abc} - \partial_c\gamma_{abd} + \gamma_{had}\gamma^h{}_{bc} - \gamma_{hac}\gamma^h{}_{bd} \\ & + \gamma_{abh}\left(\gamma^h{}_{cd} - \gamma^h{}_{dc}\right), \end{aligned} \tag{5.97}$$

$$R_{abcd}(x) = \lambda^i{}_a(x)\lambda^j{}_b(x)\lambda^k{}_c(x)\lambda^l{}_d(x)R_{ijkl}(x), \tag{5.98}$$

$$R_{ijkl}(x) = \mu^a{}_i(x)\mu^b{}_j(x)\mu^c{}_k(x)\mu^d{}_l(x)R_{abcd}(x). \tag{5.99}$$

Now, we shall investigate the algebraic identities among curvature tensor components.

Theorem 5.2.9 *The algebraic identities of the Riemann-Christoffel tensor components are furnished by the following:*

$$R_{jikl}(x) \equiv -R_{ijkl}(x), \tag{5.100}$$

$$R_{ijlk}(x) \equiv -R_{ijkl}(x), \tag{5.101}$$

$$R_{klij}(x) \equiv R_{ijkl}(x), \tag{5.102}$$

$$R_{ijkl}(x) + R_{iklj}(x) + R_{iljk}(x) \equiv 0, \tag{5.103}$$

$$R_{bacd}(x) \equiv -R_{abcd}(x), \tag{5.104}$$

$$R_{badc}(x) \equiv -R_{abcd}(x), \tag{5.105}$$

$$R_{cdab}(x) \equiv R_{abcd}(x), \tag{5.106}$$

$$R_{abcd}(x) + R_{acdb}(x) + R_{adbc}(x) \equiv 0. \tag{5.107}$$

Proof. The identities (5.100) and (5.104) follow from (5.94), (5.96), (5.97), (5.98), and (5.99). The identities (5.101) and (5.105) follow from (4.73), (5.94), (5.96), (5.97), (5.98), and (5.99). The identities (5.102) and (5.106) follow from (5.95), (5.100), and (5.101). The identities (5.103) and (5.107) follow from (4.83), (4.85), (5.94), and (5.101). ■

Corollary 5.2.10 *The number of linearly independent, non-identically zero components of $R_{ijkl}(x)$ or $R_{abcd}(x)$ in an N-dimensional manifold is $N^2(N^2-1)/12$.*

Proof. If there were no identities, the number would have been N^4. However, the identities (5.100), (5.101), (5.102), and (5.103) reduce the number considerably. By the identities (5.100) and (5.101), we find that curvature tensor components are *antisymmetric* with respect to (ij) or (kl). There exist $\widehat{N} := N(N-1)/2$ independent components of such an antisymmetric pair of indices. We can map independent components in a one-to-one manner to objects R_{AB}, where $A, B \in \{1, \ldots, \widehat{N}\}$. By the identity (5.102), $R_{BA} = R_{AB}$. Therefore, R_{AB} have $\widehat{N}(\widehat{N}+1)/2$ independent components. Examining the identity (5.103), we observe that unless the indices i, j, k, l are all distinct, that identity is already included in (5.100), (5.101), and (5.102). The number of four distinct choices from N is $\binom{N}{4} = N(N-1)(N-2)(N-3)/24$. Therefore, subtracting this number from $\widehat{N}(\widehat{N}+1)/2$, we arrive at the number of independent components of the curvature tensor as $N^2(N^2-1)/12$. ∎

The number of independent components of the curvature tensor for dimensions $N = 2, 3, 4, 5$, and 10 is $1, 6, 20, 50$, and 825, respectively.

Now, we shall state and prove Bianchi's differential identities for the curvature tensor components.

Theorem 5.2.11 *Let the Riemann-Christoffel curvature tensor field $\mathbf{R}^{\cdot}{}_{\ldots}(x)$ be of class $C^1(D \subset \mathbb{R}^N; \mathbb{R})$. Then the corresponding components satisfy the following differential identities:*

$$\nabla_j R^h{}_{ikl} + \nabla_k R^h{}_{ilj} + \nabla_l R^h{}_{ijk} \equiv 0, \tag{5.108}$$

$$\nabla_d R^e{}_{abc} + \nabla_b R^e{}_{acd} + \nabla_c R^e{}_{adb} \equiv 0. \tag{5.109}$$

Proof. Equations (5.108) and (5.109) follow directly from (4.83), (4.84), (4.86), and (5.34). ∎

Remark: The number of independent Bianchi identities in (5.108) or (5.109) is $N^2(N^2-1)(N-2)/24$.

Now, we shall state the general rule for commutators or Lie brackets of two covariant derivations. These are known as **Ricci identities**. (See problem 2 of exercises 4.3.)

Theorem 5.2.12 *Let ${}^r{}_s\mathbf{T}(x)$ be a tensor field of class C^2 in the domain $D \subset \mathbb{R}^N$. Then the following commutation relations hold:*

$$\begin{aligned}(\nabla_k \nabla_l - \nabla_l \nabla_k) & T^{i_1 \dots i_r}{}_{j_1 \dots j_s} \\ &= \sum_{\alpha=1}^{r} \left[T^{i_1 \dots i_{\alpha-1} h i_{\alpha+1} \dots i_r}{}_{j_1 \dots j_s}(x) R^{i_\alpha}{}_{hkl}(x) \right] \\ &\quad - \sum_{\beta=1}^{s} \left[T^{i_1 \dots i_r}{}_{j_1 \dots j_{\beta-1} h j_{\beta+1} \dots j_s}(x) R^h{}_{j_\beta kl}(x) \right],\end{aligned} \tag{5.110}$$

$$\begin{aligned}(\nabla_c \nabla_d - \nabla_d \nabla_c) & T^{a_1 \dots a_r}{}_{b_1 \dots b_s} \\ &= \sum_{\alpha=1}^{r} [T^{a_1 \dots a_{\alpha-1} e a_{\alpha+1} \dots a_r}{}_{b_1 \dots b_s}(x) R^{a_\alpha}{}_{ecd}(x)] \\ &\quad - \sum_{\beta=1}^{s} \left[T^{a_1 \dots a_r}{}_{b_1 \dots b_{\beta-1} e b_{\beta+1} \dots b_s}(x) R^e{}_{b_\beta cd}(x) \right].\end{aligned} \tag{5.111}$$

(The proof is left to the reader.)

Example 5.2.13 For a twice-differentiable tensor field $\mathbf{T}_{..}(x)$, the equations (5.110) and (5.111) yield

$$(\nabla_k \nabla_l - \nabla_l \nabla_k) T_{ij} = -T_{hj}(x) R^h{}_{ikl}(x) - T_{ih}(x) R^h{}_{jkl}(x),$$

$$(\nabla_c \nabla_d - \nabla_d \nabla_c) T_{ab} = -T_{eb}(x) R^e{}_{acd}(x) - T_{ae}(x) R^e{}_{bcd}(x).$$

In the case where $\mathbf{T}_{..}(x) = \mathbf{g}_{..}(x)$, the left-hand sides of the equations above are zero by (5.86), and (5.87), whereas the right-hand sides are zero by (5.100) and (5.104). □

Now, we shall define the contractions of the Riemann-Christoffel curvature tensor. The **Ricci tensor** $\mathbf{R}_{..}(x) \equiv \mathbf{R}ic(x)$ is defined by the single contraction (mentioned in (3.34)) of the curvature tensor $\mathbf{R}_{....}(x)$ given in (4.73). Thus, we define

$$\begin{aligned}\mathbf{R}ic(x) \equiv \mathbf{R}_{..}(x) &:= {}^1{}_3\mathcal{C}[\mathbf{R}^{\cdot}{}_{...}(x)] \\ &= R^k{}_{ijk}(x) dx^i \otimes dx^j =: R_{ij}(x) dx^i \otimes dx^j,\end{aligned} \tag{5.112}$$

$$\mathbf{R}_{..}(x) = R^c{}_{abc}(x) \tilde{\mathbf{e}}^a(x) \otimes \tilde{\mathbf{e}}^b(x) =: R_{ab}(x) \tilde{\mathbf{e}}^a(x) \otimes \tilde{\mathbf{e}}^b(x). \tag{5.113}$$

By (5.93), (5.94), (5.47), (5.112), and (5.113), we obtain

$$\begin{aligned} R_{ij}(x) &= \partial_j \left\{ \begin{matrix} k \\ ki \end{matrix} \right\} - \partial_k \left\{ \begin{matrix} k \\ ji \end{matrix} \right\} + \left\{ \begin{matrix} k \\ il \end{matrix} \right\} \left\{ \begin{matrix} l \\ kj \end{matrix} \right\} - \left\{ \begin{matrix} l \\ lk \end{matrix} \right\} \left\{ \begin{matrix} k \\ ji \end{matrix} \right\} \\ &= \partial_i \partial_j \ln \sqrt{|g|} - \frac{1}{\sqrt{|g|}} \partial_k \left[\sqrt{|g|} \left\{ \begin{matrix} k \\ ij \end{matrix} \right\} \right] + \left\{ \begin{matrix} k \\ il \end{matrix} \right\} \left\{ \begin{matrix} l \\ kj \end{matrix} \right\} \\ &\equiv R_{ji}(x), \qquad (5.114) \\ R_{ab}(x) &= \partial_c \gamma^c{}_{ab} - \partial_b \gamma^c{}_{ac} + \gamma^c{}_{ad} \gamma^d{}_{bc} - \gamma^c{}_{dc} \gamma^d{}_{ab}. \qquad (5.115) \end{aligned}$$

Therefore, the Ricci tensor $\mathbf{R}_{..}(x)$ is symmetric and there exist $N(N+1)/2$ linearly independent components of $R_{ij}(x)$ or $R_{ab}(x)$. In the case of general relativity, we have $N = 4$ and the number of linearly independent components of $R_{ij}(x)$ is ten.

The **curvature scalar** or **curvature invariant** is defined by

$$R(x) := R^i{}_i(x) = g^{ij}(x) R_{ij}(x) = R_j{}^j(x), \tag{5.116}$$

$$R(x) = R^a{}_a(x) = d^{ab} R_{ab}(x) = R_b{}^b(x). \tag{5.117}$$

The **Einstein tensor** field $\mathbf{G}_{..}(x)$ is defined by

$$\mathbf{G}_{..}(x) := \mathbf{R}_{..}(x) - \left(\frac{1}{2}\right) R(x) \mathbf{g}_{..}(x), \tag{5.118}$$

$$G_{ij}(x) := R_{ij}(x) - \left(\frac{1}{2}\right) R(x) g_{ij}(x) \equiv G_{ji}(x), \tag{5.119}$$

$$G_{ab}(x) := R_{ab}(x) - \left(\frac{1}{2}\right) R(x) d_{ab} \equiv G_{ba}(x). \tag{5.120}$$

The Einstein tensor is a $(0+2)$th-order symmetric tensor and possesses $N(N+1)/2$ linearly independent components. In Einstein's theory of gravitation, this tensor is important, and it has ten independent components.

There exist more differential identities involving $\mathbf{R}^{\cdot}{}_{...}(x)$, $\mathbf{R}_{..}(x)$, and $\mathbf{G}_{..}(x)$ that follow from Bianchi's identities.

Theorem 5.2.14 *Let a metric tensor field* $\mathbf{g}_{..}(x)$ *be of class* C^3 *in* $D \subset \mathbb{R}^N$. *Then the following differential identities hold:*

$$\nabla_j R^j{}_{ikl} + \nabla_k R_{il} - \nabla_l R_{ik} \equiv 0, \tag{5.121}$$

$$\nabla_d R^d{}_{abc} + \nabla_b R_{ac} - \nabla_c R_{ab} \equiv 0, \tag{5.122}$$

$$\nabla_j G^j{}_i \equiv 0, \tag{5.123}$$

$$\nabla_b G^b{}_a \equiv 0. \tag{5.124}$$

Proof. The identities (5.121) and (5.122) follow from Bianchi identities (5.108) and (5.109) by single contractions and the definitions (5.112) and (5.113). The identities (5.123) and (5.124) follow from (5.121) and (5.122) by further contractions and the definitions (5.119) and (5.120). ■

Remarks: (i) The identities (5.121) and (5.122) are called the **first contracted Bianchi identities**.

(ii) The identities (5.123) and (5.124) are called the **second contracted Bianchi identities**. These identities prove to be extremely important in Einstein's theory of gravitation. (They yield differential conservation equations.)

Example 5.2.15 Consider the spherically symmetric space-time metric field

$$\begin{aligned}
\mathbf{g}_{..}(x) &= \exp[2\alpha(x^1,x^4)]dx^1 \otimes dx^1 \\
&\quad +[Y(x^1,x^4)]^2[dx^2 \otimes dx^2 + (\sin x^2)^2 dx^3 \otimes dx^3] \\
&\quad -\exp[2\gamma(x^1,x^4)]dx^4 \otimes dx^4; \\
\vec{\mathbf{e}}_{(1)}(x) &= e^{-\alpha}\frac{\partial}{\partial x^1},\ \vec{\mathbf{e}}_{(2)}(x) = Y^{-1}\frac{\partial}{\partial x^2},\ \vec{\mathbf{e}}_{(3)}(x) = (Y\sin x^2)^{-1}\frac{\partial}{\partial x^3}, \\
&\qquad \vec{\mathbf{e}}_{(4)}(x) = e^{-\gamma}\frac{\partial}{\partial x^4}; \\
\tilde{\mathbf{e}}^{(1)}(x) &= e^{\alpha}dx^1,\ \tilde{\mathbf{e}}^{(2)}(x) = Y dx^2,\ \tilde{\mathbf{e}}^{(3)}(x) = (Y\sin x^2)dx^3, \\
&\qquad \tilde{\mathbf{e}}^{(4)}(x) = e^{\gamma}dx^4; \\
D &:= \left\{x \in \mathbb{R}^4 : r_1 < x^1 < r_2,\ 0 < x^2 < \pi,\ -\pi < x^3 < \pi,\ t_1 < x^4 < t_2\right\}.
\end{aligned} \tag{5.125}$$

By (5.41) and (5.125), the non-vanishing Christoffel symbols of the second kind are furnished by

$$\begin{aligned}
&\left\{ {1 \atop 11} \right\} = \partial_1\alpha,\ \left\{ {4 \atop 11} \right\} = e^{2(\alpha-\gamma)}\partial_4\alpha,\ \left\{ {2 \atop 12} \right\} = \left\{ {3 \atop 13} \right\} = Y^{-1}\partial_1 Y, \\
&\left\{ {1 \atop 14} \right\} = \partial_4\alpha,\ \left\{ {4 \atop 14} \right\} = \partial_1\gamma,\ \left\{ {1 \atop 22} \right\} = -e^{-2\alpha}Y\partial_1 Y,\ \left\{ {4 \atop 22} \right\} = e^{-2\gamma}Y\partial_4 Y, \\
&\left\{ {3 \atop 23} \right\} = \cot x^2,\ \left\{ {2 \atop 24} \right\} = \left\{ {3 \atop 34} \right\} = Y^{-1}\partial_4 Y, \\
&\left\{ {1 \atop 33} \right\} = -e^{-2\alpha}(\sin x^2)^2 Y\partial_1 Y,\ \left\{ {2 \atop 33} \right\} = -(\sin x^2)(\cos x^2),
\end{aligned}$$

$$\begin{aligned}
\left\{ \begin{matrix} 4 \\ 33 \end{matrix} \right\} &= e^{-2\gamma}(\sin x^2)^2 Y \partial_4 Y, \\
\left\{ \begin{matrix} 1 \\ 44 \end{matrix} \right\} &= e^{2(\gamma-\alpha)}\partial_1\gamma, \left\{ \begin{matrix} 4 \\ 44 \end{matrix} \right\} = \partial_4\gamma. \qquad (5.126)
\end{aligned}$$

Equation (4.80) yields for the torsion-free case

$$d\tilde{\mathbf{e}}^a = \tilde{\mathbf{e}}^b(x) \wedge \tilde{\mathbf{w}}^a{}_b(x). \qquad (5.127)$$

By (5.125) and (5.127), we can obtain

$$\begin{aligned}
d\tilde{\mathbf{e}}^{(1)} &= e^{-\gamma}(\partial_4\alpha)\tilde{\mathbf{e}}^{(4)} \wedge \tilde{\mathbf{e}}^{(1)},\ d\tilde{\mathbf{e}}^{(2)} = Y^{-1}(\partial_1 Y)e^{-\alpha}\tilde{\mathbf{e}}^{(1)} \wedge \tilde{\mathbf{e}}^{(2)} \\
&\quad + Y^{-1}(\partial_4 Y)e^{-\gamma}\tilde{\mathbf{e}}^{(4)} \wedge \tilde{\mathbf{e}}^{(1)}, \\
Y d\tilde{\mathbf{e}}^{(3)} &= (\partial_1 Y)e^{-\alpha}\tilde{\mathbf{e}}^{(1)} \wedge \tilde{\mathbf{e}}^{(3)} + (\partial_4 Y)e^{-\gamma}\tilde{\mathbf{e}}^{(4)} \wedge \tilde{\mathbf{e}}^{(3)} + (\cot x^2)\tilde{\mathbf{e}}^{(2)} \wedge \tilde{\mathbf{e}}^{(3)}, \\
d\tilde{\mathbf{e}}^{(4)} &= e^{-\alpha}(\partial_1\gamma)\tilde{\mathbf{e}}^{(1)} \wedge \tilde{\mathbf{e}}^{(4)}, \\
\tilde{\mathbf{w}}^{(1)}{}_{(4)} &= \tilde{\mathbf{w}}_{(1)(4)} = -\tilde{\mathbf{w}}_{(4)(1)} = \tilde{\mathbf{w}}^{(4)}{}_{(1)} = e^{-\gamma}(\partial_4\alpha)\tilde{\mathbf{e}}^{(1)} + e^{-\alpha}(\partial_1\gamma)\tilde{\mathbf{e}}^{(4)}, \\
\tilde{\mathbf{w}}^{(2)}{}_{(1)} &= \tilde{\mathbf{w}}_{(2)(1)} = -\tilde{\mathbf{w}}_{(1)(2)} = -\tilde{\mathbf{w}}^{(1)}{}_{(2)} = Y^{-1}(\partial_1 Y)e^{-\alpha}\tilde{\mathbf{e}}^{(2)}, \\
\tilde{\mathbf{w}}^{(2)}{}_{(4)} &= \tilde{\mathbf{w}}_{(2)(4)} = -\tilde{\mathbf{w}}_{(4)(2)} = \tilde{\mathbf{w}}^{(4)}{}_{(2)} = Y^{-1}(\partial_4 Y)e^{-\gamma}\tilde{\mathbf{e}}^{(2)}, \\
\tilde{\mathbf{w}}^{(3)}{}_{(1)} &= \tilde{\mathbf{w}}_{(3)(1)} = -\tilde{\mathbf{w}}_{(1)(3)} = -\tilde{\mathbf{w}}^{(1)}{}_{(3)} = Y^{-1}(\partial_1 Y)e^{-\alpha}\tilde{\mathbf{e}}^{(3)}, \qquad (5.128)
\end{aligned}$$

$$\begin{aligned}
\tilde{\mathbf{w}}^{(3)}{}_{(4)} &= \tilde{\mathbf{w}}_{(3)(4)} = -\tilde{\mathbf{w}}_{(4)(3)} = \tilde{\mathbf{w}}^{(4)}{}_{(3)} = Y^{-1}(\partial_4 Y)e^{-\gamma}\tilde{\mathbf{e}}^{(3)}, \\
\tilde{\mathbf{w}}^{(3)}{}_{(2)} &= \tilde{\mathbf{w}}_{(3)(2)} = -\tilde{\mathbf{w}}_{(2)(3)} = -\tilde{\mathbf{w}}^{(2)}{}_{(3)} = Y^{-1}(\cot x^2)\tilde{\mathbf{e}}^{(3)}.
\end{aligned}$$

By (5.56) and (5.128), we obtain for the non-vanishing Ricci rotation coefficients

$$\begin{aligned}
\gamma_{(4)(1)(1)} &= -\gamma_{(1)(4)(4)} = e^{-\gamma}\partial_4\alpha, \\
\gamma_{(4)(1)(4)} &= -\gamma_{(1)(4)(4)} = e^{-\alpha}\partial_1\gamma, \\
\gamma_{(1)(2)(2)} &= -\gamma_{(2)(1)(2)} = \gamma_{(1)(3)(3)} = -\gamma_{(3)(1)(3)} = Y^{-1}(\partial_1 Y)e^{-\alpha}, \qquad (5.129) \\
\gamma_{(4)(2)(2)} &= -\gamma_{(2)(4)(4)} = \gamma_{(4)(3)(3)} = -\gamma_{(3)(4)(4)} = Y^{-1}(\partial_4 Y)e^{-\gamma}, \\
\gamma_{(2)(3)(3)} &= -\gamma_{(3)(2)(3)} = Y^{-1}\cot x^2.
\end{aligned}$$

By (5.126), (5.93), (5.94), and (5.95), and notations $f' := \partial_1 f$ and $\dot{f} := \partial_4 f$, we obtain for the non-vanishing (coordinate) components of the Riemann-Christoffel curvature tensor

$$\begin{aligned}
R_{1212} &= Y[-Y'' + \alpha' Y' + e^{2(\alpha-\gamma)}\dot{\alpha}\dot{Y}],\\
R_{1313} &= (\sin x^2)^2 R_{1212},\\
R_{1224} &= Y[\dot{Y}' - \dot{\alpha} Y' - \gamma' \dot{Y}],\\
R_{1334} &= (\sin x^2)^2 R_{1224},\\
R_{1414} &= e^{2\alpha}[-\ddot{\alpha} - (\dot{\alpha})^2 + \dot{\alpha}\dot{\gamma}] + e^{2\gamma}[\gamma'' + (\gamma')^2 - \alpha'\gamma'],\\
R_{2424} &= Y[-\ddot{Y} + \dot{\gamma}\dot{Y} + e^{2(\gamma-\alpha)}\gamma' Y'],\\
R_{3434} &= (\sin x^2)^2 R_{2424},\\
R_{2323} &= (Y \sin x^2)^2 [1 - e^{-2\alpha}(Y')^2 + e^{-2\gamma}(\dot{Y})^2].
\end{aligned} \tag{5.130}$$

The orthonormal tetrad components of the curvature tensor are furnished by

$$\begin{aligned}
R_{(1)(2)(1)(2)} &= R_{(1)(3)(1)(3)} = Y^{-1} e^{-2\alpha}[-Y'' + \alpha' Y' + e^{2(\alpha-\gamma)}\dot{\alpha}\dot{\gamma}],\\
R_{(1)(2)(2)(4)} &= R_{(1)(3)(3)(4)} = Y^{-1} e^{-(\alpha+\gamma)}[\dot{Y}' - \dot{\alpha} Y' - \gamma' \dot{Y}],\\
R_{(2)(4)(2)(4)} &= R_{(3)(4)(3)(4)} = Y^{-1} e^{-2\gamma}[-\ddot{Y} + \dot{\gamma}\dot{Y} + e^{2(\gamma-\alpha)}\gamma' Y'],\\
R_{(2)(3)(2)(3)} &= Y^{-2}[1 - e^{-2\alpha}(Y')^2 + e^{-2\gamma}(\dot{Y})^2],\\
R_{(1)(4)(1)(4)} &= e^{-2\gamma}(-\ddot{\alpha} - (\dot{\alpha})^2 + \dot{\alpha}\dot{\gamma}) + e^{-2\alpha}(\gamma'' + (\gamma')^2 - \alpha'\gamma').
\end{aligned} \tag{5.131}$$

The **Kretschmann scalar** is provided by

$$\begin{aligned}
R_{ijkl}R^{ijkl} &= R_{abcd}R^{abcd}\\
&= 4\Big\{Y^{-4}[1 - e^{-2\alpha}(Y')^2 + e^{-2\gamma}(\dot{Y})^2]^2\\
&\quad + [e^{-2\gamma}(-\ddot{\alpha} - \dot{\alpha}^2 + \dot{\alpha}\dot{\gamma}) + e^{-2\alpha}(\gamma'' + \gamma'^2 - \alpha'\gamma')]^2\Big\}\\
&\quad + 8Y^{-2}\Big\{e^{-4\alpha}[-Y'' + \alpha' Y' + e^{2(\alpha-\gamma)}\dot{\alpha}\dot{Y}]^2\\
&\quad + e^{-4\gamma}[-\ddot{Y} + \dot{\gamma}\dot{Y} + e^{2(\gamma-\alpha)}\gamma' Y']^2\Big\}\\
&\quad - 16 Y^{-2} e^{-2(\alpha+\gamma)}[\dot{Y}' - \dot{\alpha} Y' - \gamma' \dot{Y}]^2.
\end{aligned} \tag{5.132}$$

By (5.130), (5.131), (5.114), and (5.115), the Ricci tensor components are given by

$$\begin{aligned}
R_{11} &= 2Y^{-1}Y'' + \gamma'' + (\gamma')^2 - 2\alpha' Y^{-1} Y' - \alpha'\gamma'\\
&\quad + e^{2(\alpha-\gamma)}[-\ddot{\alpha} - (\dot{\alpha})^2 - 2\dot{\alpha} Y^{-1}\dot{Y} + \dot{\alpha}\dot{\gamma}],
\end{aligned}$$

$$\begin{aligned}
R_{22} &= -1 + e^{-2\alpha}[YY'' + (Y')^2 + (\gamma' - \alpha')YY'] \\
&\quad + e^{-2\gamma}[-Y\ddot{Y} - (\dot{Y})^2 + (\dot{\gamma} - \dot{\alpha})Y\dot{Y}], \\
R_{33} &= (\sin x^2)^2 R_{22}, \\
R_{44} &= 2Y^{-1}\ddot{Y} + \ddot{\alpha} + (\dot{\alpha})^2 - \dot{\alpha}\dot{\gamma} - 2\dot{\gamma}Y^{-1}\dot{Y} \\
&\quad + e^{2(\gamma-\alpha)}[-\gamma'' - (\gamma')^2 + \alpha'\gamma' - 2\gamma' Y^{-1}Y'], \\
R_{14} &= R_{41} = 2Y^{-1}[\dot{Y}' - \dot{\alpha}Y' - \gamma'\dot{Y}],
\end{aligned} \tag{5.133}$$

$$\begin{aligned}
R_{(1)(1)} &= e^{-2\alpha}[2Y^{-1}Y'' + \gamma'' + (\gamma')^2 - 2\alpha' Y^{-1}Y' - \alpha'\gamma'] \\
&\quad + e^{2\gamma}[-\ddot{\alpha} - (\dot{\alpha})^2 - 2\dot{\alpha}Y^{-1}\dot{Y} + \dot{\alpha}\dot{\gamma}], \\
R_{(2)(2)} &= R_{(3)(3)} = -Y^{-2} + e^{-2\alpha}[Y^{-1}Y'' + (Y^{-1}Y')^2 + (\gamma' - \alpha')Y^{-1}Y'] \\
&\quad + e^{-2\gamma}[-Y^{-1}\ddot{Y} - (Y^{-1}\dot{Y})^2 + (\dot{\gamma} - \dot{\alpha})Y^{-1}\dot{Y}], \\
R_{(4)(4)} &= e^{-2\gamma}[2Y^{-1}\ddot{Y} + \ddot{\alpha} + (\dot{\alpha})^2 - \dot{\alpha}\dot{\gamma} - 2\dot{\gamma}Y^{-1}\dot{Y}] \\
&\quad + e^{-2\alpha}[-\gamma'' - (\gamma')^2 + \alpha'\gamma' - 2\gamma' Y^{-1}Y'], \\
R_{(1)(4)} &= R_{(4)(1)} = 2Y^{-1}e^{-2(\alpha+\gamma)}[\dot{Y}' - \dot{\alpha}Y' - \gamma'\dot{Y}].
\end{aligned}$$

The scalar invariant is obtained from (5.116), (5.117), and (5.133) as

$$\begin{aligned}
R(x) &= -2Y^{-2} + 2e^{-2\alpha}[2Y^{-1}Y'' + (Y^{-1}Y')^2 \\
&\quad + \gamma'' + (\gamma')^2 - \alpha'\gamma' + 2(\gamma' - \alpha')Y^{-1}Y'] \\
&\quad + 2e^{-2\gamma}[-2Y^{-1}\ddot{Y} - (Y^{-1}\dot{Y})^2 - \ddot{\alpha} - (\dot{\alpha})^2 + \dot{\alpha}\dot{\gamma} + 2(\dot{\gamma} - \dot{\alpha})Y^{-1}\dot{Y}].
\end{aligned} \tag{5.134}$$

The Einstein tensor components are furnished by (5.119), (5.120), (5.133), and (5.134) as

$$\begin{aligned}
G^1{}_1 &= G^{(1)}{}_{(1)} = Y^{-2} + e^{-2\alpha}[(Y^{-1}Y')^2 + 2\gamma' Y^{-1}Y'] \\
&\quad + e^{-2\gamma}[2Y^{-1}\ddot{Y} + (Y^{-1}\dot{Y})^2 - 2\dot{\gamma}Y^{-1}\dot{Y}], \\
G^2{}_2 &= G^{(2)}{}_{(2)} = G^3{}_3 = G^{(3)}{}_{(3)} = \\
&\quad e^{-2\alpha}[-Y^{-1}Y'' - \gamma'' - (\gamma')^2 + \alpha'\gamma' + (\alpha' - \gamma')Y^{-1}Y'] \\
&\quad + e^{-2\gamma}[Y^{-1}\ddot{Y} + \ddot{\alpha} + (\dot{\alpha})^2 - \dot{\alpha}\dot{\gamma} + (\dot{\alpha} - \dot{\gamma})Y^{-1}\dot{Y}], \\
G^4{}_4 &= G^{(4)}{}_{(4)} = Y^{-2} + e^{-2\alpha}[-2Y^{-1}Y'' - (Y^{-1}Y')^2 + 2\alpha' Y^{-1}Y'] \\
&\quad + e^{-2\gamma}[(Y^{-1}\dot{Y})^2 + 2\dot{\alpha}Y^{-1}\dot{Y}], \\
e^{2\alpha}G^1{}_4 &= -e^{2\gamma}G^4{}_1 = 2Y^{-1}[\dot{Y}' - \dot{\alpha}Y' - \gamma'\dot{Y}], \\
G^{(1)}{}_{(4)} &= -G^{(4)}{}_{(1)} = 2Y^{-1}e^{-(\alpha+\gamma)}[\dot{Y}' - \dot{\alpha}Y' - \gamma'\dot{Y}].
\end{aligned} \tag{5.135}$$

The second contracted Bianchi differential identities (5.123) and (5.124) yield by the metric in (5.125) the following identities:

$$\begin{aligned}
&\partial_1 G^1{}_1 + (2\ln|Y| + \gamma)' G^1{}_1 + \partial_4 G^4{}_1 + (\alpha + 2\ln|Y| + \gamma)^{\cdot} G^4{}_1 \\
&\quad -\gamma' G^4{}_4 - 2(\ln|Y|)' G^2{}_2 \equiv 0, \\
&\partial_1 G^1{}_4 + (\alpha + 2\ln|Y| + \gamma)' G^1{}_4 - [\dot{\alpha} G^1{}_1 + 2(\ln|Y|)^{\cdot} G^2{}_2] \\
&\quad +\partial_4 G^4{}_4 + (\alpha + 2\ln|Y|)^{\cdot} G^4{}_4 \equiv 0.
\end{aligned} \tag{5.136}$$

This example is important in the spherically symmetric gravitational fields that govern the space-time geometry of spherical stars and black holes according to Einstein's theory of general relativity. (See Das & Aruliah [9].) □

Now, we shall express Cartan's structural equations and Bianchi identities in terms of the connection 1-forms and the curvature 2-forms. By the torsion-free condition (5.32), the general Cartan structural equations (4.80) and (4.81) reduce to

$$d\tilde{\mathbf{e}}^p(x) + \tilde{\mathbf{w}}^p{}_q(x) \wedge \tilde{\mathbf{e}}^q(x) = \mathbf{O}_{..}(x), \tag{5.137}$$

$$\boldsymbol{\Omega}^p{}_{q..}(x) = d\tilde{\mathbf{w}}^p{}_q(x) + \tilde{\mathbf{w}}^p{}_u(x) \wedge \tilde{\mathbf{w}}^u{}_q(x). \tag{5.138}$$

Equation (5.127) is a consequence of (5.137). Moreover, (5.138) can be used to compute N-tuple components of the curvature tensor. (Equations (5.131) can be derived in such a manner.)

The Bianchi identities (4.83) and (4.84) reduce to

$$\boldsymbol{\Omega}^p{}_{q..}(x) \wedge \tilde{\mathbf{e}}^q(x) = \mathbf{O}_{...}(x), \tag{5.139}$$

$$d\boldsymbol{\Omega}^p{}_{q..}(x) = \boldsymbol{\Omega}^p{}_{u..}(x) \wedge \tilde{\mathbf{w}}^u{}_q(x) - \tilde{\mathbf{w}}^p{}_u(x) \wedge \boldsymbol{\Omega}^u{}_{q..}(x). \tag{5.140}$$

The identities (5.139) and (5.140) are equivalent to (5.103) and (5.108), respectively.

Example 5.2.16 Consider a thrice-differentiable two-dimensional metric field given by

$$\begin{aligned}
\mathbf{g}_{..}(x) &= \left[h_{(1)}(x)\right]^2 dx^1 \otimes dx^1 + \left[h_{(2)}(x)\right]^2 dx^2 \otimes dx^2, \\
\mathbf{g}_{..}(x) &= \delta_{ab} \tilde{\mathbf{e}}^a(x) \otimes \tilde{\mathbf{e}}^b(x).
\end{aligned}$$

Thus, we can derive the 2-forms

$$d\tilde{\mathbf{e}}^{(1)}(x) = \left[\partial_2 h_{(1)}\right] dx^2 \wedge dx^1, \quad d\tilde{\mathbf{e}}^{(2)}(x) = \left[\partial_1 h_{(2)}\right] dx^1 \wedge dx^2.$$

The non-zero connection 1-forms are furnished by

$$\begin{aligned}
\tilde{\mathbf{w}}^{(1)}{}_{(2)} &= \tilde{\mathbf{w}}_{(1)(2)} = -\tilde{\mathbf{w}}_{(2)(1)} = -\tilde{\mathbf{w}}^{(2)}{}_{(1)} \\
&= \left[\partial_{(2)} h_{(1)}\right] dx^1 - \left[\partial_{(1)} h_{(2)}\right] dx^2.
\end{aligned}$$

Now, the wedge products are

$$\begin{aligned}\tilde{\mathbf{w}}^{(1)}{}_{(2)} \wedge \tilde{\mathbf{e}}^{(2)} &= \left[h_{(2)}(x)\partial_{(2)}h_{(1)}\right] dx^1 \wedge dx^2 \\ &= \left[\partial_2 h_{(1)}\right] dx^1 \wedge dx^2 = -d\tilde{\mathbf{e}}^{(1)}(x),\end{aligned}$$

$$\begin{aligned}\tilde{\mathbf{w}}^{(2)}{}_{(1)} \wedge \tilde{\mathbf{e}}^{(1)} &= \left[h_{(1)}(x)\partial_{(1)}h_{(2)}\right] dx^2 \wedge dx^1 \\ &= \left[\partial_1 h_{(2)}\right] dx^2 \wedge dx^1 = -d\tilde{\mathbf{e}}^{(2)}(x).\end{aligned}$$

Thus, the structural equations (5.137) are explicitly verified. □

Now, we shall discuss the covariant differentiations of the relative and oriented relative tensor fields. The transformation rules are given in (5.58) and (5.59). The covariant derivative of a relative tensor field of weight w or an oriented relative tensor field of the same weight are defined by

$$\nabla_u \theta^{p_1\ldots p_r}{}_{q_1\ldots q_s} := |g|^{w/2}\nabla_u[|g|^{-w/2}\theta^{p_1\ldots p_r}{}_{q_1\ldots q_s}], \tag{5.141}$$

$$\begin{aligned}\nabla_k \theta^{i_1\ldots i_r}{}_{j_1\ldots j_s} &= \partial_k \theta^{i_1\ldots i_r}{}_{j_1\ldots j_s} + \sum_{\alpha=1}^{r}\left\{ {i_\alpha \atop kl} \right\}\theta^{i_1\ldots i_{\alpha-1} l i_{\alpha+1}\ldots i_r}{}_{j_1\ldots j_s} \\ &\quad -\sum_{\beta=1}^{s}\left\{ {l \atop kj_\beta} \right\}\theta^{i_1\ldots i_r}{}_{j_1\ldots j_{\beta-1} l j_{\beta+1}\ldots j_s} \\ &\quad -w\left\{ {l \atop kl} \right\}\theta^{i_1\ldots i_r}{}_{j_1\ldots j_s},\end{aligned} \tag{5.142}$$

$$\begin{aligned}\nabla_c \theta^{a_1\ldots a_r}{}_{b_1\ldots b_s} &= \partial_c \theta^{a_1\ldots a_r}{}_{b_1\ldots b_s} - \sum_{\alpha=1}^{r}\gamma^{a_\alpha}{}_{dc}\theta^{a_1\ldots a_{\alpha+1} d a_{\alpha+1}\ldots a_r}{}_{b_1\ldots b_s} \\ &\quad +\sum_{\beta=1}^{s}\gamma^{d}{}_{b_\beta c}\theta^{a_1\ldots a_\alpha}{}_{b_1\ldots b_{\beta-1} d b_{\beta+1}\ldots b_s}.\end{aligned} \tag{5.143}$$

(Note that $|g|^{-w/2}\theta^{p_1\ldots p_r}{}_{q_1\ldots q_s}(x)$ is an absolute or oriented absolute tensor field.)

Example 5.2.17 Now, $\operatorname{sgn}[\det g_{ij}(x)] \equiv \operatorname{sgn}(g)$ is an (absolute) invariant. Therefore,

$$\nabla_k[\operatorname{sgn}(g)] = \partial_k[\operatorname{sgn}(g)] \equiv 0. \tag{5.144}$$

Since $g = \det[g_{ij}(x)]$ is a relative scalar of weight $+2$, by (5.141),

$$\nabla_k(g) = |g|\nabla_k[|g|^{-1}g] \equiv 0. \tag{5.145}$$

Since $\sqrt{|g|}$ and $\sqrt{|\det d_{ab}|}$ is a scalar density (of weight $+1$), by (5.142) and (5.143)

$$\nabla_k\left(\sqrt{|g|}\right) = \partial_k\left(\sqrt{|g|}\right) - \left\{ {l \atop kl} \right\}\sqrt{|g|} \equiv 0, \tag{5.146}$$

$$\nabla_c\left[\sqrt{|\det d_{ab}|}\right] = \partial_c(1) \equiv 0. \tag{5.147}$$

□

Example 5.2.18 Consider the components of the totally antisymmetric, numerical tensor density $\varepsilon^{i_1 \ldots i_N}$. By the rules (5.142),

$$\nabla_k \varepsilon^{i_1 \ldots i_N} = \sum_{\alpha=1}^{N} \left\{ \begin{matrix} i_\alpha \\ kl \end{matrix} \right\} \varepsilon^{i_1 \ldots i_{\alpha-1} l i_{\alpha+1} \ldots i_N} - \left\{ \begin{matrix} l \\ kl \end{matrix} \right\} \varepsilon^{i_1 \ldots i_N}.$$

We can choose Riemann coordinates (in section 5.4) at a point such that $\left\{ \begin{smallmatrix} i \\ jk \end{smallmatrix} \right\} \equiv 0$. Therefore, in such a coordinate chart,

$$\nabla_k \varepsilon^{i_1 \ldots i_N} = 0. \tag{5.148}$$

Since this is a tensor density equation, it is valid in every coordinate system or chart. Equation (5.148) implies that

$$\nabla_b \varepsilon^{a_1 \ldots a_N} = 0. \tag{5.149}$$

Similarly, we can prove that

$$\nabla_k \varepsilon_{i_1 \ldots i_N} = 0, \tag{5.150}$$

$$\nabla_b \varepsilon_{a_1 \ldots a_N} = 0. \tag{5.151}$$

Using the Leibnitz rules of covariant derivatives and (5.146) and (5.147), we prove that

$$\begin{aligned} \nabla_k \eta_{j_1 \ldots j_N} &= \nabla_k \left[\sqrt{|g|}\, \varepsilon_{j_1 \ldots j_N}\right] \\ &= \sqrt{|g|}\, \nabla_k \varepsilon_{j_1 \ldots j_N} \equiv 0, \\ \nabla_k \eta^{i_1 \ldots i_N} &\equiv 0. \end{aligned} \tag{5.152}$$

□

Now, we shall classify differentiable manifolds with metrics and torsion-free connection into various subclasses. The signature of the metric (in (5.7)) and the curvature tensor $\mathbf{R}\ldots.(x)$ will be used for this classification. It has been mentioned previously that a domain $U \subset M$ is called **flat** provided $\mathbf{R}\ldots.(x) \equiv \mathbf{O}\ldots.(x)$ in the corresponding domain $D \subset \mathbb{R}^N$. A **non-flat** or **curved** domain is characterized by $\mathbf{R}\ldots.(x) \not\equiv \mathbf{O}\ldots.(x)$. For a positive-definite metric, $\operatorname{sgn}[\mathbf{g}..(x)] = N$. For a **Lorentz metric**, $\operatorname{sgn}[\mathbf{g}..(x)] = N - 2$. For (indefinite) **non-Lorentz metrics** $0 \leq \operatorname{sgn}[\mathbf{g}..(x)] \leq N - 4$. We represent the classification of the manifolds in the following diagram.

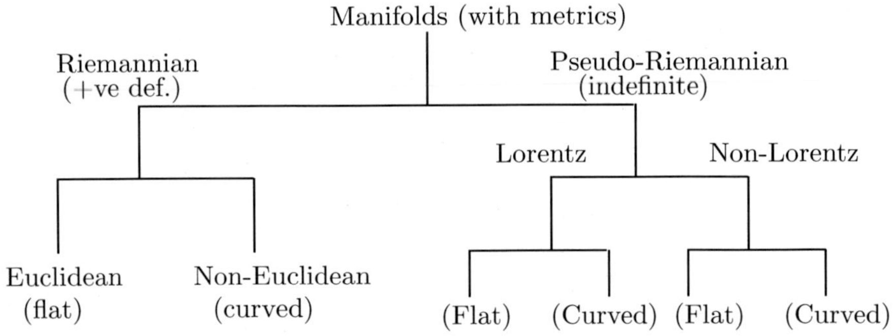

Remark: In Newtonian physics, a three-dimensional Euclidean manifold (or space) is used. In the special theory of relativity, four-dimensional flat space-time with a Lorentz metric is utilized. In Einstein's theory of gravitation, a four-dimensional curved space-time manifold with a Lorentz metric is considered.

Now, we shall discuss briefly some integral theorems. Recall the definition of an oriented volume in (2.143). In a similar approach, we define the oriented volume of a domain $D \subset \mathbb{R}^N$ by the integral

$$\begin{aligned} V(D) &:= \int_D {}_N\boldsymbol{\eta}(x) = (1/N!) \int_D \eta_{i_1 \ldots i_N}(x) dx^{i_1} \wedge \cdots \wedge dx^{i_N} \\ &= \int_D \sqrt{|g|}\varepsilon_{12\ldots N} dx^1 \wedge \cdots \wedge dx^N =: \int_D d^N v. \end{aligned} \tag{5.153}$$

Let the $(N-1)$-dimensional, differentiable, orientable, closed boundary ∂D of D be represented parametrically by (3.82) as

$$\begin{aligned} x^j &= \xi^j(u) \equiv \xi^j\left(u^1, \ldots u^{N-1}\right), \\ u &\in \mathcal{D}_{N-1} \subset \mathbb{R}^{N-1}. \end{aligned} \tag{5.154}$$

Let the Greek letters assume values from $\{1, 2, \ldots, N-1\}$. The **induced metric** on the boundary ∂D is furnished by

$$\begin{aligned} \mathbf{g}_{..}(x)_{|\partial D} &= g_{ij}\left(\xi(u)\right) \frac{\partial \xi^i(u)}{\partial u^\alpha} \frac{\partial \xi^j(u)}{\partial u^\beta} du^\alpha \otimes du^\beta \\ &=: \bar{g}_{\alpha\beta}(u) du^\alpha \otimes du^\beta. \end{aligned} \tag{5.155}$$

The $(N-1)$-dimensional volume of ∂D is given by (5.153), (5.154), and (5.155) as

$$\int_{\partial D} d^{N-1}v = \int_{\mathcal{D}_{N-1}} \sqrt{|\bar{g}|} du^1 du^2 \ldots du^{N-1}. \tag{5.156}$$

Example 5.2.19 Consider the Euclidean space $\mathbb{E}_3$ and the spherical polar coordinate chart. (See example 4.3.12.) A spherical surface of radius $a > 0$ is given by

$$S^2(a) := \left\{x \in \mathbb{R}^3 : \ x^1 = a > 0, \ 0 < x^2 < \pi, \ -\pi < x^3 < \pi\right\}.$$

It encloses the spherical domain

$$\begin{aligned} D \ &= \left\{x \in \mathbb{R}^3 : \ 0 < x^1 < a, \ 0 < x^2 < \pi, \ -\pi < x^3 < \pi\right\}, \\ \partial D &= S^2(a). \end{aligned}$$

The spherical surface can be parametrized by

$$\begin{aligned} x^1 \ &= \xi^1(u) := a, \ \ x^2 = \xi^2(u) := u^1, \ \ x^3 = \xi^3(u) := u^2, \\ \mathcal{D}_2 &:= \left\{u \in \mathbb{R}^2 : \ 0 < u^1 < \pi, \ -\pi < u^2 < \pi\right\}. \end{aligned}$$

The three-dimensional metric tensor is given by

$$\begin{aligned} \mathbf{g}_{..}(x) &= dx^1 \otimes dx^1 + (x^1)^2 \left[dx^2 \otimes dx^2 + (\sin^2 x^2) dx^3 \otimes dx^3\right], \\ \sqrt{|g|} \ &= (x^1)^2 \sin x^2 > 0. \end{aligned}$$

Therefore, the volume of the sphere D is given by (5.153) as

$$V(D) = \int_{0+}^{a-}\int_{0+}^{\pi-}\int_{-\pi+}^{\pi-} (x^1)^2 (\sin x^2) dx^1 dx^2 dx^3 = \frac{4\pi (a)^3}{3}.$$

The induced metric on the spherical surface is given by (5.155) as

$$\begin{aligned} \mathbf{g}_{..}(x)_{|_{S^2(a)}} &= \bar{\mathbf{g}}_{..}(u) = a^2 \left[du^1 \otimes du^1 + (\sin^2 u^1) du^2 \otimes du^2\right], \\ \sqrt{|\bar{g}|} \ &= a^2 \sin u^1. \end{aligned}$$

Therefore, the surface area of $S^2(a)$ by (5.156) is provided by

$$\begin{aligned} V(\partial D) =: A(S^2(a)) &= \int_{o+}^{\pi-}\int_{-\pi_+}^{\pi-} a^2 (\sin u^1) du^1 du^2 \\ &= 4\pi (a)^2. \end{aligned}$$

□

Now, we shall introduce the normal vector ν_j (*not* necessarily a unit vector) to the $(N-1)$-dimensional boundary ∂D. We define the normal components by (5.154) and with the help of $(N-1)$-forms as

$$\begin{aligned}
(N-1)!\mu_j &:= \varepsilon_{jj_2\ldots j_N} dx^{j_2} \wedge \cdots \wedge dx^{j_N}, \\
\mu_{j|_D} &= \nu_j du^1 \wedge \ldots \wedge du^{N-1}, \\
\nu_j &:= \frac{1}{(N-1)!}\varepsilon_{jj_2\ldots j_N} \frac{\partial(x^{j_2},\ldots x^{j_N})}{\partial(u^1,\ldots u^{N-1})}, \\
\nu_j \frac{\partial \xi^j(u)}{\partial u^\alpha} &= \frac{1}{(N-1)!}\varepsilon_{jj_2\ldots j_N} \frac{\partial x^j}{\partial u^\alpha}\frac{\partial(x^{j_2},\ldots x^{j_N})}{\partial(u^1,\ldots,u^{N-1})} \\
&= \frac{1}{(N-1)!}\frac{\partial(x^1,x^2,\ldots x^N)}{\partial(u^\alpha,u^1,\ldots,u^{N-1})} \equiv 0.
\end{aligned} \tag{5.157}$$

(In the Jacobian above, two of the rows coincide!) Since $\frac{\partial \xi^j(u)}{\partial u^\alpha}$ are components of a vector that is tangential to the hypersurface ∂D, the last condition indicates the orthogonality of ν_j to the hypersurface ∂D.

Now, we can prove by (5.155) and (5.156) that

$$\begin{aligned}
&\bar{g} := \det\left[\bar{g}_{\alpha\beta}(u)\right] = \left[g^{jk}\left(\xi(u)\right)\nu_j\nu_k\right]\det\left[g_{ij}\left(\xi(u)\right)\right], \\
&\sqrt{|\bar{g}|} = \sqrt{|g|}\sqrt{|g^{jk}\nu_j\nu_k|}.
\end{aligned} \tag{5.158}$$

Assuming that the boundary ∂D is nowhere null (i.e., $g^{jk}\nu_j\nu_k \neq 0$), we can define the unit, outer normal as

$$n_j := \nu_j / \sqrt{|g^{kl}\nu_k\nu_l|}. \tag{5.159}$$

Now, we shall state and prove the **generalized Gauss' theorem**.

Theorem 5.2.20 *Let D^* be a star-shaped domain in $\mathbb{R}^N$ with a continuous, piecewise-differentiable, orientable, closed, non-null boundary ∂D^* with unit normal n_j. Moreover, let $\vec{\mathbf{A}}(x)$ be a differentiable vector field in $D^* \cup \partial D^*$. Then the following integral relation holds:*

$$\int\limits_{D^*} \left(\nabla_j A^j\right) d^N v = \int\limits_{\partial D^*} A^j n_j d^{N-1} v. \tag{5.160}$$

Proof. Consider an $(N-1)$-form of weight one defined by (5.157) as

$$\begin{aligned}
\omega &:= \sqrt{|g|}A^j\mu_j, \\
d\omega &= \left(\nabla_j A^j\right)\sqrt{|g|}dx^1 \wedge \cdots \wedge dx^N.
\end{aligned} \tag{5.161}$$

Using (5.153) and (5.161), we obtain

$$\int\limits_{D^*} \left(\nabla_j A^j\right) d^N v = \int\limits_{D^*} d\omega. \tag{5.162}$$

By the Stokes' theorem 3.4.12 and (5.162), (5.161), (5.157), (5.159), and (5.156),

$$\begin{aligned}\int_{D^*} d\omega &= \int_{\partial D^*} \omega \\ &= \int_{\mathcal{D}_{N-1}} A^j\left(\xi(u)\right)\frac{\sqrt{|\bar{g}|}}{\sqrt{|g^{kl}\nu_k\nu_l|}}\nu_j du^1\wedge\cdots\wedge du^{N-1} \\ &= \int_{\partial D^*} A^j n_j d^{N-1}v.\end{aligned}$$

■

Example 5.2.21 Consider the Newtonian theory of gravitation in the Euclidean space $\mathbb{E}_3$. The gravitational potential $W(x)$, which is of class piecewise C^2, satisfies

$$\Delta W = \begin{cases} G\rho(x) \text{ in } D, \\ 0 \text{ in } \mathbb{R}^3 - D. \end{cases} \tag{5.163}$$

Here, G is the Newtonian constant of gravitation and, $\rho(x) \geq 0$ is the mass density (which is assumed to be integrable). By Gauss' theorem 5.2.20, we obtain

$$\begin{aligned}\int_D \Delta W d^3v &= \int_D (\nabla_i\nabla^i W)d^3v = \int_{\partial D} (\nabla^i W)n_i d^2v, \\ Gm := G\int_D \rho(x)d^3v &= \int_{\partial D} (\nabla^i W)n_i d^2v.\end{aligned} \tag{5.164}$$

Here, m is the total mass of the body in D. Moreover, the integral on the right-hand side is called the *total normal flux of the gravitational field across the boundary ∂D.* □

Now, we shall state and prove the **Green's theorem**.

Theorem 5.2.22 *Let a domain $D \subset \mathbb{R}^N$ and its boundary ∂D satisfy the same conditions as in theorem* 5.2.20. *Let $U, W \in C^2(D \subset \mathbb{R}^N; \mathbb{R})$. Then,*

$$\int_D [U(x)\Delta W - W(x)\Delta U]\, d^N v = \int_{\partial D} \left[U(x)\nabla^i W - W(x)\nabla^i U\right] n_i d^{N-1}v. \tag{5.165}$$

Proof. By Gauss' theorem 5.2.20, the left-hand side of (5.165) yields

$$\int_D \nabla_i\left[U(x)\nabla^i W - W(x)\nabla^i U\right] d^N v = \int_{\partial D}\left[U(x)\nabla^i W - W(x)\nabla^i U\right] n_i d^{N-1}v.$$ ■

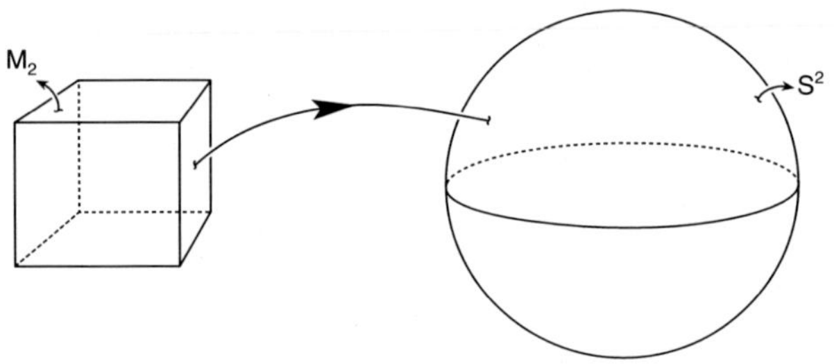

Figure 5.1: Deformation of a cubical surface into a spherical one.

The integrals over a closed, compact manifold M can yield important **topological invariants** of the manifold. (See Bocher & Yano [3].) The simplest of such results is contained in the Gauss-Bonnet theorem for a two-dimensional surface embedded in a three-dimensional space. To explain briefly the content of the theorem, consider the boundary surface M_2 of a cube in space. (See fig. 5.1.)

For the cubical surface M_2, we can count the number of vertices $v = 8$, the number of edges $e = 12$, and the number of faces $f = 6$. Therefore, the **Euler-Poincaré characteristic** $\chi(M_2) := v - e + f = 2$. The Euler-Poincaré characteristic $\chi(M_2)$ happens to be a topological invariant number. Therefore, this number does not change under a continuous and one-to-one mapping (homeomorphism). Since a cubical surface can be continuously deformed into a spherical surface, we expect $\chi(S^2) = 2$.

The Euler-Poincaré characteristic number is expressible as an integral over the surface. The following **Gauss-Bonnet theorem** precisely states that fact.

Theorem 5.2.23 *Let M_2 be a closed, compact, orientable, and piecewise-differentiable continuous surface. Moreover, let M_2 be endowed with a twice-differentiable Riemannian metric* $\mathbf{g}_{..}(x)$. *Then, the Euler-Poincaré characteristic is furnished by the integral*

$$\chi(M_2) = \frac{1}{2\pi} \int\limits_{M_2} R_{1212}(x)\,(g)^{-1/2} dx^1 \wedge dx^2. \tag{5.166}$$

For the proof, consult the book by B. O'Neill [30].

Example 5.2.24 Consider S^2, the spherical surface of unit radius. In the spherical polar coordinate chart, the metric is furnished by

$$\mathbf{g}_{..}(x) = dx^1 \otimes dx^1 + (\sin x^1)^2 dx^2 \otimes dx^2.$$

By (4.90) and (5.166), we obtain that

$$\chi(S^2) = \frac{1}{2\pi} \int_0^{\pi} \int_{-\pi}^{\pi} (\sin x^1)^2 (\sin x^1)^{-1} dx^1 dx^2$$
$$= 2.$$

Thus, the integral yields the expected number. □

Now, we shall state the generalization of the Gauss-Bonnet theorem 5.2.23 for a general hypersurface.

Theorem 5.2.25 *Let M be a closed, compact, orientable $2N$-dimensional Riemannian manifold with twice-differentiable metric field $\mathbf{g}_{..}(x)$. When the Euler-Poincaré characteristic of the manifold is $\chi(M)$, then*

$$\frac{2}{V_{2N}} \int_M K_T(x) \;{}_{2N}\eta = \chi(M). \tag{5.167}$$

Here, V_{2N} is the hyperarea of a $2N$-dimensional unit hypersphere and is explicitly given by

$$V_{2N} = \frac{2\,(\pi)^{N+1/2}}{\Gamma\left(N+\frac{1}{2}\right)} = \frac{\pi^N 2^{2N+1} N!}{(2N)!}. \tag{5.168}$$

Moreover, the 'total curvature' $K_T(x)$ is defined by

$$\begin{aligned} K_T(x) &:= \left[2^N (2N)!\right]^{-1} \eta^{i_1 \ldots i_{2N}} \eta^{j_1 \ldots j_{2N}} \times \\ &\times R_{i_1 i_2 j_1 j_2} R_{i_3 i_4 j_3 j_4} \ldots R_{i_{2N-1} i_{2N} j_{2N-1} j_{2N}}. \end{aligned} \tag{5.169}$$

(For the proof of the theorem, see Willmore's book [39].)

Exercises 5.2

1. Let $\mathbf{S}^{..}(x)$ be a differentiable, symmetric tensor field. Prove that

$$\nabla_j S^j{}_k = (1/\sqrt{|g|})[\partial_j(\sqrt{|g|} S^j{}_k)] - (1/2)(\partial_k g_{ij}) S^{ij}(x).$$

2. Consider the flat metric of special relativity furnished by

$$\mathbf{g}_{..}(x) = dx^1 \otimes dx^1 + dx^2 \otimes dx^2 + dx^3 \otimes dx^3 - dx^4 \otimes dx^4.$$

(i) Prove that for a twice-differentiable scalar field $W(x)$,

$$\Delta W = \left[\frac{\partial^2}{(\partial x^1)^2} + \frac{\partial^2}{(\partial x^2)^2} + \frac{\partial^2}{(\partial x^3)^2} - \frac{\partial^2}{(\partial x^4)^2} \right] W =: \Box W.$$

(*Remark:* The symbol $\Box$ is called the **D'Alembertian** or the **wave operator**.)

(ii) Let f and g be two *arbitrary* functions of class C^2. Moreover, let k_1, k_2, k_3 be *arbitrary* numbers with $\nu(k_1, k_2, k_3) := \sqrt{(k_1)^2 + (k_2)^2 + (k_3)^2}$. Show that the function

$$W(x) := f(k_1 x^1 + k_2 x^2 + k_3 x^3 - \nu x^4) + g(k_1 x^1 + k_2 x^2 + k_3 x^3 + \nu x^4)$$

solves the wave equation $\Box W = 0$.

3. Consider the case where $N = 2$. Show that there exist (additional) algebraic identities

(i) $R_{ijmn}(x) \equiv (1/2)R(x)\left[g_{in}g_{jm} - g_{im}g_{jn}\right]$;

(ii) $G_{ij}(x) := R_{ij}(x) - (1/2)g_{ij}(x)R(x) \equiv 0$.

4. Weyl's **conformal curvature tensor** for $N \geq 3$ is defined by

$$\begin{aligned} C^l{}_{ijk}(x) := {} & R^l{}_{ijk}(x) + (N-2)^{-1}\left[\delta^l{}_j R_{ik} - \delta^l{}_k R_{ij} + g_{ik}R^l{}_j - g_{ij}R^l{}_k\right] \\ & + (N-1)^{-1}(N-2)^{-1}R(x)\left[\delta^l{}_k g_{ij} - \delta^l_j g_{ik}\right]. \end{aligned}$$

Prove that in the case where $N = 3$, there exist (additional) algebraic identities $C^l{}_{ijk}(x) \equiv 0$.

5. The **symmetrized Riemann tensor S**.... (x) is defined by

$$\mathbf{S}....(x)(\vec{\mathbf{X}}, \vec{\mathbf{Y}}, \vec{\mathbf{U}}, \vec{\mathbf{V}}) := -(1/3)\left[\mathbf{R}....(\vec{\mathbf{X}}, \vec{\mathbf{U}}, \vec{\mathbf{Y}}, \vec{\mathbf{V}}) + \mathbf{R}....(\vec{\mathbf{X}}, \vec{\mathbf{V}}, \vec{\mathbf{Y}}, \vec{\mathbf{U}})\right]$$

for all vector fields $\vec{\mathbf{X}}(x), \vec{\mathbf{Y}}(x), \vec{\mathbf{U}}(x), \vec{\mathbf{V}}(x)$. Prove the algebraic identities

$$\begin{aligned} & S_{qpuv}(x) \equiv S_{pquv}(x), \quad S_{pqvu}(x) \equiv S_{pquv}(x), \\ & S_{uvpq}(x) \equiv S_{pquv}(x), \quad S_{pquv}(x) + S_{puvq}(x) + S_{pvqu}(x) \equiv 0. \end{aligned}$$

6. The **double Hodge dual tenser** of the four-dimensional Riemann-Christoffel curvature tensor is defined by

$$**\,R^{pquv}(x) := \left(\frac{1}{4}\right)\eta^{pqrs}(x)R_{rstw}(x)\eta^{twuv}(x).$$

(i) Show that $**R^{pquv}(x)$ satisfy exactly the same algebraic identities as $R^{pquv}(x)$ in (5.100) – (5.107).

(ii) Prove that the single contraction

$$**\, R^p{}_{qap}(x) = G_{qa}(x).$$

7. Prove that for differentiable Ricci rotation coefficients, the following differential identities hold:

$$\partial_c(\gamma^c{}_{ab} - \gamma^c{}_{ba}) + \partial_a \gamma^c{}_{bc} - \partial_b \gamma^c{}_{ac} + \gamma^c{}_{dc}(\gamma^d{}_{ba} - \gamma^d{}_{ab}) + \gamma^c{}_{ad}\gamma^d{}_{bc} - \gamma^c{}_{bd}\gamma^d{}_{ac} \equiv 0.$$

8. Suppose that the metric tensor field $\mathbf{g}_{..}(x)$ is of class C^4 in $D \subset \mathbb{R}^N$. Show that the following second-order differential identities hold:

$$\begin{aligned}\nabla_u \nabla_v R_{pqrs} - \nabla_v \nabla_u R_{pqrs} + \nabla_q \nabla_p R_{rsvu} - \nabla_p \nabla_q R_{rsvu}\\ + \nabla_s \nabla_r R_{vupq} - \nabla_r \nabla_s R_{vupq} \equiv 0.\end{aligned}$$

9. Let $\vec{\mathbf{U}}(x)$ be a differentiable vector field in $D \cup \partial D \subset \mathbb{R}^N$. Let D and ∂D satisfy conditions stated in theorem 5.2.20. Prove the **Yano's integral formula**

$$\begin{aligned}&\int_D \left[(\nabla_i U_j)(\nabla^j U^i) - (\nabla_i U^i)^2 - R_{ij}(x) U^i(x) U^j(x)\right] d^N v\\ &\quad = \int_{\partial D} \left[U^j(x) \nabla_j U^i - U^i(x) \nabla_j U^j\right] n_i d^{N-1} v.\end{aligned}$$

10. Let p be an integer such that $2 \le p+2 \le N$. Let $\partial D^*{}_{p+2}$ be the closed, orientable, differentiable boundary of a star-shaped domain $D^*{}_{p+2}$ in $\mathbb{R}^N$. Moreover, let ${}_p\mathbf{A}(x)$ be a differentiable p-form in D containing $D^*{}_{p+2}$. Prove the **integral identity**

$$\oint_{\partial D^*{}_{p+2}} \left[\nabla_u A_{q_1 \ldots q_p}\right] \tilde{\mathbf{e}}^{q_1}(x) \wedge \cdots \wedge \tilde{\mathbf{e}}^{q_p}(x) \wedge \tilde{\mathbf{e}}^u(x) \equiv 0.$$

5.3 Curves, Frenet-Serret Formulas, and Geodesics

We introduced the notion of a parametrized curve into a differentiable manifold in section 3.2. We also defined the parallel transport of tensor fields along a differentiable curve in section 4.2. In the present section, we shall deal with all these concepts in a Riemannian or a pseudo-Riemannian manifold.

Consider a non-degenerate differentiable curve $\mathcal{X}$ into $D \subset \mathbb{R}^N$ corresponding to an open subset $U \subset M$. By (3.22), (3.24), and (4.36), we write

$$\begin{aligned} x &= \mathcal{X}(t), \\ x^j &= \mathcal{X}^j(t), \\ \vec{\mathcal{X}}'(t) &= \frac{d\mathcal{X}^j(t)}{dt}\left[\frac{\partial}{\partial x^j}\right]_{|\mathcal{X}(t)} = \mathcal{X}'^a(t)\vec{\mathbf{e}}_a\left[\mathcal{X}(t)\right], \\ t &\in [a,b] \subset \mathbb{R}. \end{aligned} \tag{5.170}$$

The covariant derivative of a differentiable vector $\vec{\mathbf{V}}$ along the image Γ of $\mathcal{X}$ is given by (4.37), (4.39), and (5.85) as

$$\nabla_{\vec{\mathcal{X}}'}\left[\vec{\mathbf{V}}(\mathcal{X}(t))\right] = \frac{DV^i}{dt}(\mathcal{X}(t))\left[\frac{\partial}{\partial x^i}\right]_{|\mathcal{X}(t)} = \frac{DV^a}{dt}(\mathcal{X}(t))\vec{\mathbf{e}}_a(\mathcal{X}(t)),$$

$$\frac{DV^i}{dt}(\mathcal{X}(t)) = \left[\nabla_j V^i\right]_{|\mathcal{X}(t)}\frac{d\mathcal{X}^j(t)}{dt} = \left[\partial_j V^i + \left\{ {i \atop jk} \right\} V^k(x)\right]_{|..}\frac{d\mathcal{X}^j(t)}{dt},$$

$$\frac{DV^a}{dt}(\mathcal{X}(t)) = \left[\nabla_b V^a\right]_{|\mathcal{X}(t)}\mathcal{X}'^b(t) = \left[\partial_b V^a - \gamma^a{}_{cb}(x)V^c(x)\right]_{|..}\mathcal{X}'^b(t). \tag{5.171}$$

Example 5.3.1 Equations (5.171) yield

$$\frac{DV^i}{dt}(\mathcal{X}(t)) = \frac{dV^i}{dt}(\mathcal{X}(t)) + \left[\left\{ {i \atop jk} \right\} V^k(x)\right]_{|\mathcal{X}(t)}\frac{d\mathcal{X}^j(t)}{dt}.$$

If we choose the vector field as the tangent vector to the twice-differentiable curve, then $V^i(\mathcal{X}(t)) = \frac{d\mathcal{X}^i(t)}{dt}$ and

$$\frac{DV^i\left(\mathcal{X}(t)\right)}{dt} = \frac{d^2\mathcal{X}^i(t)}{dt^2} + \left\{ {i \atop jk} \right\}_{|\mathcal{X}(t)}\frac{d\mathcal{X}^j(t)}{dt}\frac{d\mathcal{X}^k(t)}{dt}. \tag{5.172}$$

The expression above can be considered as a component of the "**generalized acceleration**." □

The covariant derivative of a differentiable tensor field ${}^r{}_s\mathbf{T}(x)$ along a differentiable curve can be obtained from (4.49) and (5.83). It is furnished by

$$\begin{aligned} &\frac{DT^{i_1\ldots i_r}{}_{j_1\ldots j_s}}{dt}(\mathcal{X}(t)) = \left[\nabla_k T^{i_1\ldots i_r}{}_{j_1\ldots j_s}\right]_{|\mathcal{X}(t)}\frac{d\mathcal{X}^k(t)}{dt} \\ &= \frac{dT^{i_1\ldots i_r}{}_{j_1\ldots j_s}}{dt}(\mathcal{X}(t)) + \left[\sum_{\alpha=1}^{r}\left\{ {i_\alpha \atop kl} \right\} T^{i_1\ldots i_{\alpha-1} l i_{\alpha+1}\ldots i_r}{}_{j_1\ldots j_s}\right. \\ &\left. -\sum_{\beta=1}^{s}\left\{ {l \atop kj_\beta} \right\} T^{i_1\ldots i_r}{}_{j_1\ldots j_{\beta-1} l j_{\beta+1}\ldots j_s}\right]_{|\mathcal{X}(t)}\frac{d\mathcal{X}^k(t)}{dt}. \end{aligned} \tag{5.173}$$

The covariant derivatives of products of tensors obey the Leibnitz rules in (4.25). Therefore, the covariant derivative $\frac{D}{dt}$ of products of tensor components along a curve will respect the Leibnitz rules of differentiation.

Example 5.3.2 Consider a differentiable tensor field $\mathbf{g}^{\cdot\cdot}(x) \otimes \mathbf{g}_{\cdot\cdot}(x)$. The components are given by $g^{im}(x)g_{jn}(x)$. By (5.173), the covariant derivative along a curve is provided by

$$\frac{D}{dt}\left[g^{im}(\mathcal{X}(t))g_{jn}(\mathcal{X}(t))\right] = \left[\nabla_k \left(g^{im}g_{jn}\right)\right]_{|\mathcal{X}(t)} \frac{d\mathcal{X}^k(t)}{dt} \equiv 0.$$

Thus, we derive the equation

$$\frac{d}{dt}\left[g^{im}(\mathcal{X}(t))g_{jn}(\mathcal{X}(t))\right] + \left[\left(\left\{ {i \atop kl} \right\} g^{lm} + \left\{ {m \atop kl} \right\} g^{il}\right) g_{jn}\right.$$

$$\left. -g^{im}\left(\left\{ {l \atop kj} \right\} g_{ln} + \left\{ {l \atop kn} \right\} g_{jl}\right)\right]_{|\mathcal{X}(t)} \frac{d\mathcal{X}^k(t)}{dt} = 0.$$

These equations are consistent with equations (5.44), (5.45), and (5.46). □

Let us assume that the non-degenerate, differentiable curve $\mathcal{X}$ satisfies

$$\mathbf{g}_{\cdot\cdot}\left[\vec{\mathcal{X}}'(t), \vec{\mathcal{X}}'(t)\right] \neq 0.$$

Therefore, the curve is *non-null.* We shall reparametrize the curve according to (3.26) and (3.27) as

$$\begin{aligned} s &= \mathcal{S}(t), t = h(s), \\ x &= \mathcal{X}^{\#}(s) = \mathcal{X}(t), \\ \frac{d\mathcal{X}^{\#^k}(s)}{ds} &= h'(s)\frac{d\mathcal{X}^k(t)}{dt}. \end{aligned}$$

We choose a class of special reparametrizations such that

$$\frac{d\mathcal{S}(t)}{dt} = \sqrt{\left| g_{ij}(\mathcal{X}(t))\frac{d\mathcal{X}^i(t)}{dt}\frac{d\mathcal{X}^j(t)}{dt}\right|} > 0,$$

$$\sigma\left[\vec{\mathcal{X}}^{\#\prime}(s)\right] = \sqrt{\left| g_{ij}(\mathcal{X}\#(s))\frac{d\mathcal{X}^{\#^i}(s)}{ds}\frac{d\mathcal{X}^{\#^j}(s)}{ds}\right|} \tag{5.174}$$

$$= \sqrt{\left[\frac{d\mathcal{S}(t)}{dt}\frac{dh(s)}{ds}\right]^2} \equiv 1.$$

In a Riemannian manifold, such a parametrization will be called the **arc length parametrization**. In a pseudo-Riemannian manifold, s can be called an **arc separation** parameter. (See the book by O'neill[30].)

Example 5.3.3 Consider $\mathbb{E}_3$ and a Cartesian coordinate chart. A **circular helix** $\mathcal{X}$ is characterized by

$$x = \mathcal{X}(t) := (a\cos t, a\sin t, bt);\ a > 0,\ b > 0;\ t > 0.$$

The separation or length of the tangent vector $\vec{\mathcal{X}}'(t)$ is given by

$$\sigma\left(\vec{\mathcal{X}}'(t)\right) \equiv \|\vec{\mathcal{X}}'(t)\| = \sqrt{a^2 + b^2} > 0.$$

The differential equation (5.174) for the arc length parameter is given by

$$\frac{d\mathcal{S}(t)}{dt} = \|\vec{\mathcal{X}}'(t)\| = \sqrt{a^2 + b^2}.$$

Integrating the equation above, we obtain

$$s = \mathcal{S}(t) = \left(\sqrt{a^2 + b^2}\right) t + s_0.$$

Here s_0 is the arbitrary constant of integration. Putting $s_0 = 0$ for simplicity, the reparametrized curve $\mathcal{X}^{\#}(s)$ is furnished by

$$x = \mathcal{X}^{\#}(s) = \left(a\cos\left(\frac{s}{\sqrt{a^2+b^2}}\right), a\sin\left(\frac{s}{\sqrt{a^2+b^2}}\right), \frac{bs}{\sqrt{a^2+b^2}}\right) \in \mathbb{R}^3. \qquad \square$$

Subsequently, we use this parameter most often. However, *we shall drop the symbol # from $\mathcal{X}^{\#}$ in the sequel.*

Let us denote the tangent vector of a *non-null* curve $\mathcal{X}$ of differentiability class C^N by

$$\lambda^i{}_{(0)}(s) := \frac{d\mathcal{X}^i(s)}{ds}. \tag{5.175}$$

By (5.174), the vector $\vec{\lambda}_{(0)}(s)$ is a unit vector. Therefore, using (5.173), (5.174), and (5.175), we get

$$\sigma[\vec{\lambda}_{(0)}(s)] \equiv 1, \tag{5.176}$$

$$g_{ij}(\mathcal{X}(s))\lambda^i{}_{(0)}(s)\lambda^j{}_{(0)}(s) =: \varepsilon_{(0)} = \pm 1, \tag{5.177}$$

$$\frac{D}{ds}\left[g_{ij}\left(\mathcal{X}(s)\right)\lambda^i{}_{(0)}(s)\lambda^j{}_{(0)}(s)\right] \equiv 0, \tag{5.178}$$

$$g_{ij}\left(\mathcal{X}(s)\right)\lambda^i{}_{(0)}(s)\frac{D\lambda^j{}_{(0)}(s)}{ds} \equiv 0. \tag{5.179}$$

Equation (5.179) shows that the vector field $\frac{D\vec{\lambda}_{(0)}(s)}{ds}$ is orthogonal to the tangent vector field $\vec{\lambda}_{(0)}(s)$ along the image of the curve. Let us denote the unit vector field that is codirectional to $\frac{D\vec{\lambda}_{(0)}(s)}{ds}$ by $\vec{\lambda}_{(1)}(s)$. It exists in the case where $\frac{D\vec{\lambda}_{(0)}(s)}{ds}$ is *not* a null vector field. The vector field $\vec{\lambda}_{(1)}(s)$ is called the **first normal** and the separation $\sigma\left[\frac{D\vec{\lambda}_{(0)}(s)}{ds}\right]$ is called the **first or principal curvature** $\kappa_{(1)}(s)$. Assuming $\kappa_{(1)}(s) > 0$, we have

$$\frac{D\lambda^i{}_{(0)}(s)}{ds} = \kappa_{(1)}(s)\lambda^i{}_{(1)}(s), \tag{5.180}$$

$$g_{ij}(\mathcal{X}(s))\lambda^i{}_{(1)}(s)\lambda^j{}_{(1)}(s) =: \varepsilon_{(1)} = \pm 1, \tag{5.181}$$

$$g_{ij}\left(\mathcal{X}(s)\right)\lambda^i{}_{(0)}(s)\lambda^j{}_{(1)}(s) \equiv 0. \tag{5.182}$$

Suppose that

$$\sigma\left[\frac{D\vec{\lambda}_{(1)}(s)}{ds} + \varepsilon_{(0)}\varepsilon_{(1)}\kappa_{(1)}(s)\vec{\lambda}_{(0)}(s)\right] > 0.$$

In that case, we can define a positive invariant $\kappa_{(2)}(s)$ and a unit vector $\vec{\lambda}_{(2)}(s)$ by the following equations:

$$\kappa_{(2)}(s) := \sigma\left[\frac{D\vec{\lambda}_{(1)}(s)}{ds} + \varepsilon_{(0)}\varepsilon_{(1)}\kappa_{(1)}(s)\vec{\lambda}_{(0)}(s)\right], \tag{5.183}$$

$$\lambda^i{}_{(2)}(s) := \left[\kappa_{(2)}(s)\right]^{-1}\left[\frac{D\lambda^i{}_{(1)}(s)}{ds} + \varepsilon_{(0)}\varepsilon_{(1)}\kappa_{(1)}(s)\lambda^i{}_{(0)}(s)\right], \tag{5.184}$$

$$g_{ij}\left(\mathcal{X}(s)\right)\lambda^i{}_{(2)}(s)\lambda^j{}_{(2)}(s) =: \varepsilon_{(2)} = \pm 1, \tag{5.185}$$

$$g_{ij}\left(\mathcal{X}(s)\right)\lambda^i{}_{(0)}(s)\lambda^j{}_{(2)}(s) \equiv g_{ij}\left(\mathcal{X}(s)\right)\lambda^i{}_{(1)}(s)\lambda^j{}_{(2)}(s) \equiv 0. \tag{5.186}$$

The vector field $\vec{\lambda}_{(2)}(s)$ is called the **second normal** (or **binormal**), and the invariant $\kappa_{(2)}(s)$ is called the **second curvature** (or **torsion**).

In this fashion, we can continue to create more orthonormal vectors along the curve. But the process must terminate *before* or at the Nth step since the tangent space $T_{\mathcal{X}(s)}(\mathbb{R}^N)$ can admit at most N orthonormal vectors. We summarize the outcome of this process in the **Frenet-Serret formulas** in the following theorem.

Theorem 5.3.4 *Let $\mathcal{X}$ be a parametrized, non-degenerate, non-null curve of class $C^{N+1}\left([s_1, s_2] \subset \mathbb{R}; \mathbb{R}^N\right)$. Let s be the arc separation parameter for $\mathcal{X}$. Moreover, let there exist a finite sequence of N non-null vector fields, recursively defined, along the curve:*

$$\begin{aligned}
&\kappa_{(A)}(s)\lambda^i{}_{(A)}(s) := \frac{D\lambda^i{}_{(A-1)}(s)}{ds} + \varepsilon_{(A-2)}\varepsilon_{(A-1)}\kappa_{(A-1)}(s)\lambda^i{}_{(A-2)}(s),\\
&\kappa_{(0)}(s) = \kappa_{(N)}(s) \equiv 0,\\
&g_{ij}\left(\mathcal{X}(s)\right)\lambda^i{}_{(A-1)}(s)\lambda^j{}_{(A-1)}(s) =: \varepsilon_{(A-1)} = \pm 1,\\
&A \in \{1, 2, \ldots, N\}.
\end{aligned} \tag{5.187}$$

Then the set of vectors $\left\{\vec{\lambda}_{(A-1)}(s)\right\}_1^N$ *is an orthonormal basis set for* $T_{\mathcal{X}(s)}(\mathbb{R}^N)$.

The proof is left to the reader.

Caution: If any of the vectors $\frac{D\vec{\lambda}_{(A-1)}(s)}{ds} + \varepsilon_{(A-2)}\varepsilon_{(A-1)}\kappa_{(A-1)}(s)\vec{\lambda}_{(A-2)}(s)$ becomes *null*, the procedure *breaks down* in the next step.

Example 5.3.5 Let us choose $\mathbb{E}_3$ and a Cartesian coordinate chart. We denote the alternate symbols by

$$\begin{aligned} &T^i(s) := \lambda^i{}_{(0)}(s),\ N^i(s) := \lambda^i{}_{(1)}(s),\ B^i(s) := \lambda^i{}_{(2)}(s), \\ &\kappa(s) := \kappa_{(1)}(s),\ \tau(s) := \kappa_{(2)}(s). \end{aligned}$$

The Frenet-Serret formulas (5.187) yield the well-known equations

$$\begin{aligned} \frac{dT^i(s)}{ds} &= \kappa(s)N^i(s), \\ \frac{dN^i(s)}{ds} &= -\kappa(s)T^i(s) + \tau(s)B^i(s), \\ \frac{dB^i(s)}{ds} &= -\tau(s)N^i(s). \end{aligned} \tag{5.188}$$

Let us choose a circular helix furnished by

$$\begin{aligned} &\mathcal{X}(s) := \left(\cos(s/\sqrt{2}), \sin(s/\sqrt{2}), s/\sqrt{2},\right), \\ &s \in [0, \infty). \end{aligned}$$

(See fig. 5.2.) In this case, we compute explicitly

$$\begin{aligned} \vec{\mathbf{T}}(s) &= -\left(1/\sqrt{2}\right)\sin\left(s/\sqrt{2}\right)\left(\frac{\partial}{\partial x^1}\right)_{|\mathcal{X}(s)} + \left(1/\sqrt{2}\right)\cos\left(s/\sqrt{2}\right)\left(\frac{\partial}{\partial x^2}\right)_{|..} \\ &\quad + \left(1/\sqrt{2}\right)\left(\frac{\partial}{\partial x^3}\right)_{|..}, \\ \vec{\mathbf{N}}(s) &= -\cos\left(s/\sqrt{2}\right)\left(\frac{\partial}{\partial x^1}\right)_{|..} - \sin\left(s/\sqrt{2}\right)\left(\frac{\partial}{\partial x^2}\right)_{|..}, \\ \vec{\mathbf{B}}(s) &= (1/\sqrt{2})\sin\left(s/\sqrt{2}\right)\left(\frac{\partial}{\partial x^1}\right)_{|..} - (1/\sqrt{2})\cos\left(s/\sqrt{2}\right)\left(\frac{\partial}{\partial x^2}\right)_{|..} \\ &\quad + \left(1/\sqrt{2}\right)\left(\frac{\partial}{\partial x^3}\right)_{|..}, \\ \kappa(s) &\equiv (1/2), \quad \tau(s) \equiv (1/2). \end{aligned}$$

□

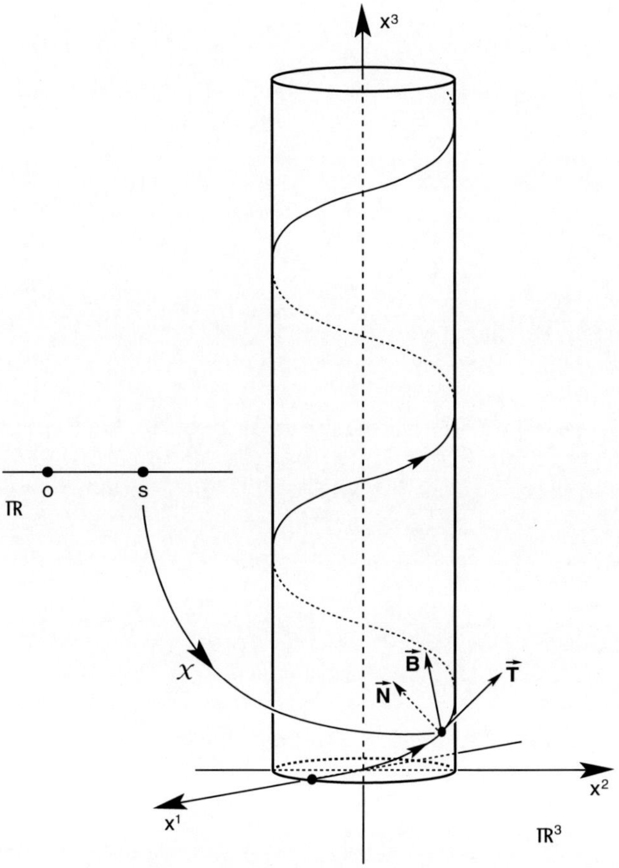

Figure 5.2: A circular helix in $\mathbb{R}^3$.

Example 5.3.6 Consider the four-dimensional space-time manifold of a Lorentz metric. Consider a timelike curve $\mathcal{X}$ of class C^5. Therefore, we must have

$$-\varepsilon_{(0)} = \varepsilon_{(1)} = \varepsilon_{(2)} = \varepsilon_{(3)} = 1.$$

The Frenet-Serret formulas (5.187) yield

$$\begin{aligned}
\frac{D\lambda^i{}_{(0)}(s)}{ds} &= \kappa_{(1)}(s)\lambda^i{}_{(1)}(s), \\
\frac{D\lambda^i{}_{(1)}(s)}{ds} &= \kappa_{(2)}(s)\lambda^i{}_{(2)}(s) + \kappa_{(1)}(s)\lambda^i{}_{(0)}(s), \\
\frac{D\lambda^i{}_{(2)}(s)}{ds} &= \kappa_{(3)}(s)\lambda^i{}_{(3)}(s) - \kappa_{(2)}(s)\lambda^i{}_{(1)}(s), \\
\frac{D\lambda^i{}_{(3)}(s)}{ds} &= -\kappa_{(3)}(s)\lambda^i{}_{(2)}(s).
\end{aligned} \tag{5.189}$$

The set of vectors $\left\{\vec{\lambda}_{(0)}(s), \vec{\lambda}_{(1)}(s), \vec{\lambda}_{(2)}(s), \vec{\lambda}_{(3)}(s)\right\}$ is the orthonormal **Frenet-Serret tetrad** along the curve.

A hyperbola of constant curvature in space-time is characterized by $\kappa_{(1)}(s) = b = \text{const.}$, $\kappa_{(0)}(s) = \kappa_{(3)}(s) \equiv 0$.

In a flat metric, using a **Minkowskian or pseudo-Cartesian coordinate system** and the Lorentz metric $g_{ij}(x) = d_{ij}$, equations (5.189) for a hyperbola with constant curvature can be integrated. The general solution is characterized by

$$x^j = \mathcal{X}^j(s) = (b)^{-1} c^j \sinh\left[b\left(s - s^j_{(0)}\right)\right] + x^j{}_{(0)}.$$

Here, c^j_j $s^j_{(0)}$, and $x^j_{(0)}$ are twelve arbitrary constants of integration. □

Now, we shall derive Frenet-Serret formulas (5.187) in an alternate form. We *change the notations*

$$\left(\vec{\lambda}_{(0)}(s), \vec{\lambda}_{(1)}(s), \ldots, \vec{\lambda}_{(N-1)}(s)\right) \text{ to } \left(\vec{\lambda}_{(1)}(s), \vec{\lambda}_{(2)}(s), \ldots, \vec{\lambda}_{(N)}(s)\right).$$

Then, (5.187) can be expressed as

$$\begin{aligned}
&\frac{D\lambda^i{}_{(a)}(s)}{ds} = \kappa_{(a)}(s)\lambda^i{}_{(a+1)}(s) - \varepsilon_{(a-1)}\varepsilon_{(a)}\kappa_{(a-1)}(s)\lambda^i{}_{(a-1)}(s), \\
&\kappa_{(0)}(s) = \kappa_{(N)}(s) \equiv 0, \\
&g_{ij}\left(\mathcal{X}(s)\right)\lambda^i{}_a(s)\lambda^j{}_b(s) = d_{ab}, \\
&\varepsilon_{(1)} := d_{11},\ \varepsilon_{(2)} := d_{22},\ \ldots,\ \varepsilon_{(N)} := d_{NN}.
\end{aligned} \tag{5.190}$$

Using (5.171), (5.54), and (5.19), we obtain

$$\begin{aligned}\frac{D\lambda^i{}_{(a)}(s)}{ds} &= \left[\nabla_k \lambda^i{}_a(s)\right] \frac{D\mathcal{X}^k(s)}{ds} = d^{bd}\gamma_{abc}(x)\left[\mu^c{}_k(x)\frac{D\mathcal{X}^k(s)}{ds}\right]\lambda^i{}_d(x) \\ &= d^{db}\gamma_{abc}(x)\mathcal{X}'^c(s)\lambda^i{}_d(x) \\ &= \sum_{d=1}^{N} \varepsilon_{(d)} \left[\gamma_{adc}(x)\mathcal{X}'^c(s)\right] \lambda^i{}_d(x). \end{aligned} \tag{5.191}$$

(Here, we have suspended the summation convention partly!) Comparing (5.191), with (5.190), we conclude that for a Frenet-Serret orthonormal basis (or frame) we must have

$$\begin{aligned}\gamma_{adc}(x)\mathcal{X}'^c(s) &\equiv 0 \quad \text{for} \quad d \neq a \pm 1, \\ \gamma_{(a)(a+1)c}(x)\mathcal{X}'^c(s) &= -\gamma_{(a+1)(a)c}(x)\mathcal{X}'^c(s) \\ &= \varepsilon_{(a+1)}\kappa_{(a)}(s). \end{aligned} \tag{5.192}$$

We shall now define the parallel transport of tensor fields along a differential curve $\mathcal{X}$. By (4.38), (4.47), (4.48), (4.49), and (5.173), the rules for parallel transport are provided by

$$\nabla_{\vec{\mathcal{X}}'} \left[{}^r{}_s\mathbf{T}(\mathcal{X}(t))\right] = {}^r{}_s\mathbf{O}(\mathcal{X}(t)), \tag{5.193}$$

$$\frac{DT^{i_1\ldots i_r}{}_{j_1\ldots j_s}}{dt}(\mathcal{X}(t)) = \left[\nabla_k T^{i_1\ldots i_r}{}_{j_1\ldots j_s}\right]_{|\mathcal{X}(t)} \frac{d\mathcal{X}^k(t)}{dt} = 0, \tag{5.194}$$

$$\frac{DT^{a_1\ldots a_r}{}_{b_1\ldots b_s}}{dt}(\mathcal{X}(t)) = \left[\nabla_c T^{a_1\ldots a_r}{}_{b_1\ldots b_s}\right]_{|\mathcal{X}(t)} \mathcal{X}'^c(t) = 0. \tag{5.195}$$

Here, the covariant derivatives ∇_k and ∇_c are provided by (5.83) and (5.84).

Example 5.3.7 By (5.172) and (5.194), a differentiable vector field $\vec{\mathbf{V}}(x)$ is parallelly propagated along a curve $\mathcal{X}$ provided

$$\frac{DV^i(\mathcal{X}(t))}{dt} = \left[\partial_j V^i + \left\{ \begin{array}{c} i \\ jk \end{array} \right\} V^k(x)\right]_{|\mathcal{X}(t)} \frac{d\mathcal{X}^j(t)}{dt} = 0, \tag{5.196}$$

$$\frac{DV^a(\mathcal{X}(t))}{dt} = \left[\partial_b V^a - \gamma^a{}_{cb}(x)V^c(x)\right]_{|\mathcal{X}(t)} \mathcal{X}'^b(t) = 0. \tag{5.197}$$

Let another vector field $\vec{\mathbf{W}}(x)$ also be parallelly propagated so that

$$\frac{DW^j(\mathcal{X}(t))}{dt} = 0.$$

Thus, by the Leibnitz rules, (5.86), (5.196), and (5.197), we get

$$\frac{D}{dt}\left[g_{ij}(x)V^i(x)W^j(x)\right]_{|\mathcal{X}(t)} = \frac{D}{dt}\left[d_{ab}V^a(x)W^b(x)\right]_{|\mathcal{X}(t)} \equiv 0, \tag{5.198}$$

$$\frac{D}{dt}\left[\sigma(\vec{\mathbf{V}}(x))\right]_{|\mathcal{X}(t)} = \frac{D}{dt}\left[\sigma(\vec{\mathbf{W}}(x))\right]_{|\mathcal{X}(t)} \equiv 0. \tag{5.199}$$

Therefore, an orthonormal basis field (or frame) $\{\vec{\mathbf{e}}_a(x)\}_1^N$, which is parallelly propagated, *remains* an orthonormal frame. □

An affine geodesic curve and an affine parameter have been defined in (4.40), (4.43), and (4.46), respectively. Thus, we are motivated to express the **geodesic equations** in terms of an *affine parameter* as

$$\frac{d^2\mathcal{X}^i(\tau)}{d\tau^2} + \left\{ \begin{matrix} i \\ jk \end{matrix} \right\}_{|\mathcal{X}(\tau)} \frac{d\mathcal{X}^j(t)}{d\tau}\frac{d\mathcal{X}^k(\tau)}{d\tau} = 0, \tag{5.200}$$

$$\mathcal{X}'^b(\tau)\left[\partial_b\mathcal{X}'^a(\tau) - \gamma^a{}_{cb}(\mathcal{X}(\tau))\mathcal{X}'^c(\tau)\right] = 0. \tag{5.201}$$

As a consequence of (5.41), (5.47), and (5.200) we obtain

$$\begin{aligned}
\frac{d}{d\tau}\left[g_{jk}(\mathcal{X}(\tau))\frac{d\mathcal{X}^j(\tau)}{d\tau}\frac{d\mathcal{X}^k(\tau)}{d\tau}\right] &= 2g_{jk}\frac{d\mathcal{X}^j}{d\tau}\frac{d^2\mathcal{X}^k}{d\tau^2} + (\partial_i g_{jk})\frac{d\mathcal{X}^i}{d\tau}\frac{d\mathcal{X}^j}{d\tau}\frac{d\mathcal{X}^k}{d\tau} \\
&= 2\left\{g_{jk}\frac{d^2\mathcal{X}^k}{d\tau^2} + [i\,k, j]\frac{d\mathcal{X}^i}{d\tau}\frac{d\mathcal{X}^k}{d\tau}\right\}\frac{d\mathcal{X}^j}{d\tau} \\
&\equiv 0.
\end{aligned}$$

Therefore, in terms of an *affine parameter* τ, the geodesic equations (5.200) and (5.201) admit a **first integral**

$$g_{jk}(\mathcal{X}(\tau))\frac{d\mathcal{X}^j(\tau)}{d\tau}\frac{d\mathcal{X}^k(\tau)}{d\tau} \equiv c = \text{const.} \tag{5.202}$$

In the case of a Riemannian manifold, $c > 0$. Thus, we can reparametrize $s = \sqrt{c}\tau$ to express the geodesic equation in terms of the arc length parameter s in (5.174). In the case of a Lorentz metric, (i) $c = 0$ implies that the geodesic is **null**, (ii) $c > 0$ implies that the geodesic is **spacelike**, and (iii) $c < 0$ implies that the geodesic is **timelike**. In the case of *non-null* geodesics, we can reparametrize $s = \sqrt{|c|}\tau$ to obtain an arc separation parameter s. In such a case, (5.200), (5.201), and (5.202) reduce to

$$\begin{aligned}
&\frac{d^2\mathcal{X}^i(s)}{ds^2} + \left\{ \begin{matrix} i \\ jk \end{matrix} \right\}_{|\mathcal{X}(s)} \frac{d\mathcal{X}^j(s)}{ds}\frac{d\mathcal{X}^k(s)}{ds} = 0, \\
&\mathcal{X}'^b(s)\left[\partial_b\mathcal{X}'^a - \gamma^a{}_{cb}(\mathcal{X}(s))\mathcal{X}'^c(s)\right] = 0, \qquad (5.203) \\
&d_{ab}\mathcal{X}'^a(s)\mathcal{X}'^b(s) = g_{jk}(\mathcal{X}(s))\frac{d\mathcal{X}^j(s)}{ds}\frac{d\mathcal{X}^k(s)}{ds} \equiv \varepsilon := \pm 1.
\end{aligned}$$

A null geodesic is characterized by (5.200), (5.201), and (5.202) as

$$\frac{d^2\mathcal{X}^i(\tau)}{d\tau^2} + \left\{ \begin{matrix} i \\ jk \end{matrix} \right\}_{|\mathcal{X}(\tau)} \frac{d\mathcal{X}^j(\tau)}{d\tau}\frac{d\mathcal{X}^k(\tau)}{d\tau} = 0,$$

$$\mathcal{X}'^b(\tau)\left[\partial_b\mathcal{X}'^a - \gamma^a{}_{cb}(\mathcal{X}(\tau))\mathcal{X}'^c(\tau)\right] = 0, \tag{5.204}$$

$$d_{ab}\mathcal{X}'^a(\tau)\mathcal{X}'^b(\tau) = g_{jk}(\mathcal{X}(\tau))\frac{d\mathcal{X}^j(\tau)}{d\tau}\frac{d\mathcal{X}^k(\tau)}{d\tau} \equiv 0.$$

If we write the unit tangent vector along a non-null geodesic as $\lambda^i{}_{(0)}(s) := \frac{d\mathcal{X}^i(s)}{ds}$, then the first equation of the Frenet-Serret formulas (5.187) and (5.196) yields for the first curvature

$$\kappa_{(1)}(s) \equiv 0. \tag{5.205}$$

The remaining Frenet-Serret formulas in (5.187) are undefined for a geodesic.

Let us consider an orthonormal basis field $\left\{\vec{\lambda}_a(x)\right\}_1^N$. Let us restrict the basis field over a non-null geodesic $\mathcal{X}$ and thus obtain the basis $\left\{\vec{\lambda}_a(\mathcal{X}(s))\right\}_1^N$ (which *need not* be the Frenet-Serret frame.) If $\lambda^i{}_c(\mathcal{X}(s)) = \frac{d\mathcal{X}^i(s)}{ds}$ is tangential to the geodesic, we obtain by (5.194), (5.197), and (5.54) that

$$0 = \frac{D\lambda^l{}_c(\mathcal{X}(s))}{ds} = \lambda^l{}_c(\mathcal{X}(s))\left[\nabla_i\lambda^l{}_c\right]$$
$$= \left[g^{jl}(x)\gamma_{cbc}(x)\mu^b{}_j(x)\right]_{|\mathcal{X}(s)}.$$

Here, the (down) index c is *not summed*, but the index b is summed! Thus, we derive the conditions for a geodesic as

$$\begin{aligned} &\gamma_{bcc}(\mathcal{X}(s)) \equiv 0, \\ &b \in \{1, 2, \ldots, N\}. \end{aligned} \tag{5.206}$$

We already discussed a geodesic in example 4.2.3. We shall provide another example presently.

Example 5.3.8 Let the metric tensor field of a two-dimensional surface embedded in $\mathbb{E}_3$ be given by

$$\begin{aligned} \mathbf{g}_{..}(u) &= E(u^1)du^1 \otimes du^1 + G(u^1)du^2 \otimes du^2, \\ &E(u^1) > 0, \;\; G(u^1) > 0, \\ &(u^1, u^2) \in D \subset \mathbb{R}^2. \end{aligned} \tag{5.207}$$

(*Remark:* A plane, sphere, helicoid, or surface of revolution is characterized by the special case of the metric in (5.207).)

By (5.207), we obtain for non-zero components of Christoffel symbols

$$\left\{ \begin{matrix} 1 \\ 11 \end{matrix} \right\} = (1/2)\partial_1(\ln E), \ \left\{ \begin{matrix} 2 \\ 12 \end{matrix} \right\} \equiv \left\{ \begin{matrix} 2 \\ 21 \end{matrix} \right\} = (1/2)\partial_1(\ln G),$$
$$\left\{ \begin{matrix} 1 \\ 22 \end{matrix} \right\} = -(1/2)\left[E(u^1)\right]^{-1} \partial_1 G. \qquad (5.208)$$

The geodesic equations (5.203) reduce to

$$\frac{d^2\mathcal{U}^1(s)}{ds^2} + \left\{ \left[2E(u^1)\right]^{-1} \left[(\partial_1 E)\left(\frac{d\mathcal{U}^1}{ds}\right)^2 - (\partial_1 G)\left(\frac{d\mathcal{U}^2}{ds}\right)^2 \right] \right\}_{|\mathcal{U}(s)} = 0, \qquad (5.209)$$

$$\begin{aligned} &\frac{d^2\mathcal{U}^2(s)}{ds^2} + (\partial_1 \ln G)_{|\mathcal{U}(s)} \frac{d\mathcal{U}^1}{ds}\frac{d\mathcal{U}^2}{ds} \\ &= [G(\mathcal{U}^1(s))]^{-1} \frac{d}{ds}\left[G(\mathcal{U}^1(s))\frac{d\mathcal{U}^2(s)}{ds}\right] = 0. \end{aligned} \qquad (5.210)$$

Integrating (5.210), we obtain a first integral

$$G\left(\mathcal{U}^1(s)\right) \frac{d\mathcal{U}^2(s)}{ds} = h = \text{const.} \qquad (5.211)$$

Instead of integrating (5.209), we can integrate the first integral in (5.203) by (5.207) to derive

$$E\left(\mathcal{U}^1(s)\right)\left[\frac{d\mathcal{U}^1(s)}{ds}\right]^2 + G\left(\mathcal{U}^1(s)\right)\left[\frac{d\mathcal{U}^2(s)}{ds}\right]^2 = 1. \qquad (5.212)$$

Eliminating $\frac{d\mathcal{U}^2(s)}{ds}$ from above by (5.211) for the case $h \neq 0$, we get

$$\left[\frac{d\mathcal{U}^1(s)}{ds} \Big/ \frac{d\mathcal{U}^2(s)}{ds}\right]^2 = \left\{G(u^1)\left(G(u^1) - h^2\right)/h^2 E(u^1)\right\}_{|\mathcal{U}^1(s)}.$$

Reparametrizing the curve, we obtain from the equation above:

$$\frac{d\mathcal{U}^{\#_2}(u^1)}{du^1} = \pm \frac{h\sqrt{E(u^1)}}{\sqrt{G(u^1)[G(u^1) - h^2]}}. \qquad (5.213)$$

Integrating the above, we get the solution

$$u^2 = \mathcal{U}^{\#_2}(u^1) = \pm h \int_{u_0}^{u^1} \frac{\sqrt{E(x)}}{\sqrt{G(x)[G(x) - h^2]}} dx. \qquad (5.214)$$

(We have tacitly *assumed* that $G(x) > (h)^2$.) □

We shall now derive the geodesic equation (5.200) by the **variational method**. The **separation** along a non-degenerate, non-null, differentiable curve $\mathcal{X}$ is defined by

$$
\begin{aligned}
s = \Sigma(\mathcal{X}) &:= \int_{t_1}^{t_2} \sigma\left[\vec{\mathcal{X}}'(t)\right] dt \\
&= \int_{t_1}^{t_2} \sqrt{\left| g_{ij}(\mathcal{X}(t)) \frac{d\mathcal{X}^i(t)}{dt} \frac{d\mathcal{X}^j(t)}{dt} \right|}\, dt \\
&=: \int_{t_1}^{t_2} L(x,u)\Big|_{\substack{x=\mathcal{X}(t) \\ u=\frac{d\mathcal{X}(t)}{dt}}} dt; \qquad (5.215) \\
L(x,u) &= \sqrt{|g_{ij}(x)u^i u^j|} > 0, \\
L(x,u)\Big|_{\substack{x=\mathcal{X}(t) \\ u=\frac{d\mathcal{X}(t)}{dt}}} &\equiv \frac{d\mathcal{S}(t)}{dt}.
\end{aligned}
$$

In a Riemannian manifold with a positive-definite metric, the separation above coincides exactly with the arc length of the curve $\mathcal{X}$. The critical values of the integral (5.215) under the variation of $\mathcal{X}$ (together with zero boundary variations) is furnished by the **Euler-Lagrange equations**

$$
\frac{\partial L(x,u)}{\partial x^k}\Big|_{\substack{x=\mathcal{X}(t) \\ u=\frac{d\mathcal{X}(t)}{dt}}} - \frac{d}{dt}\left\{ \left[\frac{\partial L(..)}{\partial u^k} \right]_{|..} \right\} = 0. \qquad (5.216)
$$

(See the book by Lovelock and Rund [26].) Now, by the **Lagrangian** L in (5.215), we have

$$
\begin{aligned}
\frac{\partial L(x,u)}{\partial x^k} &= [2L(..)]^{-1} (\partial_k g_{ij})\, u^i u^j, \\
\frac{\partial L(..)}{\partial u^k} &= [L(..)]^{-1}\, g_{kj}(x) u^j.
\end{aligned} \qquad (5.217)
$$

By (5.217) and (5.215), the Euler-Lagrange equations (5.216) yield

$$
0 = \left[2\frac{d\mathcal{S}(t)}{dt}\right]^{-1} \left[(\partial_k g_{ij}) u^i u^j\right]_{|..} - \frac{d}{dt}\left\{ \left[\frac{d\mathcal{S}(t)}{dt}\right]^{-1} \left[g_{kj} u^j\right]_{|..} \right\},
$$

or (5.218)

$$
\frac{d^2\mathcal{X}^k(t)}{dt^2} + \left\{ \begin{matrix} k \\ ij \end{matrix} \right\}_{|x=\mathcal{X}(t)} \frac{d\mathcal{X}^i(t)}{dt} \frac{d\mathcal{X}^j(t)}{dt} = \frac{d\mathcal{X}^k(t)}{dt} \frac{d}{dt}\left\{ \ln\left| \frac{d\mathcal{S}(t)}{dt} \right| \right\}.
$$

Choosing the parameter $t = s$, the arc separation parameter, we get back equation (5.203). Let us consider the *alternate* Lagrangian

$$\begin{aligned}
&L_2(x,u) := (1/2)\left[g_{ij}(x)u^i u^j\right],\\
&\frac{\partial L_2(x,u)}{\partial x^k} = (1/2)\,(\partial_k g_{ij})\,u^i u^j,\\
&\frac{\partial L_2(x,u)}{\partial u^k} = g_{kj}(x)u^j,\\
&\frac{\partial^2}{\partial u^i \partial u^j}[L_2(x,u)] = g_{ij}(x).
\end{aligned} \tag{5.219}$$

The Euler-Lagrangian equations (5.216) from the alternate Lagrangian (5.171) yield again the geodesic equation (involving an *affine parameter*)

$$\frac{d^2\mathcal{X}^k(\tau)}{d\tau^2} + \left\{ {k \atop ij} \right\}_{|x=\mathcal{X}(\tau)} \frac{d\mathcal{X}^i(\tau)}{d\tau}\frac{d\mathcal{X}^j(\tau)}{d\tau} = 0.$$

Remark: The Euler-Lagrange equations from the alternate Lagrangian (5.219) can be used to *compute* Christoffel symbols $\left\{ {k \atop ij} \right\}$.

Since the geodesic equation was derived by the variational methods from the Lagrangian (5.215) and (5.217), we can conclude that the separation along the geodesic $\mathcal{X}$ is **critical-valued**. In other words, for a geodesic $\mathcal{X}$ joining two points $\mathcal{X}(\tau_1)$ and $\mathcal{X}(\tau_2)$, the separation is **maximum**, **minimum**, or **stationary** compared with that of any neighboring non-geodesic curve joining the same two points.

From the equation

$$\left[\frac{dS(\tau)}{d\tau}\right]^2 = \left| g_{ij}(\mathcal{X}(\tau))\frac{d\mathcal{X}^i(\tau)}{d\tau}\frac{d\mathcal{X}^j(\tau)}{d\tau}\right|$$

in (5.215), the **first fundamental form**, **metric form**, or **line element**

$$ds^2 = g_{ij}dx^i dx^j \tag{5.220}$$

has been invented.

Example 5.3.9 Let us choose a *flat* manifold and a Cartesian, or a pseudo-Cartesian, chart. The metric tensor field and the Christoffel symbols are given by

$$\begin{aligned}
\mathbf{g}_{..}(x) &= d_{ij}dx^i \otimes dx^j,\\
\left\{ {i \atop jk} \right\} &\equiv 0.
\end{aligned}$$

Equation (5.203) for a non-null geodesic implies that

$$\begin{aligned} \frac{d^2\mathcal{X}^k(s)}{ds^2} &= 0, \\ x^k &= \mathcal{X}^k(s) = c^k s + x^k{}_0. \end{aligned} \tag{5.221}$$

Here c^k and x_0^k are $2N$ constants of integration. Thus, the geodesics are **straight lines** in $\mathbb{R}^N$.

Equations (5.203) and (5.221) yield

$$\left|d_{ij}c^i c^j\right| = 1.$$

Therefore, there exists one second-degree constraint on N constant c^i's.

The separation $\Sigma(\mathcal{X})$ in (5.215) along the geodesic in (5.221) is given by

$$\begin{aligned} \Sigma(\mathcal{X}) &= \int_{\tau_1}^{\tau_2} \frac{dS(\tau)}{d\tau} d\tau = \int_{s_1}^{s_2} ds = (s_2 - s_1) = \sqrt{\left|d_{ij}c^i c^j\right|}\,(s_2 - s_1) \\ &= \sqrt{\left|d_{ij}[c^i(s_2 - s_1)]\,[c^j(s_2 - s_1)]\right|} \\ &= \sqrt{\left|d_{ij}(x^i{}_2 - x^i{}_1)\,(x^j{}_2 - x^j{}_1)\right|}, \\ s^2 = [\Sigma(\mathcal{X})]^2 &= \left|d_{ij}(x^i{}_2 - x^i{}_1)\,(x^j{}_2 - x^j{}_1)\right|. \end{aligned} \tag{5.222}$$

The equation above is the generalization of the usual **Pythagorean theorem.** □

Now we shall derive the equations for the **geodesic deviation**. Consider a two-dimensional parametric surface ξ into $D \in \mathbb{R}^N$ given by

$$\begin{aligned} & x = \xi(\tau, v) \\ & \xi \in C^3(\mathcal{D} \in \mathbb{R}^2; \mathbb{R}^N), \\ & \text{Rank}\left[\frac{\partial \xi^i}{\partial \tau}, \frac{\partial \xi^i}{\partial v}\right] = 2. \end{aligned} \tag{5.223}$$

(See figure 5.3.)

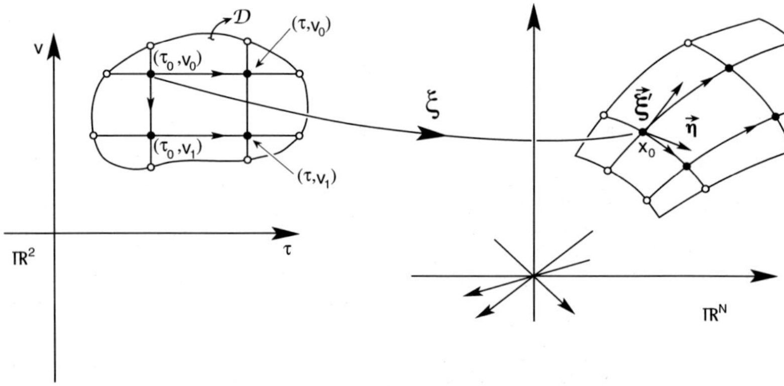

Figure 5.3: Two-dimensional surface generated by geodesics.

Let a typical parametric curve $\xi(\tau, v_0)$ be a *geodesic* in $D \subset \mathbb{R}^N$ with τ as an *affine parameter*. Therefore, we have

$$
\begin{aligned}
&\xi'^i(\tau, v) := \frac{\partial \xi^i}{\partial \tau}(\tau, v), \\
&\vec{\xi}'(\tau, v) := \xi'^i(\tau, v)\left(\frac{\partial}{\partial x^i}\right)_{|\xi(\tau,v)} = \xi'^a(\tau, v)\vec{\mathbf{e}}_a(\xi(\tau, v)), \\
&\frac{D\xi'^i}{\partial \tau}(\tau, v) = \frac{\partial}{\partial \tau}\xi'^i(\tau, v) + \left\{ \begin{array}{c} i \\ jk \end{array} \right\}_{|\xi(\tau,v)} \xi'^j(\tau, v)\xi'^k(\tau, v) = 0, \\
&g_{ij}(\xi(\tau, v))\xi'^i(\tau, v)\xi'^j(\tau, v) = d_{ab}\xi'^a(\tau, v)\xi'^b(\tau, v) \equiv c = \text{const.}
\end{aligned}
\tag{5.224}
$$

The tangent vector $\vec{\xi}'(\tau, v_0)$ is parallelly transported along the geodesic $x = \xi(\tau, v_0)$. Let us consider the other coordinate curve, $x = \xi(\tau_0, v)$. This curve is non-degenerate, differentiable, and need *not* be a geodesic. Thus, $\xi(\tau, v)$ defines a one-parameter congruence of geodesics spanning a two-dimensional surface in

$D \subset \mathbb{R}^N$. The tangent vector along $\xi(\tau_0, v)$ is given by

$$
\begin{aligned}
\eta^i(\tau, v) &:= \frac{\partial \xi^i}{\partial v}(\tau, v), \\
\vec{\eta}(\tau, v) &:= \eta^i(\tau, v)\left(\frac{\partial}{\partial x^i}\right)_{|\xi(\tau,v)} = \eta^a(\tau, v)\vec{\mathbf{e}}_a(\xi(\tau, v)), \\
\xi^i(\tau_0, v_1) - \xi^i(\tau_0, v_0) &= (v_1 - v_0)\eta^i(\tau_0, v_0) + (1/2)\left[\frac{\partial \eta^i}{\partial v}\right]_{|v^\#}(v_1 - v_0)^2, \\
v^\# &:= v_0 + \theta(v_1 - v_0), \\
0 < \theta < 1. &
\end{aligned}
\tag{5.225}
$$

Therefore, $|v_1 - v_0|\sigma[\vec{\eta}(\tau_0, v_0)]$ is the first-order approximation of the *separation* between two neighboring geodesics, $\xi(\tau, v_0)$ and $\xi(\tau, v_1)$, at τ_0. We can deduce from (5.224) and (5.225) that

$$
\begin{aligned}
\frac{\partial}{\partial \tau}\eta^i(\tau, v) &= \frac{\partial^2 \xi^i(\tau, v)}{\partial \tau \partial v} = \frac{\partial^2 \xi^i(\tau, v)}{\partial v \partial \tau} = \frac{\partial}{\partial v}\xi'^i(\tau, v), \\
\frac{D\eta^i}{\partial \tau}(\tau, v) &= \frac{\partial \eta^i}{\partial \tau}(\tau, v) + \left\{ \begin{matrix} i \\ jk \end{matrix} \right\}_{|\xi(\tau,v)} \eta^j(\tau, v)\xi'^k(\tau, v), \\
\frac{D\xi'^i}{\partial v}(\tau, v) &= \frac{\partial \xi'^i(\tau, v)}{\partial v} + \left\{ \begin{matrix} i \\ jk \end{matrix} \right\}_{|\xi(\tau,v)} \xi'^j(\tau, v)\eta^k(\tau, v) = \frac{D\eta^i(\tau, v)}{\partial \tau}.
\end{aligned}
\tag{5.226}
$$

The vector field $(v_1 - v_0)\frac{D\vec{\eta}}{\partial \tau}(\tau, v_0)$ is the approximate "relative velocity" between two neighboring geodesics, $\xi(\tau, v_0)$ and $\xi(\tau, v_1)$. Similarly, the approximate "relative acceleration" between the same two geodesics is furnished by $(v_1 - v_0)\frac{D^2\vec{\eta}}{\partial \tau^2}(\tau, v_0)$. We would like to investigate the "relative acceleration." Now, by (5.110), (5.111), and (5.173), we obtain for a twice-differentiable vector field $\vec{\mathbf{T}}(x)$ restricted to a smooth surface the following commutation relation:

$$
\begin{aligned}
&\frac{D^2 T^i}{\partial \tau \partial v}(\xi(\tau, v)) - \frac{D^2 T^i}{\partial v \partial \tau}(\xi(\tau, v)) \\
&\quad = R^i{}_{lkj}(\xi(\tau, v))T^\ell(\xi(\tau, v))\xi'^k(\tau, v)\eta^j(\tau, v).
\end{aligned}
\tag{5.227}
$$

By (5.224), (5.226), and (5.227) we can derive that

$$
\begin{aligned}
\frac{D^2\eta^i}{\partial \tau^2}(\tau, v) = \frac{D^2\xi'^i}{\partial \tau \partial v}(\tau, v) &= \frac{D^2\xi'^i}{\partial v \partial \tau}(\tau, v) + R^i{}_{lkj}(\xi(\tau, v))\xi'^\ell(\tau, v)\xi'^k(\tau, v)\eta^j(\tau, v) \\
&= R^i{}_{lkj}(\xi(\tau, v))\xi'^\ell(\tau, v)\xi'^k(\tau, v)\eta^j(\tau, v).
\end{aligned}
$$

The equation above is *exact*. We have essentially proved the following theorem due to Synge. (See Synge & Schild [37].)

Theorem 5.3.10 *Let $D \subset \mathbb{R}^N$ correspond to an open subset of a C^3-differentiable manifold with a metric. Let $\xi \in C^3(\mathcal{D} \subset \mathbb{R}^2; \mathbb{R}^N)$ be a two-dimensional, non-degenerate parametric surface into $\mathbb{R}^N$. Let the parametric surface ξ be generated by the one-parameter geodesic congruence $x = \xi(\tau, v)$ (with $\frac{D\xi^i(\tau,v)}{\partial\tau} = 0$). Then, the geodesic deviation vector components $\eta^i(\tau, v) := \frac{\partial\xi^i}{\partial v}(\tau, v)$ must satisfy*

$$\frac{D^2\eta^i(\tau, v)}{(\partial\tau)^2}(\xi(\tau, v)) + R^i{}_{ljk}(\xi(\tau, v))\xi'^{\ell}(\tau, v)\eta^j(\tau, v)\xi'^k(\tau, v) = 0, \qquad (5.228)$$

$$\xi'^c(\tau, v)\left[\nabla_c(\xi'^b\nabla_b\eta^a)\right] + R^a{}_{bcd}(\xi(\tau, v))\xi'^b(\tau, v)\eta^c(\tau, v)\xi'^d(\tau, v) = 0. \qquad (5.229)$$

Remark: The equations (5.228) and (5.229) are called the **geodesic deviation equations**.

Example 5.3.11 Consider a locally flat manifold and a Cartesian, or a pseudo-Cartesian, chart with $R^i{}_{jkl}(x) \equiv 0$. By (5.221), a one-parameter family of straight lines (or geodesics) spanning a **ruled surface** is provided by

$$\begin{aligned}
&x^i = \xi^i(\tau, v) = \tau F^i(v) + G^i(v),\\
&\sum_{i=1}^{N}\left[F^i(v)\right]^2 > 0,\ d_{ij}F^i(v)F^j(v) \equiv \text{const.},\\
&\xi'^i(\tau, v) = F^i(v), \eta^i(\tau, v) = \tau F'^i(v) + G'^i(v),\\
&\frac{D\eta^i}{\partial\tau}(\tau, v) = F'^i(v),\ \frac{D^2\eta^i}{\partial\tau^2}(\tau, v) \equiv 0.
\end{aligned} \qquad (5.230)$$

Thus, the geodesic deviation equations (5.228) are identically satisfied. □

Example 5.3.12 Consider the surface S^2 of the unit sphere embedded in $\mathbb{R}^3$. (See examples 4.3.12, 4.3.13, 5.1.8, and 5.2.19.)

The metric tensor, the non-zero Christoffel symbols, and the non-zero Riemann tensor components are furnished by

$$\begin{aligned}
\mathbf{g}_{\cdot\cdot}(x) &= dx^1 \otimes dx^1 + (\sin x^1)^2 dx^2 \otimes dx^2\\
&\equiv d\theta \otimes d\theta + (\sin\theta)^2 d\phi \otimes d\phi,\\
\left\{ {1 \atop 22} \right\} &= -\sin\theta\cos\theta, \left\{ {2 \atop 12} \right\} = \left\{ {2 \atop 21} \right\} = \cot\theta,\\
R^1{}_{212} &= -R^1{}_{221} = \sin^2\theta,\\
R^2{}_{112} &= -R^2{}_{121} = -1.
\end{aligned}$$

Consider the one-parameter family of geodesic congruence (or great circles) provided by the longitudes as

$$\begin{aligned}
\xi(\theta,\phi_0) &:= (\theta,\phi_0), && (\theta,\phi)\in(0,\pi)\times(-\pi,\pi)\subset\mathbb{R}^2;\\
\xi'^i(\theta,\phi) := \frac{\partial\xi^i}{\partial\theta}(\theta,\phi) &= \delta^i{}_{(1)}, && \eta^i(\theta,\phi) := \frac{\partial\xi^i(\theta,\phi)}{\partial\phi} = \delta^i{}_{(2)},\\
g_{ij}(\xi(\theta,\phi))\xi'^i(\theta,\phi)\xi'^j(\theta,\phi) &\equiv 1, && g_{ij}(\xi(\theta,\phi))\eta^i(\theta,\phi)\eta^j(\theta,\phi) = \sin^2\theta,\\
g_{ij}(\xi(\theta,\phi))\xi'^i(\theta,\phi)\eta^j(\theta,\phi) &\equiv 0.
\end{aligned}$$

(See fig. 5.4.) Obviously, θ is the arc length parameter along a longitude.

The geodesic conditions (5.224) can be checked explicitly as

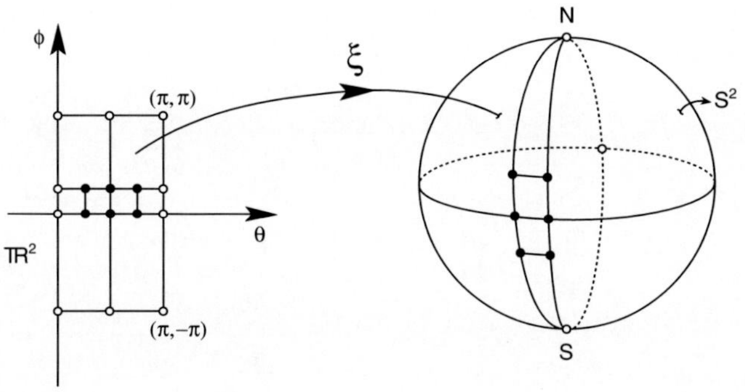

Figure 5.4: Geodesic deviation between two neighboring longitudes.

$$\begin{aligned}
\frac{D\xi'^1}{\partial\theta}(\theta,\phi) &= 0 + \left\{ \begin{array}{c} 1 \\ 22 \end{array} \right\} \left[\delta^2{}_{(1)}\right]^2 \equiv 0,\\
\frac{D\xi'^2}{\partial\theta}(\theta,\phi) &= 0 + 2\left\{ \begin{array}{c} 2 \\ 12 \end{array} \right\} \left[\delta^2{}_{(1)}\right]^2 \equiv 0.
\end{aligned}$$

The geodesic deviation equations (5.228) reduce to

$$\begin{aligned}
\frac{D^2\delta^i{}_{(2)}}{\partial\theta^2} + R^i{}_{121} &= \frac{D}{\partial\theta}\left[\left\{ \begin{array}{c} i \\ 12 \end{array} \right\}\right] + R^i{}_{121} = \nabla_1\left[(\cot\theta)\delta^i{}_{(2)}\right] + R^i{}_{121}\\
&= \left[\partial_1(\cot\theta)\right]\delta^i{}_{(2)} + \left\{ \begin{array}{c} i \\ 1k \end{array} \right\} \delta^k{}_{(2)}\cot\theta + R^2{}_{121}\delta^i{}_{(2)}\\
&= \delta^i{}_{(2)}\left[-\operatorname{cosec}^2\theta + \cot^2\theta + 1\right] \equiv 0.
\end{aligned}$$

Thus, the geodesic deviation equations are identically satisfied by $\eta^i(\theta, \phi)$. In this example,

$$\|\vec{\boldsymbol{\eta}}(\theta,\phi)\| := \sigma(\vec{\boldsymbol{\eta}}(\theta,\phi)) = \sin\theta, \quad \lim_{\theta\to 0_+} \|\vec{\boldsymbol{\eta}}(\theta,\phi)\| = \lim_{\theta\to\pi_-} \|\vec{\boldsymbol{\eta}}(\theta,\phi)\| = 0.$$

(Longitudinal geodesics *do meet* at the north and south poles, which are not covered by the coordinate chart.) □

The solution $\eta^i(\tau, v_0)$ of (5.228) along the geodesic $\xi^i(\tau, v_0)$ is also called a **Jacobi field** (from the viewpoint of the second variation). (See the book by Hawking and Ellis [18].) If the Jacobi field $\vec{\eta}(\tau, v_0)$ is not an identically zero vector field but vanishes at two points ($\eta^i(\tau_1, v_0) = \eta^i(\tau_2, v_0) = 0$), the points $\xi(\tau_1, v_0)$ and $\xi(\tau_2, v_0)$ are called **conjugate points** on the geodesic $\xi(\tau, v_0)$. In example 5.3.12, the north and south poles are conjugate points on every longitudinal geodesic on S^2.

Now we shall discuss briefly **geodesic completeness** of a differentiable manifold. Recall that the **completeness** of the set of real numbers $\mathbb{R}$ is one of the corresponding axioms. It is equivalent to the condition that every **Cauchy sequence** converges. (Completeness is also one of the axioms for the Hilbert space that is applicable in quantum mechanics.) Consider a Riemannian manifold (endowed with a positive-definite metric). In a coordinate chart, the distance between two points q_1 and q_2 in M along the image of a curve γ is given by (5.215) as

$$\begin{aligned} l(\gamma) := \Sigma(\mathcal{X}) &= \int_{t_1}^{t_2} \sqrt{g_{ij}(\mathcal{X}(t))\frac{d\mathcal{X}^i(t)}{dt}\frac{d\mathcal{X}^j(t)}{dt}}\, dt, \\ x_1 = \mathcal{X}(t_1),&\ \ x_2 = \mathcal{X}(t_2), \\ q_1 = \chi^{-1}(x_1),&\ \ q_2 = \chi^{-1}(x_2). \end{aligned} \tag{5.231}$$

We define the **distance function** $\rho(q_2, q_1)$ by the *greatest of lower bounds* for lengths of all C^1-curves joining q_1 and q_2. The distance function $\rho(q_2, q_1)$ satisfies the axioms of a metric space. Consider a sequence of points $\{q_n\}_1^\infty$ in M. This sequence is Cauchy provided that for every $\varepsilon > 0$ there exists an integer $N(\varepsilon) > 0$ such that $\rho(q_m, q_n) < \varepsilon$ for all $m, n > N(\varepsilon)$. In the case where every Cauchy sequence in M converges, the manifold M is called **complete**. As an example, consider the Euclidean plane $\mathbb{E}_2$ and a Cartesian chart. We choose the usual distance function

$$d(q_1, q_2) = d\left[\chi^{-1}(x_1), \chi^{-1}(x_2)\right] := \sqrt{(x_1^1 - x_2^1)^2 + (x_1^2 - x_2^2)^2}.$$

By the completeness of real numbers, we can prove that every Cauchy sequence $\{x_n\}_1^\infty$ corresponding to $\{q_n\}_1^\infty$ converges. Thus, the space $\mathbb{E}_2$ is complete. Now, let us *excise* one point (corresponding to the origin) from $\mathbb{E}_2$. Consider the sequence $\{q_n\}_1^\infty$ corresponding to $\{x_n\}_1^\infty := \{(0, 1/n)\}_1^\infty$. It is a Cauchy sequence that does *not* converge in $\mathbb{R}^2 - \{(0,0)\}$. Therefore, such a differentiable manifold is *not* complete. Moreover, in this case, two points, (a, a) and $(-a, -a)$, in $\mathbb{R}^2$ *cannot* be joined by a straight line (or a geodesic) lying entirely in $\mathbb{R}^2 - \{(0,0)\}$. (See fig. 5.5.)

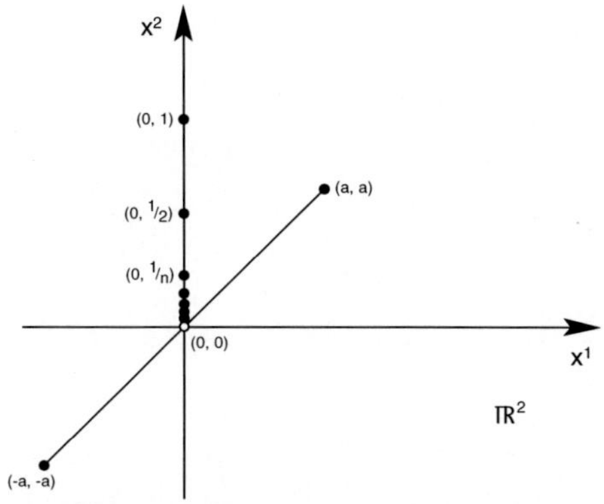

Figure 5.5: An incomplete manifold $\mathbb{R}^2 - \{(0,0)\}$.

We can use for the positive-definite metric either of the following two equivalent criteria for the completeness.

(i) Every Cauchy sequence of points $\{q_n\}_1^\infty$ in M is convergent.

(ii) Every geodesic in M can be extended for all real values of an affine parameter.

The second criterion is called the **geodesic completeness** for a Riemannian manifold. In a pseudo-Riemannian manifold, the concept of geodesic completeness is more subtle because of the existence of timelike, spacelike, and null geodesics. Nevertheless, such concepts are useful in the study of **singularity theorems** in general relativity.

Another relevant concept associated with geodesics is the idea of a **world function** Ω (introduced by Synge [38]). Let a twice-differentiable geodesic $\mathcal{X}$ in a coordinate chart be expressed in terms of an affine parameter τ as

$$x = \mathcal{X}(\tau) \in D \subset \mathbb{R}^N,$$
$$\tau \in [\tau_1, \tau_2] \subset \mathbb{R}.$$

The world function Ω is defined by

$$\Omega(\mathcal{X}(\tau_2), \mathcal{X}(\tau_1)) := (1/2)(\tau_2 - \tau_1) \int_{\tau_1}^{\tau_2} g_{ij}(\mathcal{X}(\tau)) \frac{d\mathcal{X}^i(\tau)}{d\tau} \frac{d\mathcal{X}^j(\tau)}{d\tau} d\tau. \qquad (5.232)$$

Since $g_{ij}(\mathcal{X}(\tau)) \frac{d\mathcal{X}^i(\tau)}{d\tau} \frac{d\mathcal{X}^j(\tau)}{d\tau}$ is *constant-valued* along the geodesic, the world function Ω in (5.232) can also be expressed as

$$\begin{aligned} \Omega(\mathcal{X}(\tau_2), \mathcal{X}(\tau_1)) &= (1/2)(\tau_2 - \tau_1)^2 g_{ij}(\mathcal{X}(\tau)) \frac{d\mathcal{X}^i(\tau)}{d\tau} \frac{d\mathcal{X}^j(\tau)}{d\tau}, \\ \tau &\in [\tau_1, \tau_2]. \end{aligned} \qquad (5.233)$$

Caution: The function Ω above is *not* to be confused with the curvature 2-form $\mathbf{\Omega}$...

Using (5.174), (5.215), (5.232), and (5.233), we obtain

$$\begin{aligned} \Omega(x_2, x_1) = \Omega(\mathcal{X}(\tau_2), \mathcal{X}(\tau_1)) &= (1/2) \left[\frac{d\mathcal{S}(\tau)}{d\tau} (\tau_2 - \tau_1) \right]^2 \\ &= (1/2) \left[\Sigma(\mathcal{X}) \right]^2. \end{aligned} \qquad (5.234)$$

Here, $\Sigma(\mathcal{X})$ is the arc separation along the (unique) geodesic joining the points $x_1 = \mathcal{X}(\tau_1)$ and $x_2 = \mathcal{X}(\tau_2)$.

Example 5.3.13 Let M be an N-dimensional flat manifold. In a pseudo-Cartesian chart, the metric tensor is expressed as

$$\mathbf{g}_{..}(x) = d_{ij} dx^i \otimes dx^j.$$

Equations (5.222) and (5.234) yield

$$\begin{aligned} \Omega(x_2, x_1) &= (1/2) d_{ij} (x_2^i - x_1^i)(x_2^j - x_1^j), \\ \frac{\partial \Omega(x_2, x_1)}{\partial x_2^i} &= d_{ij}(x_2^j - x_1^j), \\ \frac{\partial \Omega(x_2, x_1)}{\partial x_1^i} &= -d_{ij}(x_2^j - x_1^j), \\ \lim_{x_1 \to x_2} \Omega(x_2, x_1) &= 0, \\ \lim_{x_1 \to x_2} \frac{\partial \Omega(x_2, x_1)}{\partial x_2^i} &= \lim_{x_1 \to x_2} \frac{\partial \Omega(x_2, x_1)}{\partial x_1^i} = 0. \end{aligned}$$

□

Exercises 5.3

1. Consider S^2, the surface of a unit sphere and the spherical polar coordinate chart characterized by

$$\mathbf{g}..(\theta, \phi) = d\theta \otimes d\theta + \sin^2\theta \, d\phi \otimes d\phi.$$

Integrate the corresponding (non-degenerate) geodesic equations to obtain

$$c_1 \sin\theta \cos\phi + c_2 \sin\theta \sin\phi + c_3 \cos\theta = 0,$$

where the constants satisfy $c_1^2 + c_2^2 + c_3^2 > 0$.

2. Consider a three-dimensional pseudo-Riemannian manifold and the metric field

$$\mathbf{g}..(x) = \left[f(x^3)\right]^2 \left[dx^1 \otimes dx^1 + dx^2 \otimes dx^2\right] - dx^3 \otimes dx^3.$$

Here $f \in C^1(D \subset \mathbb{R}; \mathbb{R})$ and $f(x^3) \neq 0$. Integrate the equations for a non-degenerate null geodesic.

3. Let a four-dimensional metric field be furnished by

$$\mathbf{g}..(x) = \delta_{\alpha\beta} dx^\alpha \otimes dx^\beta - (1 + x^1)^2 dx^4 \otimes dx^4.$$

Show that the solutions of timelike geodesic equations with initial conditions $\mathcal{X}^\alpha(0) = 0$, $\frac{d\mathcal{X}^\alpha}{dx^4}\big|_{x^4=0} = 0$, are provided by

$$\mathcal{X}^1(x^4) = \text{ sech } (x^4) - 1, \quad \mathcal{X}^2(x^4) = \mathcal{X}^3(x^4) \equiv 0.$$

4. Prove that the arc separation function

$$\Sigma(\mathcal{X}) := \int_{t_2}^{t_2} \sqrt{\left| g_{ij}(\mathcal{X}(t)) \frac{d\mathcal{X}^i(t)}{dt} \frac{d\mathcal{X}^j(t)}{dt} \right|} dt$$

for a differentiable curve $\mathcal{X}$ into $\mathbb{R}^N$ is *invariant* under a reparametrization.

5. Consider the Euclidean space $\mathbb{E}_3$ and a Cartesian coordinate chart. Let $\mathcal{X}$ be a thrice-differentiatle curve from the arc length parameter s into $\mathbb{R}^3$. Assuming $\kappa(s) \equiv \kappa_{(1)}(s) > 0$, prove that $\mathcal{X}$ is a plane curve if and only if $\tau(s) \equiv \kappa_{(2)}(s) \equiv 0$.

6. Consider a flat four-dimensional pseudo-Riemannian manifold and a pseudo-Cartesian (or Minkowskian) coordinate chart. A timelike helix satisfies the following Frenet-Serret equations:

$$\begin{aligned}
\frac{d\lambda^i_{(0)}(s)}{ds} &= c_{(1)}\lambda^i_{(1)}(s), \\
\frac{d\lambda^i_{(1)}(s)}{ds} &= c_{(2)}\lambda^i_{(2)}(s) + c_{(1)}\lambda^i_{(0)}(s), \\
\frac{d\lambda^i_{(2)}(s)}{ds} &= -c_{(2)}\lambda^i_{(1)}(s), \\
\frac{d\lambda^i_{(3)}(s)}{ds} &\equiv 0.
\end{aligned}$$

Here, the constants satisfy $c_{(1)} > 0$ and $c_{(2)} \neq 0$. Prove that the curve (with $a > 0,\ s > 0$) defined by

$$\mathcal{X}(s) := (\sin[(\sinh a)s], \cos[(\sinh a)s], 0, (\cosh a)s)$$

is a timelike helix.

7. Recall the symmetrized Riemann tensor $\mathbf{S}....(x)$ in problem 5 of exercises 5.2. Prove that the geodesic deviation equation (5.228) can be cast into the form

$$\frac{D^2\eta^i(\tau,v)}{\partial\tau^2} + 3S^i{}_{klj}\left(\xi(\tau,v)\right)\xi'^k(\tau,v)\xi'^l(\tau,v)\eta^j(\tau,v) = 0.$$

8. Investigate the spacelike geodesics in an **Anti-De Sitter space** characterized by

$$\begin{aligned}
\mathbf{g}..(x) &= e^{-2t}\left[\delta_{\alpha\beta}dx^\alpha \otimes dx^\beta\right] - dt \otimes dt, \\
&\alpha, \beta \in \{1, 2, \cdots, N-1\}.
\end{aligned}$$

(i) Deduce that the first integrals

$$\begin{aligned}
\exp[-2T(s)]\delta_{\alpha\beta}\frac{d\mathcal{X}^\beta(s)}{ds} &= k_\alpha = \text{const.}, \\
\exp[2T(s)](\delta^{\mu\nu}k_\mu k_\nu) - \left[\frac{dT(s)}{ds}\right]^2 &= 1,
\end{aligned}$$

exist locally.

(ii) Prove that the world function is furnished by

$$\begin{aligned}
\Omega(x, x_0) = (1/2)&\left[\text{Arctan}\sqrt{e^{2(t-t_0)} - 1}\right]^2 \times \\
&\left\{\left[e^{2(t-t_0)}/(e^{2t} - e^{2t_0})\right]\delta_{\mu\nu}(x^\mu - x^\mu_0)(x^\nu - x^\nu_0) - e^{2(t-t_0)} + 1\right\}, \\
-\frac{\pi}{2} &< \text{Arctan}\, x < \frac{\pi}{2}, \\
0 < s_0 &< s < \frac{\pi}{2} + s_0.
\end{aligned}$$

5.4 Special Coordinate Charts

Consider a particular point $p_0 \in U \subset M$. In a coordinate chart (χ, U), the corresponding image is $x_0 = \chi(p_0) \in D \subset \mathbb{R}^N$. The metric field at that point is given by

$$\mathbf{g}_{..}(x_0) = g_{ij}(x_0)dx^i \otimes dx^j =: s_{ij}dx^i \otimes dx^j. \tag{5.235}$$

Here, the $N \times N$ matrix $[S] := [s_{ij}]$ is *symmetric* with p positive eigenvalues and n negative eigenvalues. Let us consider a coordinate transformation by (3.2) and the tensor transformation rules (3.48) to obtain

$$\begin{aligned} \widehat{s}_{ij} &:= \widehat{g}_{ij}(\widehat{x}_0) = \left[\frac{\partial X^k(\widehat{x})}{\partial \widehat{x}^i}\frac{\partial X^l(\widehat{x})}{\partial \widehat{x}^j}\right]_{|\widehat{x}_0} g_{kl}(x_0) =: \lambda^k{}_i \lambda^l{}_j s_{kl}, \\ [\widehat{S}] &= [\lambda]^T[S][\lambda]. \end{aligned} \tag{5.236}$$

In linear algebra, there is a theorem asserting the existence of an $N \times N$ matrix $[\lambda]$ such that

$$[\widehat{S}] = [D] = [d_{ij}]. \tag{5.237}$$

(See the book by Hoffman and Kunze [21].) Therefore, we seek a coordinate transformation satisfying first-order differential equations

$$\begin{aligned} \frac{\partial \widehat{X}^k(x_0)}{\partial x^i} &= \mu^k{}_i, \\ [\mu] &:= [\lambda]^{-1}. \end{aligned} \tag{5.238}$$

Thus, we can have coordinates so that at a particular point $\widehat{x}_0$,

$$\widehat{g}_{ij}(\widehat{x}_0)d\widehat{x}^i \otimes d\widehat{x}^j = d_{ij}d\widehat{x}^i \otimes d\widehat{x}^j.$$

Example 5.4.1 Consider the metric tensor field in a two-dimensional manifold given by

$$\mathbf{g}_{..}(x) = e^{-x^2}dx^1 \otimes dx^1 - dx^2 \otimes dx^2.$$

At the particular point $x_0 = (0, 1) \in D \subset \mathbb{R}^2$, we have

$$\begin{aligned} \mathbf{g}_{..}(x_0) &\equiv \mathbf{g}_{..}(0,1) = e^{-1}dx^1 \otimes dx^1 - dx^2 \otimes dx^2, \\ [S] &\equiv [s_{ij}] = \begin{bmatrix} e^{-1}, & 0 \\ 0, & -1 \end{bmatrix}, \\ [\lambda^i{}_j] &= \begin{bmatrix} e^{1/2}, & 0 \\ 0, & 1 \end{bmatrix}, \quad [\mu^i{}_j] = \begin{bmatrix} e^{-1/2}, & 0 \\ 0, & 1 \end{bmatrix}. \end{aligned}$$

Therefore, the differential equations (5.238) reduce to

$$\frac{\partial \widehat{X}^1(x)}{\partial x^1}_{|(0,1)} = e^{-1/2}, \quad \frac{\partial \widehat{X}^1(x)}{\partial x^2}_{|(0,1)} = 0,$$

$$\frac{\partial \widehat{X}^2(x)}{\partial x^1}_{|(0,1)} = 0, \quad \frac{\partial \widehat{X}^2(x)}{\partial x^2}_{|(0,1)} = 1.$$

A general class of solutions is furnished by

$$\begin{aligned} \widehat{X}^1(x) &= e^{-1/2}x^1 + (x^1)^2 f(x), \\ \widehat{X}^2(x) &= (1/2)(x^2)^2 + (x^2-1)^2 g(x). \end{aligned}$$

Here, f and g are *arbitrary* functions of class C^1. □

Remarks: Note that in the construction above, simplification of the metric tensor occurs at a *single point.* Moreover, the transformation is *not* unique.

We would now like to simplify the metric tensor and also the connection coefficients at a point. For that purpose, we introduce the **exponential mapping**. Consider the geodesic equations (5.200) in terms of an affine parameter τ. The equivalent first-order system of ordinary differential equations is provided by

$$\begin{aligned} \frac{d\mathcal{X}^i(\tau)}{d\tau} &= V^i(\tau), \\ \frac{dV^i(\tau)}{d\tau} &= -\left\{ \begin{matrix} i \\ jk \end{matrix} \right\}_{|\mathcal{X}(\tau)} V^j(\tau)V^k(\tau). \end{aligned} \tag{5.239}$$

The system of equations above possesses unique solutions to an initial-value problem provided the coefficients $\left\{ \begin{smallmatrix} i \\ jk \end{smallmatrix} \right\}$ satisfy **Lipschitz conditions**. Let the initial values be presented by and the corresponding geodesic be furnished by

$$\begin{aligned} &\mathcal{X}^i(0) = x^i{}_0, \quad \frac{d\mathcal{X}^i(\tau)}{d\tau}_{|\tau=0} = W^i, \\ &x^i = \mathcal{X}^i{}_W(\tau), \quad \tau \in [-\delta, \delta]; \\ &\mathcal{X}_W : [-\delta, \delta] \longrightarrow \mathbb{R}^N, \\ &\mathcal{X}_W(0) = x_0, \quad \vec{\mathcal{X}}'{}_W(0) = \vec{\mathbf{W}}_{x_0} = W^i \frac{\partial}{\partial x^i}_{|x_0} \in T_{x_0}(\mathbb{R}^N). \end{aligned} \tag{5.240}$$

The geodesic above may or may not be extendible for the interval $\tau \in [-1, 1]$. In a geodesically complete manifold, every geodesic passing through

x_0 can be extended for all $\tau \in \mathbb{R}$. In the present context, we assume that the unique geodesic in (5.240) exists for the whole interval $\tau \in [-1, 1]$. We define the exponential mapping $\exp_{x_0} : T_{x_0}(\mathbb{R}^N) \to \mathbb{R}^N$ by the rule $\exp_{x_0}[\vec{\mathbf{W}}_{x_0}] := \mathcal{X}_W(1)$. (See fig. 5.6.)

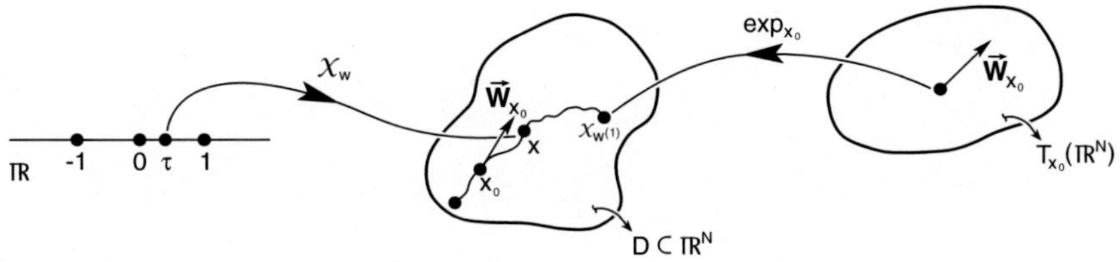

Figure 5.6: The exponential mapping.

Now, τ appearing in (5.240) is an affine parameter. Therefore, by (4.46), $x = \mathcal{X}_W(\kappa\tau)$ (with $\kappa \neq 0$) yields another geodesic having a common segment with the original geodesic $x = \mathcal{X}_W(\tau)$. The new geodesic has the initial values $\mathcal{X}_W(\kappa\tau)_{|\tau=0} = x_0$ and $\frac{d\mathcal{X}^i{}_W(\kappa\tau)}{d\tau}_{|\tau=0} = \left[\frac{d(\kappa\tau)}{d\tau}\frac{d\mathcal{X}^i{}_W(\kappa\tau)}{d(\kappa\tau)}\right]_{|\tau=0} = \kappa W^i$. Thus, we conclude that

$$\begin{aligned}
&\mathcal{X}_W(\kappa\tau) = \mathcal{X}_{\kappa W}(\tau),\\
&\exp_{x_0}\left[\kappa\vec{\mathbf{W}}_{x_0}\right] = \mathcal{X}_{\kappa W}(1) = \mathcal{X}_W(\kappa),\\
&\exp_{x_0}\left[\tau\vec{\mathbf{W}}_{x_0}\right] = \mathcal{X}_W(\tau).
\end{aligned} \tag{5.241}$$

Example 5.4.2 We choose a flat N-dimensional manifold with

$$\begin{aligned}
&\mathbf{g}_{..}(x) = d_{ij}dx^i \otimes dx^j,\\
&\left\{ {k \atop ij} \right\} \equiv 0.
\end{aligned}$$

The geodesic equations (5.239) yield

$$\begin{aligned}
&V^i(\tau) = A^i,\\
&\mathcal{X}^i(\tau) = A^i\tau + B^i,\\
&\tau \in \mathbb{R}.
\end{aligned}$$

Here, A^i's and B^i's are arbitrary constants of integration (with $\sum_{i=1}^{N} |A^i| > 0$).

The initial values can be incorporated by

$$x^i{}_0 \equiv B^i, \quad \frac{d\mathcal{X}^i(\tau)}{d\tau}_{|\tau=0} \equiv A^i,$$

$$\vec{\mathcal{X}}'(0) = A^i \frac{\partial}{\partial x^i{}_0} = \vec{\mathbf{W}}_{x0}.$$

Therefore, (5.240) yields

$$\mathcal{X}^i{}_W(\tau) = A^i\tau + B^i = W^i\tau + x^i{}_0,$$
$$\pi^i \circ \exp_{x_0}\left[\vec{\mathbf{W}}_{x0}\right] = \mathcal{X}^i{}_W(1) = W^i + x^i{}_0,$$
$$\pi^i \circ \exp_{x_0}\left[\tau\vec{\mathbf{W}}_{x0}\right] = \tau W^i + x^i{}_0 = \mathcal{X}^i{}_W(\tau). \qquad \square$$

We have already tacitly assumed for the geodesic that $\mathcal{X}_W \in C^2([-1,1] \subset \mathbb{R}; \mathbb{R}^N)$. Let us make a much *stronger* assumption that $\mathcal{X}^i = \pi^i \circ \mathcal{X}$ are real-analytic functions. In that case, we obtain the Taylor series

$$\begin{aligned} x^i &= \pi^i \circ \exp\left[\tau\vec{\mathbf{W}}_{x0}\right] = \mathcal{X}^i{}_W(\tau) \\ &= \sum_{n=0}^{\infty}(1/n!)\left[\frac{d^n\mathcal{X}^i{}_W(\tau)}{d\tau^n}\right]_{|\tau=0}\tau^n \\ &= x^i{}_0 + \tau W^i - (1/2)\tau^2\left\{\begin{matrix} i \\ jk \end{matrix}\right\}_{|x_0} W^jW^k + \tau^3 c^i{}_{jkl}W^jW^kW^l + \ldots; \\ & |\tau| < 1, \quad 0 < \delta_{ij}W^iW^j < r_0^2. \end{aligned} \tag{5.242}$$

By the repeated use of the geodesic equations, each term in the series above is of the type $(\tau)^{j_i+..+j_l}c^i{}_{j_i+..+j_l}W^{j_i}..W^{j_l}$. We now introduce a coordinate transformation in a convex neighborhood of x_0 by *imitating* the series in (5.242). The transformation is given by

$$\begin{aligned} x^i &= X^i(\widehat{x}) := x^i{}_0 + \widehat{x}^i - (1/2)\left\{\begin{matrix} i \\ jk \end{matrix}\right\}_{|x_0}\widehat{x}^j\widehat{x}^k + c^i{}_{jkl}\widehat{x}^j\widehat{x}^k\widehat{x}^l + \ldots, \\ \frac{\partial X^i(\widehat{x})}{\partial\widehat{x}^j} &= \delta^i{}_j - \left\{\begin{matrix} i \\ jk \end{matrix}\right\}_{|x_0}\widehat{x}^k + O(\widehat{x}^2), \\ \det&\left[\frac{\partial X^i(\widehat{x})}{\partial\widehat{x}^j}_{|\widehat{x}=0}\right] = 1. \end{aligned} \tag{5.243}$$

We conclude from (5.243) that the transformation is one-to-one in the neighborhood of the origin $\widehat{x}_0 \equiv (0,\ldots,0)$. Therefore, by (5.243), we get

$$X(\widehat{x}_0) := X(0,\ldots,0) = x_0. \tag{5.244}$$

The hatted coordinate system is called the **Riemann normal coordinate chart**.

We notice that the curve

$$\widehat{x} = \widehat{\mathcal{X}}(\tau) := W^i \tau \tag{5.245}$$

has for the pre-image (by (5.242)) the geodesic

$$x = \mathcal{X}_W(\tau).$$

Since the geodesic equation (5.239) is a vector equation, the coordinate transformation (5.243) yields the corresponding geodesic equation (5.245) in the Riemann normal coordinates. However, this geodesic is a straight line passing through the origin $\widehat{x}_0 = (0, \ldots, 0)$. Moreover, on every point of this geodesic, we have

$$\begin{aligned} &\frac{d^2 \widehat{\mathcal{X}}^i(\tau)}{d\tau^2} \equiv 0, \\ &\widehat{\left\{ \begin{matrix} i \\ jk \end{matrix} \right\}}_{|\hat{x}(\tau)} \frac{d\widehat{\mathcal{X}}^j(\tau)}{d\tau} \frac{d\widehat{\mathcal{X}}^k(\tau)}{d\tau} \equiv 0, \\ &\widehat{\left\{ \begin{matrix} i \\ jk \end{matrix} \right\}}_{|(0,\ldots,0)} W^j W^k = 0. \end{aligned} \tag{5.246}$$

At a point $\widehat{x} = \widehat{\mathcal{X}}(\tau)$ away from the origin, other geodesics (not passing through the origin) will intersect the particular geodesic in (5.245). For *other* geodesics, equations $\frac{d^2 \widehat{\mathcal{X}}^i(\tau)}{d\tau^2} = 0$ and $\widehat{\left\{ \begin{smallmatrix} i \\ jk \end{smallmatrix} \right\}}_{|X(\tau)} \frac{d\widehat{\mathcal{X}}^j(\tau)}{d\tau} \frac{d\widehat{\mathcal{X}}^k(\tau)}{d\tau} = 0$ may *not* hold. However, *every* straight-line geodesic in (5.245) *intersecting at the origin* must satisfy

$$\begin{aligned} &\widehat{\left\{ \begin{matrix} i \\ jk \end{matrix} \right\}}_{|(0,\ldots,0)} W^j W^k = 0, \\ &\left\{ (W^1, \ldots, W^N) \in \mathbb{R}^N; \quad 0 < \delta_{ij} W^i W^j < r_0^2 \right\}. \end{aligned} \tag{5.247}$$

Therefore, we obtain

$$\begin{aligned} (1/2) \frac{\partial^2}{\partial W^j \partial W^k} \left[\widehat{\left\{ \begin{matrix} i \\ lm \end{matrix} \right\}}_{|(0,\ldots,0)} W^l W^m \right] = \widehat{\left\{ \begin{matrix} i \\ jk \end{matrix} \right\}}_{|(0,\ldots,0)} = 0, \\ \widehat{[jk, i]}_{|(0,\ldots,0)} = 0, \\ \frac{\partial \widehat{g}_{jk}(\widehat{x})}{\partial \widehat{x}^i}_{|(0,0\ldots,0)} = 0. \end{aligned} \tag{5.248}$$

Thus, at the origin of the Riemann normal coordinates, we can replace covariant derivatives by partial derivatives. Moreover, by (5.93) and (5.95), we have

$$
\begin{aligned}
\widehat{R}^i{}_{jkl}(0,\ldots,0) &= \left[\partial_k \left\{ \begin{matrix} i \\ lj \end{matrix} \right\} - \partial_l \left\{ \begin{matrix} i \\ kj \end{matrix} \right\}\right]_{|(0,\ldots,0)}, \\
\widehat{R}_{ijkl}(0,\ldots,0) &= \frac{1}{2}\left[\frac{\partial^2 \widehat{g}_{il}(\widehat{x})}{\partial \widehat{x}^j \partial \widehat{x}^k} + \frac{\partial^2 \widehat{g}_{jk}(\widehat{x})}{\partial \widehat{x}^i \partial \widehat{x}^l} - \frac{\partial^2 \widehat{g}_{ik}(\widehat{x})}{\partial \widehat{x}^j \partial \widehat{x}^l} - \frac{\partial^2 \widehat{g}_{jl}(\widehat{x})}{\partial \widehat{x}^i \partial \widehat{x}^k}\right]_{|(0,\ldots,0)}.
\end{aligned}
\tag{5.249}
$$

The equations above can be utilized to prove the algebraic identities the and Bianchi's identities of the Riemann-Christoffel curvature tensor in (5.108), and (5.109), (5.100) – (5.107).

Example 5.4.3 Consider a metric field given by

$$
\begin{aligned}
&\mathbf{g}_{..}(x) = (x^1)^4 dx^1 \otimes dx^1 + (x^2)^4 dx^2 \otimes dx^2, \\
&D := \mathbb{R}^2 - \{(0,0)\}\,.
\end{aligned}
$$

The non-zero Christoffel symbols are furnished by

$$
\left\{ \begin{matrix} 1 \\ 11 \end{matrix} \right\} = \frac{2}{x^1}, \quad \left\{ \begin{matrix} 2 \\ 22 \end{matrix} \right\} = \frac{2}{x^2}.
$$

We choose the initial point $x_0 := (1,1)$. The unique geodesic with

$$
x_0 = \mathcal{X}_W(0) = (1,1), \quad \vec{\mathcal{X}}'_W(0) = \vec{\mathbf{W}}_{x_0} = W^i \frac{\partial}{\partial x^i}_{|(1,1)},
$$

is provided by

$$
\mathcal{X}^i{}_W(s) = \pi^i \circ \mathcal{X}_W(s) = (1 + 3W^i s)^{1/3}.
$$

The functions above are real-analytic and admit Taylor's expansions

$$
\begin{aligned}
\mathcal{X}^1{}_W(s) &= 1 + sW^1 - (sW^1)^2 + \ldots \\
&= 1 + sW^1 - \frac{s^2}{2}\left\{ \begin{matrix} 1 \\ 11 \end{matrix} \right\}_{(1,1)} (W^1)^2 + \ldots, \\
\mathcal{X}^2{}_W(s) &= 1 + sW^2 - (sW^2)^2 + \ldots \\
&= 1 + sW^2 - \frac{s^2}{2}\left\{ \begin{matrix} 2 \\ 22 \end{matrix} \right\}_{(1,1)} (W^2)^2 + \ldots, \\
|sW^i| &< \frac{1}{3}.
\end{aligned}
$$

The Riemann normal coordinates are provided by

$$
\begin{aligned}
x^i &= (1+3\widehat{x}^i)^{1/3}, \\
\widehat{x}^i &= (1/3)\left[(x^i)^3 - 1\right], \\
\widehat{x}_0 &= (0,0), \\
\widehat{\mathbf{g}}..(\widehat{x}) &= \delta_{ij} d\widehat{x}^i \otimes d\widehat{x}^j, \quad \widehat{D}_s := \mathbb{R}^2 - \left\{\left(-\frac{1}{3}, -\frac{1}{3}\right)\right\}, \\
\widehat{\left\{ \begin{matrix} i \\ jk \end{matrix} \right\}}_{|(0,0)} &\equiv 0. \qquad \square
\end{aligned}
$$

Let a function $f : D \subset \mathbb{R}^N$ be differentiable so that we can obtain a 1-form by (3.15),

$$df(x) = (\partial_j f)dx^j = (\partial_a f)\widehat{\mathbf{e}}^a(x). \tag{5.250}$$

We *assume* that

$$
\begin{aligned}
&df(x) \neq \tilde{\mathbf{o}}(x), \\
&\sum_{i=1}^{N} (\partial_i f)^2 > 0.
\end{aligned}
\tag{5.251}
$$

The separation of the 1-form can be defined from (2.137) as

$$\sigma(df) := \sqrt{|\mathbf{g}^{\cdot\cdot}(x)(df, df)|}. \tag{5.252}$$

For a Riemannian manifold, (5.251) and (5.252) yield that $\sigma(df) > 0$. However, for a pseudo-Riemannian manifold, $\mathbf{g}^{\cdot\cdot}\,(x)(df, df)$ can have positive, negative, or zero values. Now, consider an equation of the type

$$
\begin{aligned}
&f(x) \equiv f(x^1, \ldots, x^N) = c = \text{const.}, \\
&x \in D \subset \mathbb{R}^N.
\end{aligned}
\tag{5.253}
$$

An equation like (5.253) may have no solution, or infinitely many solutions constituting one or more $(N-1)$-dimensional hypersurfaces. For example, consider $\mathbb{E}_3$ and a Cartesian coordinate chart. An equation $f(x) := (x^1)^2 + (x^2)^2 + (x^3)^2 + 2 = 0$ has *no* solution. An equation $f(x) := x^1 + x^2 + x^3 = 0$ yields one two-dimensional plane passing through the origin.

Another example is given by $f(x) := (x^1 + x^2 - x^3)(x^1 - x^2 + x^3) = 0$. Solutions lie on two planes $x^1 + x^2 - x^3 = 0$ and $x^1 - x^2 + x^3 = 0$. Thus planes intersect on the line given by $x^1 = 0$ and $x^2 = x^3$. The solution $x_0 := (1, 1, 2)$ is on the plane $x^1 + x^2 - x^3 = 0$ but not on the other plane $x^1 - x^2 + x^3 = 0$. There is another example in $f(x) := (x^1 x^2 + x^3)^2 - (x^1 x^2)^2 - 2x^1 x^2 x^3 - (x^3)^2 \equiv 0$. This is an *identity* and is satisfied by *every* $(x^1, x^2, x^3) \in \mathbb{R}^3$. We *assume* that (5.252) yields only *one* $(N-1)$-dimensional **hypersurface** $D_{N-1} \subset D \subset \mathbb{R}^N$ and the point $x_0 \in D_{N-1}$. We classify the hypersurface D_{N-1} according to

$$\begin{aligned} \sigma(df)_{|x_0} &> 0 \longrightarrow \text{ spacelike or timelike,} \\ \sigma(df)_{|x_0} &= 0 \longrightarrow \text{ null.} \end{aligned} \tag{5.254}$$

A null hypersurface D_{N-1} can also be expressed as

$$\begin{aligned} g^{ij}(x)(\partial_i f)(\partial_j f) = 0, \\ d^{ab}(\partial_a f)(\partial_b f) = 0. \end{aligned} \tag{5.255}$$

Consider two differentiable non-null hypersurfaces given by $f(x) = c$ and $h(x) = k$. In the case where these hypersurfaces intersect orthogonally at x_0, we must have

$$\begin{aligned} \mathbf{g}^{\cdot\cdot}(x)(df, dh)_{|x_0} &= 0, \\ \left[g^{ij}(x)(\partial_i f)(\partial_j h)\right]_{|x_0} &= 0, \\ d^{ab}\left[(\partial_a f)(\partial_b h)\right]_{|x_0} &= 0. \end{aligned} \tag{5.256}$$

The first-order partial differential equation

$$g^{ij}(x)(\partial_i f)(\partial_j h) = 0$$

for a *prescribed* differentiable function f (with $\sigma(df) > 0$) locally admits $N-1$ independent solutions $h_1(x), \ldots, h_{N-1}(x)$. We make a coordinate transformation

$$\begin{aligned} \widehat{x}^\alpha &= \widehat{X}^\alpha(x) := h_\alpha(x), \\ \widehat{x}^N &= \widehat{X}^N(x) := f(x), \\ \alpha &\in \{1, \ldots, N-1\}. \end{aligned} \tag{5.257}$$

The transformed components of the contravariant metric tensor satisfy

$$\begin{aligned} \widehat{g}^{N\alpha}(\widehat{x}) &= \frac{\partial f(x)}{\partial x^i} \frac{\partial h_\alpha(x)}{\partial x^j} g^{ij}(x) = \mathbf{g}^{\cdot\cdot}(x)\,(df, dh_\alpha) = 0, \\ \left|\widehat{g}^{NN}(\widehat{x})\right| &= \left|\frac{\partial f(x)}{\partial x^i} \frac{\partial f(x)}{\partial x^j} g^{ij}(x)\right| = [\sigma(df)]^2 > 0. \end{aligned} \tag{5.258}$$

The equations above yield for the metric field

$$\begin{aligned} \widehat{\mathbf{g}}_{\cdot\cdot}(\widehat{x}) &= \widehat{g}_{\alpha\beta}(\widehat{x}) dx^\alpha \otimes dx^\beta + \widehat{g}_{NN}(\widehat{x}) d\widehat{x}^N \otimes d\widehat{x}^N, \\ \widehat{g}_{NN}(\widehat{x}) &\neq 0. \end{aligned} \tag{5.259}$$

Here, the coordinate $\widehat{x}^N$ is called a **normal** or **hypersurface orthogonal** coordinate.

The geometrical interpretation of these results is that the image of the coordinate curve for $\widehat{x}^N$ intersects *orthogonally* each of the other coordinate curves for $\widehat{x}^\alpha$ contained in a hypersurface D_{N-1}. (See fig. 5.7.)

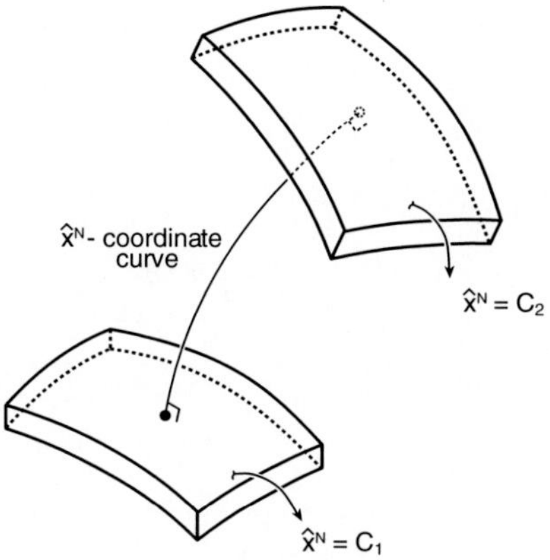

Figure 5.7: A normal coordinate $\widehat{x}^N$.

Example 5.4.4 Consider a two-dimensional differentiable manifold such that the conjugate metric field is furnished by

$$\mathbf{g}^{\cdot\cdot}(x) := \frac{\partial}{\partial x^1} \otimes \frac{\partial}{\partial x^1} + \left\{\exp\left[(x^1)^2 + x^2\right]\right\} \left(\frac{\partial}{\partial x^1} \otimes \frac{\partial}{\partial x^2} + \frac{\partial}{\partial x^2} \otimes \frac{\partial}{\partial x^1}\right)$$
$$+ \left[\exp(x^1)^2\right] \frac{\partial}{\partial x^2} \otimes \frac{\partial}{\partial x^2},$$
$$D := \left\{(x^1, x^2) \in \mathbb{R}^2 : (x^1)^2 + 2x^2 < 0\right\}.$$

Let a differentiable function be chosen as $f(x) := 2x^2$. We verify the condition

$$[\sigma(df)]^2 = 4[\exp(x^1)^2] > 0.$$

The partial differential equation in (5.256) reduces to

$$[\exp(x^2)]\partial_1 h + \partial_2 h = 0.$$

A particular solution is provided by

$$h(x^1, x^2) = x^1 - \exp(x^2).$$

We make a coordinate transformation

$$\widehat{x}^1 = x^1 - e^{x^2}, \quad \widehat{x}^2 = 2x^2.$$

The metric tensor components in the hatted coordinates are given by

$$\begin{aligned}\widehat{\mathbf{g}}..(\widehat{x}) &= \left\{1 - \widehat{x}^2 + \exp\left[\left(\widehat{x}^1 + e^{\widehat{x}^2/2}\right)^2\right]\right\}^{-1} d\widehat{x}^1 \otimes d\widehat{x}^1 \\ &\quad +(1/4)\exp\left[-\left(\widehat{x}^1 + e^{\widehat{x}^2/2}\right)^2\right] d\widehat{x}^2 \otimes d\widehat{x}^2.\end{aligned}$$

In the metric tensor above, the coordinate $\widehat{x}^2$ is a normal or hypersurface orthogonal. □

We now drop hats from a normal coordinate chart. Consider an $(N-1)$-dimensional hypersurface given by

$$D_{N-1} := \{x \in D \subset \mathbb{R}^N : x^N = c\} \tag{5.260}$$

in the normal coordinate chart. The metric field in (5.259) (with dropped hats) restricted to the hypersurface D_{N-1} can be expressed as

$$\begin{aligned}&\mathbf{g}..\,(x)_{|D_{N-1}} = g_{\mu\nu}(\mathbf{x}, c)dx^\mu \otimes dx^\nu =: \bar{g}_{\mu\nu}(\mathbf{x})dx^\mu \otimes dx^\nu, \\ &x \equiv (x^1, \ldots, x^{N-1}, x^N) =: (\mathbf{x}, x^N), \;\; \mathbf{x} := (x^1, \ldots, x^{N-1}), \\ &\mu, \nu \in \{1, \ldots, N-1\}.\end{aligned} \tag{5.261}$$

Caution: The bar does *not* indicate the complex conjugation!

By (5.41), (5.259), and (5.261), we derive that

$$\begin{aligned}&\left\{ \begin{matrix} \rho \\ \mu\nu \end{matrix} \right\}_{|x^N=c} = \overline{\left\{ \begin{matrix} \rho \\ \mu\nu \end{matrix} \right\}}, \\ &\left\{ \begin{matrix} \rho \\ N\sigma \end{matrix} \right\} = (1/2)g^{\rho\mu}\partial_N g_{\mu\sigma}, \quad \left\{ \begin{matrix} N \\ \rho\sigma \end{matrix} \right\} = -\left(1/2g_{NN}\right)\partial_N g_{\rho\sigma}, \\ &\left\{ \begin{matrix} N \\ N\rho \end{matrix} \right\} = \left(1/2g_{NN}\right)\partial_\rho g_{NN}, \quad \left\{ \begin{matrix} \rho \\ NN \end{matrix} \right\} = -(1/2)g^{\rho\mu}(x)\partial_\rho g_{NN}, \\ &\left\{ \begin{matrix} N \\ NN \end{matrix} \right\} = \left(1/2g_{NN}\right)\partial_N g_{NN}.\end{aligned} \tag{5.262}$$

Consider components $T^{i_1 \ldots i_r}{}_{j_1,\ldots,j_s}(x)$ of an $(r+s)$th-order tensor field relative to a *normal coordinate* chart. We introduce a coordinate transformation to another normal coordinate chart by

$$
\begin{aligned}
&x^{\#_\alpha} = X^{\#_\alpha}(x^1, x^2, \ldots, x^{N-1}) \equiv X^{\#_\alpha}(\mathbf{x}),\\
&x^{\#_N} = X^{\#_N}(x) := x^N,\\
&\frac{\partial X^{\#_\alpha}(\mathbf{x})}{\partial x^N} \equiv 0, \qquad \frac{\partial X^{\#_N}(x)}{\partial x^N} \equiv 1,\\
&\det\left[\frac{\partial x^{\#_\alpha}(\mathbf{x})}{\partial x^\beta}\right] \neq 0.
\end{aligned}
\tag{5.263}
$$

By the transformation rules (3.40), (3.49), and (5.263), we obtain

$$
\begin{aligned}
T^{\#_{\alpha_1 \ldots \alpha_p N \ldots N}}{}_{\beta_1 \ldots \beta_q N \ldots N}(\mathbf{x}^{\#}, x^{\#_N}) &= \frac{\partial X^{\#_{\alpha_1}}}{\partial x^{\gamma_1}} \cdots \frac{\partial X^{\#_{\alpha_p}}}{\partial x^{\gamma_p}} \frac{\partial X^{\#_{\mu_1}}}{\partial x^{\#_{\beta_1}}}\\
\cdots \frac{\partial X^{\#_{\mu_q}}}{\partial x^{\#_{\beta_q}}} T^{\gamma_1 \ldots \gamma_p N \ldots N}{}_{\mu_1 \ldots \mu_q N \ldots N}(\mathbf{x}, x^N).
\end{aligned}
\tag{5.264}
$$

This is a transformation rule for a **subtensor** field of order $(p+q)$ defined on the hypersurface D_{N-1} with $x^N = c$ under the restricted coordinate transformation in (5.263). We can generate *new subtensors* in D_{N-1} by covariant derivatives with Christoffel symbols in (5.262). For example, let us compute the covariant derivative of a 1-form $T_i(x)dx^i = T_\alpha(x)dx^\alpha + T_N(x)dx^N$. We obtain from (5.83) and (5.262)

$$
\nabla_\alpha T_{\beta}{}_{|_{x^N=c}} = \left[\bar{\nabla}_\alpha T_\beta + \left(\frac{1}{2}\right)(\partial_N g_{\alpha\beta})\left(g^{NN} T_N\right)\right]_{|_{x^N=c}}, \tag{5.265}
$$

$$
\begin{aligned}
&\nabla_\alpha T_{N}{}_{|_{x^N=c}} =\\
&\left[\partial_\alpha T_N - \left(\frac{1}{2}\right)(\partial_N g_{\mu\alpha})\left(g^{\mu\beta} T_\beta\right) - \left(\frac{1}{2}\right)(\partial g_{NN})\left(g^{NN} T_N\right)\right]_{|_{x^N=c}},
\end{aligned}
\tag{5.266}
$$

$$
\begin{aligned}
&\nabla_N T_{\alpha}{}_{|_{x^N=c}} =\\
&\left[\partial_N T_\alpha - \left(\frac{1}{2}\right)(\partial_N g_{\alpha\mu})\left(g^{\mu\beta} T_\beta\right) - \left(\frac{1}{2}\right)(\partial_\alpha g_{NN})\left(g_{NN} T_N\right)\right]_{|_{x^N=c}},
\end{aligned}
\tag{5.267}
$$

$$
\begin{aligned}
&\nabla_N T_{N}{}_{|_{x^N=c}} =\\
&\left[\partial_N T_N + \frac{1}{2}(\partial_\beta g_{NN})\left(g^{\mu\beta} T_\mu\right) - \left(\frac{1}{2}\right)(\partial_N g_{NN})\left(g^{NN} T_N\right)\right]_{|_{x^N=c}}.
\end{aligned}
\tag{5.268}
$$

The expressions on the right-hand side of (5.265)–(5.268) yield one subtensor, two subcovectors, and one subinvariant, respectively, in D_{N-1} (under the restricted transformation (5.263)).

Let us go back to the metric form for a normal coordinate chart in (5.259) (*with dropped hats*). Corresponding orthonormal N-tuples must satisfy by (5.18)

$$\begin{aligned}
\vec{\mathbf{e}}_\alpha(x) &= \lambda^i{}_{(\alpha)}(x)\frac{\partial}{\partial x^i} = \lambda^\beta_{(\alpha)}(x)\frac{\partial}{\partial x^\beta},\\
\lambda^N_{(\alpha)}(x) &\equiv 0,\\
\vec{\mathbf{e}}_N(x) &= \lambda^i{}_{(N)}(x)\frac{\partial}{\partial x^i} = \pm\sqrt{d_{NN}g^{NN}(x)}\,\delta^i_{(N)}\frac{\partial}{\partial x^i}\\
&\qquad = \pm\sqrt{\varepsilon_N g^{NN}(x)}\,\frac{\partial}{\partial x^N},\\
\lambda^\alpha_{(N)}(x) &\equiv 0,\\
\bar{g}_{\mu\nu}(\mathbf{x})\left[\lambda^\mu{}_{(\alpha)}(x)\lambda^\nu{}_{(\beta)}(x)\right]_{|x^N=c} &= d_{(\alpha)(\beta)}.
\end{aligned} \tag{5.269}$$

We can derive necessary conditions for Ricci rotation coefficients in a normal coordinate chart. Suppose that x^N is a normal or hypersurface orthogonal coordinate. Therefore, (5.269) holds. By (5.52) with $j = N$, we obtain that

$$\begin{aligned}
0 &= \lambda^i_{(\alpha)}(x)\left[\partial_i\lambda^N_{(\beta)}\right] - \lambda^k_{(\beta)}(x)\left[\partial_k\lambda^N_{(\alpha)}\right]\\
&= \left[\gamma^N_{\alpha\beta}(x) - \gamma^N_{\beta\alpha}(x)\right]\lambda^N_{(N)}(x),
\end{aligned} \tag{5.270}$$

or

$$\gamma_{N\alpha\beta}(x) = \gamma_{N\beta\alpha}(x).$$

There exists a special type of normal coordinate chart, namely a **geodesic normal coordinate chart** (or a **Gaussian normal coordinate chart**). The corresponding metric field is furnished by

$$\begin{aligned}
\mathbf{g}_{..}(x) &= g_{\alpha\beta}(x)dx^\alpha\otimes dx^\beta + d_{NN}dx^N\otimes dx^N\\
&\equiv g_{\alpha\beta}(x)dx^\alpha\otimes dx^\beta + \varepsilon_N dx^N\otimes dx^N\\
\mathbf{g}_{..}(x)_{|x^N=c} &= g_{\alpha\beta}(\mathbf{x},c)dx^\alpha\otimes dx^\beta =: \bar{g}_{\alpha\beta}(\mathbf{x})dx^\alpha\otimes dx^\beta.
\end{aligned} \tag{5.271}$$

By (5.271) and (5.262), we can obtain non-zero Christoffel symbols

$$\begin{aligned}
\left\{\begin{matrix}\rho\\ \mu\nu\end{matrix}\right\}_{|x^N=c} &= \overline{\left\{\begin{matrix}\rho\\ \mu\nu\end{matrix}\right\}},\ \left\{\begin{matrix}\rho\\ N\sigma\end{matrix}\right\} = (1/2)g^{\rho\mu}(x)\left(\partial_N g_{\mu\sigma}\right),\\
\left\{\begin{matrix}N\\ \rho\sigma\end{matrix}\right\} &= (\varepsilon_N/2)\partial_N g_{\rho\sigma}.
\end{aligned} \tag{5.272}$$

An x^N-coordinate curve

$$\begin{aligned}
x^\alpha &= \mathcal{X}^\alpha(\tau) := \kappa^\alpha = \text{const.},\\
x^N &= \mathcal{X}^N(\tau) := \tau,
\end{aligned}$$

satisfies the geodesic equation (5.200). Thus, each of the x^N-coordinate curves is a non-null geodesic. Moreover, the $(N-1)$-dimensional hypersurfaces $x^N = c_1$ and $x^N = c_2$ are called **geodesically parallel**.

We can compute curvature tensor components and related tensor components restricted to the $(N-1)$-dimensional hypersurface $x^N = c$ from (5.271), (5.95), and (5.272). The results are exhibited in the following equations:

$$\dot{g}_{\alpha\beta}(\mathbf{x}) := \partial_N g_{\alpha\beta_{|_{x^N=c}}}, \quad \ddot{g}_{\alpha\beta}(\mathbf{x}) := \partial_N \partial_N g_{\alpha\beta_{|_{x^N=c}}}, \tag{5.273}$$

$$R_{\lambda\rho\mu\nu}(x)_{|_{x^N=c}} = \overline{R}_{\lambda\rho\mu\nu}(\mathbf{x}) + (\varepsilon_N/4)\left[\dot{g}_{\rho\mu}\dot{g}_{\lambda\nu} - \dot{g}_{\rho\nu}\dot{g}_{\lambda\mu}\right], \tag{5.274}$$

$$2R_{N\rho\mu\nu}(\mathbf{x}, c) = \overline{\nabla}_\nu \dot{g}_{\rho\mu} - \overline{\nabla}_\mu \dot{g}_{\rho\nu}, \tag{5.275}$$

$$R_{N\rho\mu N}(\mathbf{x}, c) = (1/2)\ddot{g}_{\rho\mu}(\mathbf{x}) - (1/4)\overline{g}^{\alpha\beta}(\mathbf{x})\dot{g}_{\alpha\rho}\dot{g}_{\beta\mu}, \tag{5.276}$$

$$\begin{aligned} G_{\mu\nu}(x)_{|_{x^N=c}} = {} & \overline{G}_{\mu\nu}(\mathbf{x}) + (\varepsilon_N/2)\ddot{g}_{\mu\nu}(\mathbf{x}) + (\varepsilon_N/4)\overline{g}^{\rho\sigma}(\mathbf{x})\dot{g}_{\rho\sigma}(\mathbf{x})\dot{g}_{\mu\nu}(\mathbf{x}) \\ & -(\varepsilon_N/2)\overline{g}^{\alpha\beta}(\mathbf{x})\dot{g}_{\mu\alpha}(\mathbf{x})\dot{g}_{\nu\beta}(\mathbf{x}) - (\varepsilon_N/2)\overline{g}_{\mu\nu}(\mathbf{x}) \\ & \times \Big[(1/4)\left(\overline{g}^{\rho\sigma}(\mathbf{x})\dot{g}_{\rho\sigma}(\mathbf{x})\right)^2 - (3/4)\overline{g}^{\rho\sigma}(\mathbf{x})\overline{g}^{\alpha\lambda}(\mathbf{x})\dot{g}_{\alpha\rho}(\mathbf{x})\dot{g}_{\lambda\sigma}(\mathbf{x}) \\ & +\overline{g}^{\rho\sigma}(\mathbf{x})\ddot{g}_{\rho\sigma}(\mathbf{x})\Big], \end{aligned} \tag{5.277}$$

$$G_{\mu N}(x)_{|_{x^N=c}} = R_{\mu N}(\mathbf{x}) = (1/2)\overline{g}^{\rho\sigma}(\mathbf{x})\left[\overline{\nabla}_\mu \dot{g}_{\rho\sigma} - \overline{\nabla}_\rho \dot{g}_{\sigma\mu}\right], \tag{5.278}$$

$$\begin{aligned} G_{NN}(x)_{|_{x^N=c}} = {} & -(\varepsilon_N/2)\overline{R}(\mathbf{x}) - (1/8)\left[\overline{g}^{\mu\nu}(\mathbf{x})\dot{g}_{\mu\nu}(\mathbf{x})\right]^2 \\ & +(1/8)\overline{g}^{\mu\rho}(\mathbf{x})\overline{g}^{\nu\lambda}\dot{g}_{\rho\nu}(\mathbf{x})\dot{g}_{\lambda\mu}(\mathbf{x}). \end{aligned} \tag{5.279}$$

Example 5.4.5 Consider the four-dimensional space-time manifold with the metric field

$$\begin{aligned} & \mathbf{g}_{..}(x) = [A(\tau)]^2 g^{\#}{}_{\alpha\beta}(\mathbf{x}) dx^\alpha \otimes dx^\beta - d\tau \otimes d\tau, \\ & \alpha, \beta \in \{1, 2, 3\}, \\ & \mathbf{x} := (x^1, x^2, x^3), \quad x^4 \equiv \tau, \\ & x \in D \subset R^4; \quad A(\tau) > 0, \quad \dot{A}(\tau) := \frac{dA(\tau)}{d\tau}, \\ & \operatorname{sgn}[\mathbf{g}_{..}(x)] = +2. \end{aligned} \tag{5.280}$$

Here, we have

$$\begin{aligned} g_{\alpha\beta}(x) &= [A(\tau)]^2 g^{\#}_{\alpha\beta}(\mathbf{x}), \ g_{\alpha 4}(x) \equiv 0, \quad g_{44}(x) = d_{44} \equiv \varepsilon_4 = -1, \\ g_{\alpha\beta}(x)_{|\tau=c} &= [A(c)]^2 g^{\#}_{\alpha\beta}(\mathbf{x}) =: \overline{g}_{\alpha\beta}(\mathbf{x}), \ \dot{g}_{\alpha\beta}(\mathbf{x}) := \partial_4 g_{\alpha\beta_{|_{\tau=c}}} \\ & \quad = 2[A(\tau)\dot{A}(\tau)]_{|_{\tau=c}} g^{\#}{}_{\alpha\beta}(\mathbf{x}), \\ \overline{\left\{ \begin{matrix} \alpha \\ \beta\gamma \end{matrix} \right\}} &= \left\{ \begin{matrix} \alpha \\ \beta\gamma \end{matrix} \right\}^{\#}. \end{aligned} \tag{5.281}$$

By (5.273)–(5.279), we obtain that

$$\begin{aligned}
G_{\mu\nu}(x)_{|\tau=c} &= \overline{G}_{\mu\nu}(\mathbf{x}) + \left[\frac{2A(\tau)\ddot{A}(\tau) + (\dot{A}(\tau))^2}{(A(\tau))^2}\right]_{|\tau=c} \overline{g}_{\mu\nu}(\mathbf{x}), \\
G_{\mu\nu}(x)_{|\tau=c} &= G^{\#}_{\mu\nu}(\mathbf{x}) + \left[2A(\tau)\ddot{A}(\tau) + (\dot{A}(\tau))^2\right]_{|\tau=c} g^{\#}_{\mu\nu}(\mathbf{x}), \\
G_{\mu 4}(x) &\equiv 0, \\
G_{44}(x)_{|\tau=c} &= (1/2)\overline{R}(\mathbf{x}) - 3\left[\dot{A}(\tau)/A(\tau)\right]^2_{|\tau=c}, \\
G_{44}(x)_{|\tau=c} &= [A(\tau)]^{-2}\left[(1/2)R^{\#}(\mathbf{x}) - 3(\dot{A}(\tau))^2\right]_{|\tau=c}.
\end{aligned} \tag{5.282}$$

The equations above are relevant to the relativistic cosmology. The physical terminology for the chart inherent in (5.280) is the **comoving, synchronous coordinate system.** □

Now, we shall introduce the **orthogonal coordinate chart**. Suppose that there exist N differentiable functions $f_{(i)}(x)$ such that

$$\begin{aligned}
\mathbf{g}^{\cdot\cdot}(x)\left[df_{(i)}, df_{(j)}\right] &= 0 \quad \text{for } i \neq j, \\
g^{kl}(x)\left(\partial_k f_{(i)}\right)\left(\partial_l f_{(j)}\right) &= 0 \quad \text{for } i \neq j, \\
g^{kl}(x)\left(\partial_k f_{(i)}\right)\left(\partial_l f_{(i)}\right) &\neq 0.
\end{aligned} \tag{5.283}$$

The system above comprises $N(N+1)/2$ coupled, non-linear, first-order partial differential equations. The solution of this overdetermined system may *not* exist for $N > 3$ cases. If the solution exists locally, the non-null hypersurface $f_{(i)}(x) = c_{(i)}$ and $f_{(j)}(x) = c_{(j)}$ intersect orthogonally for $i \neq j$. We make a coordinate transformation

$$\widehat{x}^i = \widehat{X}^i(x) := f_{(i)}(x).$$

By (5.283), we obtain for $i \neq j$

$$\begin{aligned}
\widehat{g}^{ij}(\widehat{x}) &= \frac{\partial \widehat{X}^i(x)}{\partial x^k}\frac{\partial \widehat{X}^j(x)}{\partial x^l} g^{kl}(x) = g^{kl}(x)(\partial_k f_{(i)})(\partial_l f_{(j)}) \equiv 0, \\
\widehat{g}^{11}(\widehat{x}) &\neq 0,\ \widehat{g}^{22}(\widehat{x}) \neq 0, \ldots, \widehat{g}^{NN}(\widehat{x}) \neq 0.
\end{aligned} \tag{5.284}$$

Therefore, the metric field reduces to

$$\begin{aligned}
&\widehat{\mathbf{g}}_{\cdot\cdot}(\widehat{x}) = \widehat{g}_{11}(\widehat{x})d\widehat{x}^1 \otimes d\widehat{x}^1 + \widehat{g}_{22}(\widehat{x})d\widehat{x}^2 \otimes d\widehat{x}^2 + \cdots + \widehat{g}_{NN}(\widehat{x})d\widehat{x}^N \otimes d\widehat{x}^N, \\
&\widehat{g}_{11}(\widehat{x}) \neq 0,\ \widehat{g}_{22}(\widehat{x}) \neq 0, \ldots, \widehat{g}_{NN}(\widehat{x}) \neq 0.
\end{aligned} \tag{5.285}$$

Every coordinate curve $\widehat{x}^i$ intersects the other $(N-1)$ coordinate curves of $\widehat{x}^j (i \neq j)$ orthogonally. Let us *drop* the hats from the orthogonal coordinates in the sequel. We also *suspend* the summation convention for the orthogonal coordinates. With this understanding, we rewrite (5.285) as

$$\begin{aligned}
\mathbf{g}_{..}(x) &= \sum_{i=1}^{N}\sum_{j=1}^{N}\left[h_{(i)}(x)\right]^2 d_{ij}dx^i \otimes dx^j = \sum_{i=1}^{N} \varepsilon_{(i)}\left[h_{(i)}(x)\right]^2 dx^i \otimes dx^i, \\
h_{(i)}(x) &> 0, \\
g_{ij}(x) &= \left[h_{(i)}(x)\right]^2 d_{ij}, \quad g^{ij}(x) = \left[h_{(i)}(x)\right]^{-2} d^{ij}, \\
g &\equiv \det\left[g_{ij}(x)\right] = \left\{\prod_{i=1}^{N}\left[h_{(i)}(x)\right]^2\right\} \det\left[d_{ij}\right], \\
\sqrt{|g|} &= \prod_{i=1}^{N}\left[h_{(i)}(x)\right].
\end{aligned} \tag{5.286}$$

For a corresponding orthonormal basis (or orthonormal frame), we can deduce from (5.286) that

$$\begin{aligned}
\tilde{\mathbf{e}}^a(x) &= \sum_i h_{(i)}(x)\delta^a{}_i\, dx^i, \quad \vec{\mathbf{e}}_a(x) = \sum_i \left[h_{(i)}(x)\right]^{-1}\delta^i{}_a \frac{\partial}{\partial x^i}, \\
\lambda^i{}_a(x) &= \left[h_{(i)}(x)\right]^{-1}\delta^i{}_a, \quad \mu^a{}_i(x) = h_{(i)}(x)\delta^a{}_i.
\end{aligned} \tag{5.287}$$

Denoting by i, j, k only *distinct* indices for $N \geq 3$, the Christoffel symbols from (5.286), (5.40), and (5.41) are computed as

$$\begin{aligned}
&[ij,k] \equiv 0, \quad [ij,i] = -[ii,j] = d_{ii}h_{(i)}(x)\partial_j h_{(i)}, \\
&[ii,i] = d_{ii}h_{(i)}(x)\partial_i h_{(i)}, \\
&\left\{ {i \atop jk} \right\} \equiv 0, \quad \left\{ {j \atop ii} \right\} = -d_{ii}d^{jj}h_{(i)}(x)\left[h_{(j)}(x)\right]^{-2}\partial_j h_{(i)}, \\
&\left\{ {i \atop ij} \right\} = \partial_j \ln h_{(i)}(x), \quad \left\{ {i \atop ii} \right\} = \partial_i \ln h_{(i)}(x).
\end{aligned} \tag{5.288}$$

The Ricci rotation coefficients are provided by (5.287) and (5.53) (*with a, b, c as distinct indices*) as

$$\begin{aligned}
\gamma_{abc}(x) &\equiv 0, \\
\gamma_{aba}(x) &= -d_{aa}\left[h_b(x)\right]^{-1}\frac{\partial}{\partial x^b}\left[\ln h_{(a)}(x)\right] \equiv -d_{aa}\partial_b \ln\left[h_a(x)\right] \equiv -\gamma_{baa}(x), \\
\gamma_{aaa}(x) &\equiv 0.
\end{aligned} \tag{5.289}$$

The connection 1-forms from (5.55) and (5.289) are furnished as

$$\begin{aligned}
\tilde{\mathbf{w}}_{ab}(x) &= d_{aa}\left[\partial_b h_{(a)}\right]dx^a - d_{bb}\left[\partial_a h_{(b)}\right]dx^b \\
&= d_{aa}\left[\partial_b \ln h_{(a)}\right]\tilde{\mathbf{e}}^a(x) - d_{bb}\left[\partial_a \ln h_{(b)}\right]\tilde{\mathbf{e}}^b(x) \equiv -\tilde{\mathbf{w}}_{ba}(x).
\end{aligned} \tag{5.290}$$

The covariant divergence of a differentiable vector field provided by (5.85) and (5.286) is

$$\sum_{i=1}^{N} \nabla_i A^i = \left[h_{(i)} h_{(i)} \cdots h_{(N)}\right]^{-1} \frac{\partial}{\partial x^i} \left\{ \left[h_{(1)} h_{(2)} \cdots h_{(N)}\right] A^i \right\} \equiv \sum_{b=1}^{N} \nabla_b A^b. \quad (5.291)$$

The invariant Laplacian is given by (5.90), (5.91), (5.286), and (5.289) as

$$\Delta V = \sum_i \sum_j \left\{ \left[h_{(1)} h_{(2)} \cdots h_{(N)}\right]^{-1} \partial_i \left[\left(h_{(i)} \cdots h_{(N)}\right) \cdot h_{(j)}^{-2} d^{ij} \partial_j V \right] \right\}. \quad (5.292)$$

Now, we shall compute the Riemann tensor components from (5.95), (5.286), and (5.288). (We restrict the dimensions to $N \geq 4$ and denote by h, i, j, k only *distinct* indices.) Long calculations yield

$$\begin{aligned}
R_{hijk}(x) &\equiv 0, \\
R_{hiik}(x) &= d_{ii} h_{(i)}(x) \left\{ \partial_h \partial_k h_{(i)} - \left[\partial_h h_{(i)}\right] \left[\partial_k \ln h_{(h)}\right] \right. \\
&\quad \left. - \left[\partial_k h_{(i)}\right] \left[\partial_h \ln h_{(k)}\right] \right\}, \\
R_{kiik}(x) &= d_{ii} d_{kk} h_{(i)}(x) h_{(k)}(x) \left\{ d^{ii} \partial_i \left[h_{(i)}^{-1} \partial_i h_{(k)} \right] + d^{kk} \partial_k \left[h_{(k)}^{-1} \partial_k h_{(i)} \right] \right. \\
&\quad \left. + \sum_{l \neq i,k}{}' d^{ll} (h_l)^{-2} \left[\partial_l h_{(i)}\right] \left[\partial_l h_{(k)}\right] \right\}. \qquad (5.293)
\end{aligned}$$

Relative to an orthonormal basis, the curvature components computed from (5.97), (5.287), and (5.289) are given by $(a, b, c, d \neq)$:

$$\begin{aligned}
R_{abcd}(x) &\equiv 0, \\
R_{abbd}(x) &= d_{bb} \left[h_{(b)}(x)\right]^{-1} \left\{ \partial_a \partial_d h_{(b)} - \left[\partial_a h_{(b)}\right] \left[\partial_d \ln h_{(a)}\right] \right\}, \\
R_{abba}(x) &= d_{aa} d_{bb} \left\{ d^{bb} \left[h_{(a)}(x)\right]^{-1} \partial_b \partial_b h_{(a)} + d^{aa} \left[h_{(b)}(x)\right]^{-1} \partial_a \partial_a h_{(b)} \right. \\
&\quad \left. + \sum_{c \neq a,b}{}' d^{cc} \left[\partial_c \ln h_{(a)}\right] \left[\partial_c \ln h_{(b)}\right] \right\}. \qquad (5.294)
\end{aligned}$$

The Ricci tensor components relative to a coordinate basis are given by (5.114) and (5.293) for $l \neq k$ as

$$\begin{aligned}
R_{lk}(x) &= \sum_{i \neq k,l}{}' \left[h_{(i)}(x)\right]^{-1} \left\{ \partial_l \partial_k h_{(i)} - \left[\partial_l h_{(i)}\right] \left[\partial_k \ln h_{(l)}\right] \right. \\
&\quad \left. - \left[\partial_k h_{(i)}\right] \left[\partial_l \ln h_{(k)}\right] \right\}, \\
R_{kk}(x) &= d_{kk} h_{(k)}(x) \sum_{i \neq k}{}' \left[h_{(i)}(x)\right]^{-1} \left\{ d^{ii} \partial_i \left[h_{(i)}^{-1} \partial_i h_{(k)} \right] + d^{kk} \partial_k \left[h_{(k)}^{-1} \partial_k h_{(i)} \right] \right. \\
&\quad \left. + \sum_{l \neq i,k}{}' d^{ll} \left[h_{(l)}^{-2} \left(\partial_l h_{(i)}\right) \left(\partial_l h_{(k)}\right) \right] \right\}. \qquad (5.295)
\end{aligned}$$

The Ricci tensor components relative to an orthonormal basis are provided by (5.115) and (5.294) for $a \neq b$ as

$$R_{ab}(x) = \sum_{c\neq a,b}{}' \left[h_{(c)}(x)\right]^{-1}\{\partial_a\partial_b h_{(c)} - \left[\partial_a h_{(c)}\right]\left[\partial_b \ln h_{(a)}\right] - \left[\partial_b h_{(c)}\right]\left[\partial_a \ln h_{(b)}\right]\},$$

$$\begin{aligned} R_{aa}(x) = d_{aa} \sum_{b\neq a}{}' & \left\{d^{bb}\left[h_{(a)}(x)\right]^{-1}\partial_b\partial_b h_{(a)} + d^{aa}\left[h_{(b)}(x)\right]^{-1}\partial_a\partial_a h_{(b)}\right. \\ & \left. + \sum_{c\neq a,b}{}' d^{cc}\left[\partial_c \ln h_{(b)}\right]\left[\partial_c \ln h_{(a)}\right]\right\}. \end{aligned} \tag{5.296}$$

The curvature scalar by (5.116), (5.117), (5.295), and (5.296) is furnished by

$$\begin{aligned} R(x) &= \sum_k \left[h_{(k)}(x)\right]^{-1} \sum_{i\neq k}{}' \left\{d^{ii}\partial_i\left[h_{(i)}^{-1}\partial_i h_{(k)}\right] + d^{kk}\partial_k\left[h_{(k)}^{-1}\partial_k h_{(i)}\right]\right. \\ &\quad \left. + \sum_{l\neq i,k}{}' d^{ll} h_{(l)}^{-2}\left[\partial_l h_{(i)}\right]\left[\partial_l h_{(k)}\right]\right\} \\ &= \sum_a \sum_{b\neq a}{}' \left\{d^{bb} h_{(a)}^{-1}\partial_b\partial_b h_{(a)} + d^{aa} h_{(b)}^{-1}\partial_a\partial_a h_{(b)}\right. \\ &\quad \left. + \sum_{c\neq a,b}{}' d^{cc}\left[\partial_c \ln h_{(b)}\right]\left[\partial_c \ln h_{(a)}\right]\right\}. \end{aligned} \tag{5.297}$$

Remark: These equations are important in continuum mechanics as well as general relativity.

Exercises 5.4

1. Consider the two-dimensional, real-analytic metric field given by

$$\mathbf{g}_{..}(x) = \left[1 + (1/4)({x^1}^2 + {x^2}^2)\right]^{-2}\left[dx^1 \otimes dx^1 + dx^2 \otimes dx^2\right],$$
$$x \in \mathbb{R}^2.$$

Determine whether or not the coordinate chart above is Riemann normal.

2. Suppose that the N-dimensional, real-analytic metric field in a *Riemann normal coordinate chart* is provided by

$$\mathbf{g}_{..}(x) = g_{ij}(x)dx^i \otimes dx^j,$$
$$x \in D \subset \mathbb{R}^N; \quad \left\{ \begin{matrix} i \\ jk \end{matrix} \right\}_{|0} \equiv 0.$$

Prove that

$$\left[\partial_l\left\{ {i \atop jk} \right\} - 2\left\{ {i \atop mj} \right\}\left\{ {m \atop kl} \right\} + \partial_j\left\{ {i \atop kl} \right\} - 2\left\{ {i \atop mk} \right\}\left\{ {m \atop lj} \right\} \right.$$
$$\left. +\partial_k\left\{ {i \atop lj} \right\} - 2\left\{ {i \atop ml} \right\}\left\{ {m \atop jk} \right\}\right]_{|0} \equiv 0.$$

3. Consider a 1-form $\tilde{\mathbf{A}}(x) = A_i(x)dx^i$ such that $d\tilde{\mathbf{A}}(x)$ is a continuous 2-form. Let the oriented tensor field equation $A_{i_1}(x)\left[\partial_{i_2}A_{i_3} - \partial_{i_3}A_{i_2}\right]\eta^{i_1 i_2 i_3 \ldots i_N}$ $(x) = 0$ hold in $D \subset \mathbb{R}^N$. Prove that there exist functions λ and f such that $\tilde{\mathbf{A}}(x) = \lambda(x)df(x)$.

(*Remark:* Such a 1-form field is called **hypersurface orthogonal**.)

4. Consider a normal coordinate chart. Show that the corresponding Christoffel symbols $\left\{ {\alpha \atop \beta N} \right\}$, $\left\{ {N \atop \alpha\beta} \right\}$, $\left\{ {\alpha \atop NN} \right\}$, $\left\{ {N \atop \alpha N} \right\}$, and $\left\{ {N \atop NN} \right\}$ transform as *subtensors* under the restricted coordinate transformation $x^{\#\alpha} = X^{\#\alpha}(x^1, \ldots, x^{N-1})$, $x^{\#_N} = x^N$.

5. Consider a geodesic normal coordinate chart characterized by (5.271). Prove that the corresponding N-tuple components $\lambda^i{}_{(N)}(x)$ must satisfy

$$\nabla_j \lambda_{(N)i} = \nabla_i \lambda_{(N)j}\,.$$

6. Consider a class of twice-differentiable geodesic normal coordinate charts characterized by

$$\begin{aligned}
&\mathbf{g}_{..}(x) = g_{ij}(x)dx^i \otimes dx^j = \left[A(x^N)\right]^2 g^{\#}{}_{\alpha\beta}(\mathbf{x})dx^\alpha \otimes dx^\beta + \varepsilon_N (dx^N)^2,\\
&\alpha, \beta \in \{1, 2, \ldots, N-1\},\\
&\mathbf{x} := (x^1, \ldots, x^{N-1}) \in \mathbf{D} \subset \mathbb{R}^{N-1},\\
&x^N \in (\tau_1, \tau_2) \in \mathbb{R}.
\end{aligned}$$

Assume furthermore that $g^{\#}{}_{\alpha\beta}(\mathbf{x})$ yield $(N-1)$-dimensional *positive definite* metric components. Prove that the invariant eigenvalues from the equation $\det[G_{ij} - \lambda(x)g_{ij}(x)] = 0$ are all real.

7. Consider a *three-dimensional* manifold and a twice-differentiable orthogonal coordinate chart characterized by

$$\mathbf{g}_{..}(x) = \sum_{i=1}^{3} \varepsilon_{(i)} \left[h_{(i)}(x)\right]^2 dx^i \otimes dx^i.$$

(In this problem, the summation convention is *suspended.*) Prove that for distinct indices i and k, the Riemann-Christoffel tensor components must satisfy

$$R_{kiik}(x) - \varepsilon_{(k)} \left[h_{(k)}(x)\right]^2 R_{ii}(x) - \varepsilon_{(i)} \left[h_{(i)}(x)\right]^2 R_{kk}(x)$$
$$+(1/2)R(x)\varepsilon_{(i)}\varepsilon_{(k)} \left[h_{(i)}(x)h_{(k)}(x)\right]^2 \equiv 0.$$

Chapter 6

Special Riemannian and Pseudo-Riemannian Manifolds

6.1 Flat Manifolds

It was mentioned in section 5.2 that a domain $D \subset \mathbb{R}^N$ corresponding to the domain $U \subset M_N$ is *flat* provided

$$\begin{aligned} \mathbf{R}^{\cdot}{}_{\cdots}(x) &\equiv \mathbf{O}^{\cdot}{}_{\cdots}(x), \\ R^{i}{}_{jkl}(x) &\equiv 0, \end{aligned} \tag{6.1}$$

for all $x \in D$.

Eisenhart, in his book on Riemannian geometry [10], proves that a metric tensor component $g_{ij}(x)$ of a prescribed signature (corresponding to that of the flat metric component d_{ij}) solves the non-linear equations (6.1) if and only if

$$g_{ij}(x) = d_{kl} \frac{\partial f^k(x)}{\partial x^i} \frac{\partial f^l(x)}{\partial x^j}. \tag{6.2}$$

Here, N functions $f^k(x)$ are of class C^3 in D and otherwise *arbitrary*.

If we make a C^3-coordinate transformation

$$\widehat{x}^i = f^i(x),$$

then (6.2) yields

$$\mathbf{g}_{\cdot\cdot}(x) = g_{ij}(x) dx^i \otimes dx^j, \quad \widehat{\mathbf{g}}_{\cdot\cdot}(\widehat{x}) = d_{kl} d\widehat{x}^k \otimes d\widehat{x}^l. \tag{6.3}$$

In the above, the metric tensor components d_{kl} are diagonal as well as constant-valued. The corresponding $\widehat{x}$-coordinate system (or chart) surely deserves a special name. If the metric tensor $\mathbf{g}_{..}(x)$ is positive-definite (thus $d_{ij} = \delta_{ij}$), the special coordinate system is called a **Cartesian coordinate chart**. If $\mathbf{g}_{..}(x)$ is *not* positive-definite, the chart is called **pseudo-Cartesian**. In the case of dimension $N = 4$, and $\mathbf{g}_{..}(x)$ is of Lorentz signature, the chart is known as **Minkowskian** (in the theory of relativity).

The coordinate transformation from one Cartesian (or pseudo-Cartesian) chart to another is governed by the equations

$$\begin{aligned} x^{\#i} &= X^{\#i}(\widehat{x}), \\ d_{ij} &= d_{kl}\frac{\partial X^{\#k}(\widehat{x})}{\partial \widehat{x}^{i}}\frac{\partial X^{\#l}(\widehat{x})}{\partial \widehat{x}^{j}}. \end{aligned} \tag{6.4}$$

The system of first-order, non-linear partial differential equations in (6.4) can be solved. The general solution, comprising $N(N+1)/2$ arbitrary constants (or parameters), is furnished by

$$\begin{aligned} & x^{\#i} = X^{\#i}(\widehat{x}) = c^{i} + l^{i}{}_{j}\widehat{x}^{j}, \\ & l^{i}{}_{m} d_{ij} l^{j}{}_{n} = d_{mn}, \\ & [L] := \left[l^{i}{}_{j}\right], \ [D] := [d_{ij}], \\ & [L]^{T}[D][L] = [D], \\ & \det[D] = \pm 1. \end{aligned} \tag{6.5}$$

(Compare the equations above with (5.17).) The proof for the general solution in (6.5) is available in books by Synge [38] or Das [7] on the special theory of relativity.

The set of non-homogeneous linear transformations in (6.5) constitutes an $N(N+1)/2$ parameter continuous group. This group is called the **generalized Poincaré group**, denoted by $\mathcal{IO}(p, n; \mathbb{R})$.

The Cartesian (or pseudo-Cartesian) tensor components transform under a generalized Poincaré transformation (6.5) (by (3.48)) as

$$\begin{aligned} T^{\#k_1\ldots k_r}{}_{l_1\ldots l_s}(x^{\#}) &= l^{k_1}{}_{i_1}\ldots l^{k_r}{}_{i_r} a^{j_1}{}_{l_1}\ldots a^{j_s}{}_{l_s} \widehat{T}^{i_1\ldots i_r}{}_{j_1\ldots j_s}(\widehat{x}), \\ [A] &:= \left[a^{i}{}_{j}\right] = [L]^{-1}. \end{aligned} \tag{6.6}$$

Under the transformation (6.6), we notice the following consequences:

(1) There is no difference between the Cartesian (or pseudo-Cartesian) components and the corresponding orthonormal components.

(2) In a Euclidean space $\mathbb{E}_N$, relative to a Cartesian chart, tensor components satisfy $\widehat{T}^{i_1\ldots i_r j_1\ldots j_s}(\widehat{x}) \equiv \widehat{T}^{i_1\ldots i_r}{}_{j_1\ldots j_s}(\widehat{x}) \equiv \widehat{T}_{i_1\ldots i_r j_1\ldots j_s}(\widehat{x})$. Thus, the contravariant components and covariant components are identical.

(3) The flat metric tensor components d_{ij} and the corresponding tensor-product components $d_{i_1 j_1} d_{i_2 j_2} \ldots d_{i_r j_r}$ transform under (6.6) as *numerical* tensor components (retaining the values intact).

(4) The totally antisymmetric permutation symbol $\widehat{\varepsilon}_{i_1 \ldots i_N} \equiv \widehat{\eta}_{i_1 \ldots i_N}$.

(5) The Christoffel symbol components $\widehat{\left\{ {i \atop jk} \right\}} \equiv 0$ relative to a Cartesian (or pseudo-Cartesian) system. Therefore, covariant derivatives *reduce* to partial derivatives.

Example 6.1.1 Consider the three-dimensional Euclidean space $\mathbb{E}_3$ and a Cartesian coordinate chart. Let three vector fields

$$\vec{\mathbf{A}}(\mathbf{x}) = a^i(\mathbf{x})\frac{\partial}{\partial x^i}, \quad \vec{\mathbf{B}}(\mathbf{x}) = b^i(\mathbf{x})\frac{\partial}{\partial x^i}, \quad \vec{\mathbf{C}}(\mathbf{x}) = c^i(\mathbf{x})\frac{\partial}{\partial x^i},$$

be defined in a domain $D \subset \mathbb{R}^3$. (Here, $\mathbf{x} := (x^1, x^2, x^3)$.) Recall the usual vector cross product given by

$$\vec{\mathbf{A}}(\mathbf{x}) \times \vec{\mathbf{B}}(\mathbf{x}) = \left[\delta^{li}\varepsilon_{ijk}a^j(\mathbf{x})b^k(\mathbf{x})\right]\frac{\partial}{\partial x^l}.$$

By (2.96), we express

$$\delta^{in}\varepsilon_{nlm}\varepsilon_{ijk} = \delta_{lj}\delta_{mk} - \delta_{lk}\delta_{mj}.$$

Therefore,

$$\begin{aligned}
\vec{\mathbf{A}}(\mathbf{x}) \times \left[\vec{\mathbf{B}}(\mathbf{x}) \times \vec{\mathbf{C}}(\mathbf{x})\right] &= \left[\delta^{hn}\delta^{li}\varepsilon_{nml}\varepsilon_{ijk}a^m(\mathbf{x})b^j(\mathbf{x})c^k(\mathbf{x})\right]\frac{\partial}{\partial x^h} \\
&= \left\{\left[\delta_{mk}a^m(\mathbf{x})c^k(\mathbf{x})\right]b^h(\mathbf{x}) - \left[\delta_{mj}a^m(\mathbf{x})b^j(\mathbf{x})\right]c^h(\mathbf{x})\right\}\frac{\partial}{\partial x^h} \\
&= \left[\vec{\mathbf{A}}(\mathbf{x}) \cdot \vec{\mathbf{C}}(\mathbf{x})\right]\vec{\mathbf{B}}(\mathbf{x}) - \left[\vec{\mathbf{A}}(\mathbf{x}) \cdot \vec{\mathbf{B}}(\mathbf{x})\right]\vec{\mathbf{C}}(\mathbf{x}).
\end{aligned}$$

Thus, we recover an equation from the elementary vector calculus. □

Example 6.1.2 Consider the flat product manifold $\mathbb{E}_3 \times \mathbb{R}$. Here, the Euclidean space $\mathbb{E}_3$ represents the physical space according to Newton. The set $\mathbb{R}$ represents the absolute time.

A Cartesian coordinate system is used for a domain of $\mathbb{E}_3$. Tensors are considered in the corresponding domain D of $\mathbb{R}^3 \times \mathbb{R}$. A typical point in D is denoted by $(\mathbf{x}, t)$.

An **elastic deformation** of a material body is mathematically characterized by a C^2-diffeomorphism of a domain of $\mathbb{E}_3 \longrightarrow \mathbb{E}_3$ (at a given instant).

The spatial gradient of a *small* deformation, u_i, yields the symmetric **strain tensor**

$$\sigma_{ij}(\mathbf{x},t) := (1/2)\left(\partial_i u_j + \partial_j u_i\right). \tag{6.7}$$

The **generalized Hooke's law** states that the symmetric **stress tensor** $\tau_{ij}(\mathbf{x},t)$ is linearly related to the strain tensor by the following equations:

$$\begin{aligned} \tau_{ij}(\mathbf{x},t) &= c_{ijkl}\sigma^{kl}(\mathbf{x},t), \\ c_{ijlk} &\equiv c_{ijkl}, \\ c_{klij} &\equiv c_{lkij}. \end{aligned} \tag{6.8}$$

Here, c_{ijkl} are components of the constant-valued first **elasticity tensor**.

The elasticity tensor of an isotropic material body is furnished by

$$c_{ijkl} = \lambda\delta_{ij}\delta_{kl} + \mu\left(\delta_{ik}\delta_{jl} + \delta_{il}\delta_{jk}\right). \tag{6.9}$$

Here, the constants λ and μ are called **Lame's constant** and **rigidity**, respectively.

The equations of motion for a particle inside an isotropic elastic body (with sufficiently small velocity) are governed by the equations of motion

$$\rho(\mathbf{x},t)\frac{\partial^2 U_i(\mathbf{x},t)}{\partial t^2} = \rho(\mathbf{x},t)F_i(\mathbf{x},t) + (\lambda+\mu)\partial_i\left[\delta^{kl}\sigma_{kl}\right] + \mu\Delta u_i. \tag{6.10}$$

Here, ρ is the mass density, F_i is the body force density, Δ is the three-dimensional Laplacian, and $u_i = U_i(\mathbf{x},t)$ are deformation components. (See the book by Green & Zerna [16].) □

Example 6.1.3 We consider again the flat product manifold $\mathbb{E}_3 \times \mathbb{R}$ of the preceding example. Let us investigate fluid flow in a domain $D \subset \mathbb{R}^3 \times \mathbb{R}$ using a Cartesian coordinate chart for $\mathbb{R}^3$. We denote the fluid velocity field by the components

$$\vec{\mathbf{v}} = v^i\frac{\partial}{\partial x^i}, \quad v^i = V^i(\mathbf{x},t), \quad \frac{d\mathcal{X}^i(t)}{dt} = V^i\left[\mathcal{X}(t),t\right]. \tag{6.11}$$

The acceleration vector of a fluid particle is given by

$$\left[\frac{dV^i(\mathbf{x},t)}{dt}\right]_{|x=\mathcal{X}(t)} = \left[\frac{\partial V^i(\mathbf{x},t)}{\partial t} + \left(\partial_j v^i\right)V^j(\mathbf{x},t)\right]_{|x=\mathcal{X}(t)}. \tag{6.12}$$

By the principle of conservation of mass, the **continuity equation**

$$\frac{\partial}{\partial t}\left[\rho(\mathbf{x},t)\right] + \partial_j\left[\rho(\mathbf{x},t)V^j(\mathbf{x},t)\right] = 0 \tag{6.13}$$

follows. Here, ρ is the fluid mass density.

The **vorticity tensor** components and the vorticity 2-form at a point in a fluid are defined by

$$\begin{aligned}\omega_{ij} &:= (1/2)\left(\partial_i v_j - \partial_j v_i\right) \equiv -\omega_{ji},\\ \boldsymbol{\omega}_{..} &:= \omega_{ij} dx^i \wedge dx^j = d\left[v_i dx^i\right].\end{aligned} \tag{6.14}$$

(Here, we have utilized (3.65).)

In the case of a perfect fluid, the density of pressure $p(\mathbf{x}, t)$ plays an important part. The equations of motion of a perfect fluid particle are governed by

$$\rho(\mathbf{x}, t)\left[\frac{\partial V^i(\mathbf{x}, t)}{\partial t} + \left(\partial_j v^i\right) V^j(\mathbf{x}, t)\right] = \rho(\mathbf{x}, t) F^i(\mathbf{x}, t) - \delta^{ij}\partial_j p. \tag{6.15}$$

Here, $F^i(\mathbf{x}, t)$ is the density of body force.

In the case where there is an **equation of state**, the mass density and pressure are functionally related. We can express the relationship as

$$\rho(\mathbf{x}, t) = \mathcal{R}[p(\mathbf{x}, t)] > 0. \tag{6.16}$$

Here, $\mathcal{R}$ is a differentiable function of p. Thus, we can define another C^2 function by

$$\begin{aligned}\mathcal{P}(p) &:= \int \frac{dp}{\mathcal{R}(p)} \equiv \int \rho^{-1} dp,\\ \partial_i \mathcal{P} &= \rho^{-1}\partial_i p.\end{aligned} \tag{6.17}$$

When the body force $F^i(\mathbf{x}, t)$ is **conservative**, it is derivable from a potential $U(\mathbf{x}, t)$ so that

$$F^i(\mathbf{x}, t) = -\delta^{ij}\partial_j U. \tag{6.18}$$

In the case of an **irrotational motion**, the vorticity tensor vanishes. Therefore, by (6.14),

$$\boldsymbol{\omega}_{..}(\mathbf{x}, t) \equiv \mathbf{O}_{..} \equiv d\left[v_i dx^i\right]. \tag{6.19}$$

By theorem 3.4.10, there exists a function $\phi(\mathbf{x}, t)$ of class C^2 such that

$$v_i = -\partial_i \phi. \tag{6.20}$$

(The function ϕ is called the **velocity potential**.) □

Exercises 6.1

1. Consider $\mathbb{E}_3$ and a Cartesian coordinate chart. Let a differentiable vector field $\vec{\mathbf{U}}(\mathbf{x}) = U^i(\mathbf{x})\frac{\partial}{\partial x^i}$ and a twice-differentiable vectorfield $\vec{\mathbf{V}}(x) = V^i(\mathbf{x})\frac{\partial}{\partial x^i}$

be defined in a domain $D \subset \mathbb{R}^3$. Using the Cartesian components of the vector fields, prove the following vector identities:

$$\text{(i)}\ \nabla \times \left[\nabla \times \vec{\mathbf{V}}(\mathbf{x})\right] \equiv \nabla \left[\nabla \cdot \vec{\mathbf{V}}(\mathbf{x})\right] - \nabla \cdot \left[\nabla \vec{\mathbf{V}}(\mathbf{x})\right].$$

$$\begin{aligned}\text{(ii)}\ \nabla \times \left[\vec{\mathbf{U}}(\mathbf{x}) \times \vec{\mathbf{V}}(\mathbf{x})\right] \equiv\ & \vec{\mathbf{V}}(\mathbf{x}) \cdot \left[\nabla \vec{\mathbf{U}}(\mathbf{x})\right] - \vec{\mathbf{U}}(\mathbf{x}) \cdot \left[\nabla \vec{\mathbf{V}}(\mathbf{x})\right] \\ & + \vec{\mathbf{U}}(\mathbf{x}) \left[\nabla \cdot \vec{\mathbf{V}}(\mathbf{x})\right] - \vec{\mathbf{V}}(\mathbf{x}) \left[\nabla \cdot \vec{\mathbf{U}}(\mathbf{x})\right].\end{aligned}$$

2. Consider the equations (6.5) with $c^i \equiv 0$. (Compare this with (5.17).) Prove that the resulting set of transformations, called $O(p, n; \mathbb{R})$, constitutes an $N(N-1)/2$ parameter subgroup of the generalized Poincaré group.

3. Prove that in the absence of a body force $(F_i(x,t) \equiv 0)$, and equilibrium $\left(\frac{\partial U_i}{\partial t} \equiv 0\right)$, the equations (6.7), (6.9), and (6.10) yield

$$\Delta \left[\delta^{kl} \sigma_{kl}(\mathbf{x})\right] = 0.$$

4. Consider a perfect fluid endowed with an equation of state following an irrotational motion under a conservative body force. Prove that the equations of motion (6.15), (6.16), (6.18), and (6.20) admit the **Bernoulli integral**

$$-\frac{\partial \phi}{\partial t}(\mathbf{x}, t) + \frac{1}{2} \delta^{kl} \partial_k \phi \cdot \partial_l \phi + \mathcal{R}\left[p(\mathbf{x}, t)\right] + U(\mathbf{x}, t) = f(t).$$

(Here, f is an arbitrary function of integration.)

6.2 The Space of Constant Curvature

Consider a smooth surface M_2 embedded in $\mathbb{R}^3$. (See fig. 6.1.) Let us use a Cartesian coordinate chart in $\mathbb{R}^3$. By (5.154), the equations of the surface are provided by

$$\begin{aligned} & x^i = \xi^i(u), \\ & \text{Rank}\left[\frac{\partial \xi^i(u)}{\partial u^\alpha}\right] = 2, \\ & u \in \mathcal{D}_2 \subset \mathbb{R}^2. \end{aligned} \tag{6.21}$$

Let a plane P be generated by the tangent vector $\vec{\mathbf{t}}(\xi(u))$ on the surface M_2 and the unit normal $\vec{\mathbf{n}}(\xi(u))$ to M_2. Let the plane P intersect the surface along the image of the curve $\xi \circ \mathcal{U}$.

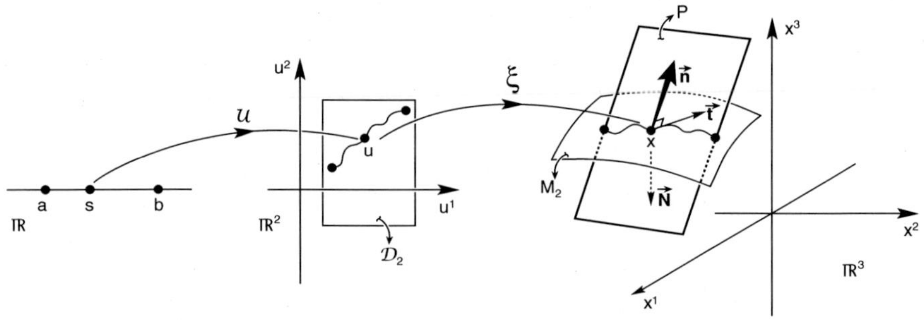

Figure 6.1: The normal section of M_2 along $\vec{\mathbf{t}}(x)$.

The first normal and the first (or principal) curvature of the curve $\xi \circ \mathcal{U}$ are denoted by $\vec{\mathbf{N}}(\xi(u))$ and $k(u) := \pm\kappa(\xi(u))$. (See the Frenet-Serret equations (5.188).) As the tangent vector $\vec{\mathbf{t}}$, the plane P, and the (consequent) normal section of $\xi \circ \mathcal{U}$ change along the curve, the value of the curvature $k(u)$ also changes. We can prove that there exist a maximum value $k_{(1)}(u)$ and a minimum value $k_{(2)}(u)$. These correspond to the principal tangential directions $\vec{\mathbf{e}}_1(\xi(u))$ and $\vec{\mathbf{e}}_2(\xi(u))$. The **Gaussian curvature** and the **mean curvature** of the surface at the point $\xi(u) \in M_2$ are defined respectively by

$$K(u) := k_{(1)}(u) \cdot k_{(2)}(u), \tag{6.22}$$

$$\mu(u) := (1/2)\left[k_{(1)}(u) + k_{(2)}(u)\right]. \tag{6.23}$$

Pictorial depictions for three simple surfaces are exhibited in fig. 6.2 below.

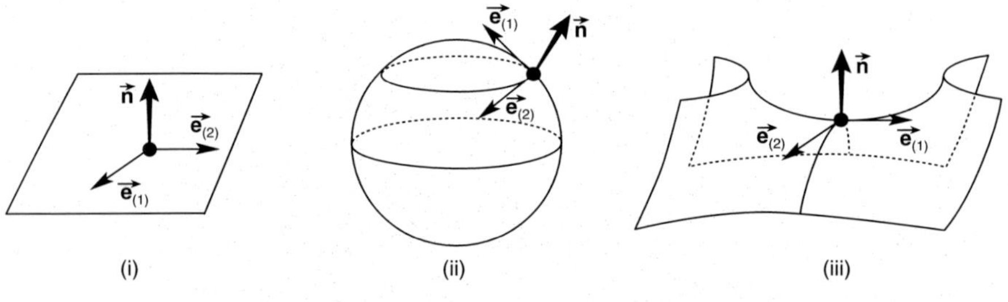

Figure 6.2: (i) A plane, (ii) a sphere, and (iii) a saddle-shaped surface.

It is intuitively evident that for (i) a plane, $k_{(1)}(u) = k_{(2)}(u) = K(u) = \mu(u) \equiv 0$. (ii) For a sphere, $k_{(1)}(u) \equiv k_{(2)}(u) < 0$, $K(u) > 0$,and $\mu(u) < 0$.

(iii) Moreover, for a saddle-shaped surface, $k_{(1)}(u) > 0$, $k_{(2)}(u) < 0$, and $K(u) < 0$.

The metric tensor field (or the **first fundamental form**) of M_2 can be derived from (6.21) and (3.49) as

$$\begin{aligned} {}_2(\boldsymbol{\xi}^{-1})'(\delta_{ij}dx^i \otimes dx^j{}_{|\xi(u)}) &= \delta_{ij}\left[\frac{\partial \xi^i(u)}{\partial u^\alpha}du^\alpha\right] \otimes \left[\frac{\partial \xi^j(u)}{\partial u^\beta}du^\beta\right] \\ &= \left[\delta_{ij}\frac{\partial \xi^i(u)}{\partial u^\alpha}\frac{\partial \xi^j(u)}{\partial u^\beta}\right] du^\alpha \otimes du^\beta \\ &=: g_{\alpha\beta}(u)du^\alpha \otimes du^\beta. \end{aligned} \tag{6.24}$$

Gauss, in his famous "**Theorema Egregium**," proved that the curvature in (6.22) can be expressed as

$$K(u) = R_{1212}(u)/\left[g_{11}(u)g_{22}(u) - (g_{12}(u))^2\right]. \tag{6.25}$$

Here, $R_{1212}(u)$ stands for the Riemann-Christoffel tensor component derived from $g_{\alpha\beta}(u)$. By the right-hand side of (6.25), it has been demonstrated that the Gaussian curvature of a surface M_2 embedded in the Euclidean space $\mathbb{E}_3$ can be expressed in terms of the *intrinsic* geometrical properties of the surface M_2 *alone* without reference to the three-dimensional space in which it is embedded.

We would like to generalize now the Gaussian curvature $K(x)$ for higher-dimensional Riemannian or pseudo-Riemannian manifolds. Consider a two-dimensional "plane" Π_2 **embedded** in M_N. It can be characterized by

$$\begin{aligned} &x^i = \xi^i(u^1, u^2) := t^i u^1 + v^i u^2, \\ &(u^1, u^2) \in \mathcal{D}_2 \subset \mathbb{R}^2, \\ &\vec{\mathbf{t}}(x) := t^i \frac{\partial}{\partial x^i} \in T_x\left(\mathbb{R}^N\right), \\ &\vec{\mathbf{v}}(x) := v^i \frac{\partial}{\partial x^i} \in T_x\left(\mathbb{R}^N\right). \end{aligned} \tag{6.26}$$

(Here, we have assumed that $\vec{\mathbf{t}}(x)$ and $\vec{\mathbf{v}}(x)$ are *linearly independent.*) We consider all geodesics locally passing through the point $x \in \Pi_2$ that are *tangential* to Π_2. The collection of all such geodesics generates a two-dimensional **geodesic surface** M_2. The Gaussian curvature of the geodesic surface M_2 at the point $x \in \Pi_2 \cap M_2$ is expressed from (6.25) as

$$K(x, \vec{\mathbf{t}}, \vec{\mathbf{v}}) := \frac{R_{lijk}(x)t^l v^i t^j v^k}{\left[g_{hm}(x)g_{in}(x) - g_{hn}(x)g_{im}(x)\right] t^h v^i t^m v^n}. \tag{6.27}$$

The $K(x, \vec{\mathbf{t}}, \vec{\mathbf{v}})$ above is called the (**Riemannian**) **sectional curvature** of the N-dimensional manifold M_N *relative* to the plane Π_2 (spanned by $\vec{\mathbf{t}}$ and $\vec{\mathbf{v}}$) at the point $x \in \Pi_2 \cap M_N$.

In the case $K(x, \vec{\mathbf{t}}, \vec{\mathbf{v}})$ assumes the *same* value $k(x)$ for every pair $\vec{\mathbf{t}}(x), \vec{\mathbf{v}}(x) \in T_x(\mathbb{R}^N)$, we call the point $x \in M_N$ **isotropic**. We shall now derive the necessary and sufficient conditions for isotropy. At a point of isotropy, by (6.27) we obtain the identity

$$\begin{aligned} &T_{lijk}(x) t^l v^i t^j v^k \equiv 0, \\ &T_{lijk}(x) := R_{lijk}(x) - k(x)\left[g_{lj}(x) g_{ik}(x) - g_{lk}(x) g_{ij}(x)\right]. \end{aligned} \tag{6.28}$$

The identities (6.28) for arbitrary values of t^i, v^j, with the help of identities (5.100)–(5.107), yield

$$\begin{aligned} &T_{lijk}(x) \equiv -T_{iljk}(x) \equiv -T_{likj}(x) \equiv T_{jkli}(x), \\ &T_{lijk}(x) + T_{ljki}(x) + T_{lkij}(x) \equiv 0, \\ &0 \equiv \frac{\partial^4}{\partial t^h \partial v^m \partial t^g \partial v^n}\left[T_{lijk} t^l v^i t^j v^k\right] \\ &\quad = T_{gnhm}(x) + T_{hngm}(x) + T_{gmhn}(x) + T_{hmgn}(x). \end{aligned} \tag{6.29}$$

We can conclude from (6.28) and (6.29) that

$$\begin{aligned} &T_{lijk}(x) \equiv 0, \\ &R_{lijk}(x) \equiv k(x)\left[g_{lj}(x) g_{ik}(x) - g_{lk}(x) g_{ij}(x)\right]. \end{aligned} \tag{6.30}$$

Now, we come to an interesting theorem due to Schur.

Theorem 6.2.1 *Let M_N be a differentiable manifold with $N > 2$. Moreover, let the metric field $\mathbf{g}_{..}(x)$ be thrice differentiable and every $x \in D \subset \mathbb{R}^N$ be isotropic. Then, the sectional curvature $k(x)$ must be constant-valued.*

Proof. By (6.29) and the Bianchi identities (5.108), we derive that

$$\begin{aligned} &(\partial_m k)(g_{lj} g_{ik} - g_{lk} g_{ij}) + (\partial_j k)(g_{lk} g_{im} - g_{lm} g_{ik}) \\ &\quad + (\partial_k k)(g_{lm} g_{ij} - g_{lj} g_{im}) \equiv 0. \end{aligned}$$

By the multiplication of $g^{lj}(x) g^{ik}(x)$ and the contractions, the identities above yield

$$(N-1)(N-2)(\partial_m k) \equiv 0.$$

Thus, for $N > 2$, $\partial_m k \equiv 0$ and $k(x)$ is constant-valued. ■

Remark: The constancy of the sectional curvatures is also called **homogeneity**. The three-dimensional spatial submanifold of our cosmological universe is assumed to be isotropic and homogeneous.

A Riemannian or pseudo-Riemannian manifold with *both* isotropy and homogeneity is called a manifold or **space of constant curvature** ($k(x) = K_0 =$ *constant*). By (6.30), for a space of constant curvature, we obtain

$$\begin{aligned}
R_{lijk}(x) &= K_0 \left[g_{lj}(x)g_{ik}(x) - g_{lk}(x)g_{ij}(x)\right], \\
R_{ij}(x) &= -(N-1)K_0 g_{ij}(x), \\
R(x) &= -N(N-1)K_0, \\
G_{ij}(x) &= \left(\frac{1}{2}\right)(N-1)(N-2)K_0 g_{ij}(x), \\
R_{abcd}(x) &= K_0\left(d_{ac}d_{bd} - d_{ad}d_{bc}\right), \\
R_{ab} &= -(N-1)K_0 d_{ab}, \\
G_{ab}(x) &= \left(\frac{1}{2}\right)(N-1)(N-2)K_0 d_{ab}.
\end{aligned} \tag{6.31}$$

Note that orthonormal components of the curvature tensor, etc., are all *constant-valued.*

Example 6.2.2 In a two-dimensional Riemannian surface of constant curvature, (6.31) reduces in an orthogonal coordinate chart to the following:

$$\partial_1 \left[\frac{\partial_1 g_{22}}{\sqrt{g}}\right] + \partial_2 \left[\frac{\partial_2 g_{11}}{\sqrt{g}}\right] + 2K_0\sqrt{g} = 0. \tag{6.32}$$

Assuming a geodesic normal coordinate chart (5.271) in this case, (6.32) reduces to

$$\begin{aligned}
\mathbf{g}_{..}(x) &= dx^1 \otimes dx^1 + \left[h(x^1, x^2)\right]^2 dx^2 \otimes dx^2, \\
&\frac{\partial^2}{(\partial x^1)^2} h(x^1, x^2) + K_0 h(x^1, x^2) = 0, \\
&(x^1, x^2) \in D \subset \mathbb{R}^2.
\end{aligned} \tag{6.33}$$

We impose initial values

$$h(0, x^2) \equiv 1, \quad \frac{\partial}{(\partial x^1)} h(x^1, x^2)_{|x^1=0} \equiv 0. \tag{6.34}$$

We shall now derive solutions of (6.33) and (6.34) in three distinct cases. □

Example 6.2.3 For the flat case $K_0 = 0$, the general solution of (6.33) is furnished by

$$h(x^1, x^2) = f(x^2)x^1 + g(x^2),$$

where f and g are arbitrary functions of integration. Solving the initial values in (6.34), we derive that

$$\mathbf{g}_{..}(x) = dx^1 \otimes dx^1 + dx^2 \otimes dx^2.$$

Therefore, the domain of the flat surface M_2 is locally isometric to a plane. □

Example 6.2.4 In the case where $K_0 > 0$, the integration of (6.33) yields the general solution

$$h(x^1, x^2) = A(x^2)\sin(\sqrt{K_0}x^1) + B(x^2)\cos(\sqrt{K_0}x^1).$$

Solving the initial values in (6.34), we deduce that

$$\mathbf{g}_{..}(x) = dx^1 \otimes dx^1 + \left[\cos(\sqrt{K_0}x^1)\right]^2 dx^2 \otimes dx^2.$$

Making a coordinate transformation

$$x^1 = \left(1/\sqrt{K_0}\right)\left[(\pi/2) - \widehat{x}^1\right], \quad x^2 = \left(1/\sqrt{K_0}\right)\widehat{x}^2,$$

we finally obtain

$$\widehat{\mathbf{g}}_{..}(\widehat{x}) = (1/K_0)\left[d\widehat{x}^1 \otimes d\widehat{x}^1 + \left(\sin \widehat{x}^1\right)^2 d\widehat{x}^2 \otimes d\widehat{x}^2\right].$$

The metric above shows that the domain of a surface M_2 of the constant positive curvature K_0 is locally isometric to the domain of a spherical surface of radius $(1/\sqrt{K_0})$. □

Example 6.2.5 Assume that $K_0 < 0$. The general solution of (6.33) is given by

$$h(x^1, x^2) = A(x^2)\exp\left[\sqrt{-K_0}x^1\right] + B(x^2)\exp\left[-\sqrt{-K_0}x^1\right].$$

The solution of the initial-value problem (6.34) yields

$$\mathbf{g}_{..}(x) = dx^1 \otimes dx^1 + \left[\cosh\left(\sqrt{-K_0}x^1\right)^2\right]^2 dx^2 \otimes dx^2.$$

In the case where we do not use the geodesic normal coordinates, we can produce *another* special solution of (6.32),

$$\mathbf{g}_{..}(x) = (-1/K_0)\left[(\cot x^1)^2 dx^1 \otimes dx^1 + (\sin x^1)^2 dx^2 \otimes dx^2\right],$$
$$(x^1, x^2) \in (0, \pi/2) \times (-\pi, \pi) \subset \mathbb{R}^2.$$

This is the surface of revolution in the shape of a *bugle*, and it is a *geodesically incomplete* surface. □

Now we shall derive the canonical form of the metric tensor for a space of constant curvature K_0.

Theorem 6.2.6 *Let M_N be a manifold (or space) of constant curvature K_0 of the differentiability class C^2. Then there exists locally a coordinate chart such that*

$$\begin{aligned}\mathbf{g}_{..}(x) &= \left[1 + (K_0/4)(d_{kl}x^k x^l)\right]^{-2}\left[d_{ij}dx^i \otimes dx^j\right]; \\ x &\in D \subset \mathbb{R}^N.\end{aligned} \tag{6.35}$$

Proof. For $N \geq 2$, we *assume* that there exists a coordinate chart such that

$$\begin{aligned} \mathbf{g}_{..}(x) &= [U(x)]^{-1} d_{ij} dx^i \otimes dx^j = [U(x)]^{-1} \sum_{j=1}^{N} \varepsilon_{(j)} dx^j \otimes dx^j; \\ U(x) &\neq 0. \end{aligned} \tag{6.36}$$

(This assumption will be *validated* in section 6.4.) *Suspending the summation convention,* we obtain from (6.36), (6.31), and (5.293) for *distinct indices* the following equations:

$$\begin{aligned} R_{hiik}(x) &= -\varepsilon_{(i)} U^{-3}(x) \partial_h \partial_k U = 0, \\ R_{kiik}(x) &= -\varepsilon_{(i)}\varepsilon_{(k)} U^{-3}(x) \left\{ \left[\varepsilon_{(i)} \partial_i \partial_i U + \varepsilon_{(k)} \partial_k \partial_k U\right] \right. \\ &\qquad \left. - U^{-1}(x) \sum_{j=1}^{N} \varepsilon_{(j)} \left(\partial_j U\right)^2 \right\} \\ &= -\varepsilon_{(i)}\varepsilon_{(k)} K_0 U^{-4}(x). \end{aligned}$$

Simplifying the equations above, we get

$$\frac{\partial^2 U(x)}{\partial x^h \partial x^k} = 0, \tag{6.37}$$

$$U(x) \left[\varepsilon_{(k)} \left(\partial_i^2 U\right) + \varepsilon_{(i)} \left(\partial_k^2 U\right)\right] = \varepsilon_{(i)}\varepsilon_{(k)} \left[K_0 + \sum_j \varepsilon_{(j)} \left(\partial_j U\right)^2\right]. \tag{6.38}$$

The general solution of (6.37) in a convex subdomain of $D \subset \mathbb{R}^N$ is furnished by

$$U(x) = F^1(x^1) + F^2(x^2) + \cdots + F^N(x^N) = \sum_{j=1}^{N} F^j(x^j). \tag{6.39}$$

Here, each F^j is an arbitrary C^2-function of the *single* variable x^j. Substituting (6.39) into (6.38), we get

$$\begin{aligned} U(x) \left[\varepsilon_{(i)}\varepsilon_{(k)} \frac{d^2 F^i(x^i)}{(dx^i)^2} + \frac{d^2 F^k(x^k)}{(dx^k)^2}\right] &= \varepsilon_{(k)} \left[K_0 + \sum_j \varepsilon_{(j)} \left(\partial_j U\right)^2\right] \\ &= U(x) \left[\varepsilon_{(l)}\varepsilon_{(k)} \frac{d^2 F^l(x^l)}{(dx^l)^2} + \frac{d^2 F^k(x^k)}{(dx^k)^2}\right]. \end{aligned}$$

Therefore, for each pair of indices i and l, we must have

$$\varepsilon_{(i)} \frac{d^2 F^i(x^i)}{(dx^i)^2} = \varepsilon_{(l)} \frac{d^2 F^l(x^l)}{(dx^l)^2} = 2a = \text{const.} \tag{6.40}$$

Here, "a" is a constant of separation. Integrating N ordinary differential equations in (6.40), we deduce that

$$\begin{aligned} F^i(x^i) &= \varepsilon_{(i)}\left[a(x^i)^2 + 2b^{(i)}x^i + c^{(i)}\right], \\ \partial_j U &= 2\varepsilon_{(j)}\left[ax^j + b^{(j)}\right]. \end{aligned} \tag{6.41}$$

Substituting (6.41) into (6.38), we derive that

$$K_0 = 4\sum_j \varepsilon_{(j)}\left[ac^{(j)} - b^{(j)}\right], \tag{6.42}$$

$$4aU(x) = K_0 + 4\sum_j \left(ax^j + b^{(j)}\right)^2. \tag{6.43}$$

In the case where $K_0 = 0$, the proof of (6.35) can be obtained from (6.31) and the existence of a Cartesian (or pseudo-Cartesian) coordinate chart for a flat manifold.

In the sequel of this proof, we choose $K_0 \neq 0$ and $a \neq 0$. By (6.43) and (6.36), we have

$$\mathbf{g}_{..}(x) = (4a)^2\left[K_o + 4\sum_j \varepsilon_{(j)}\left(ax^j + b^{(j)}\right)^2\right]^{-2} d_{ij}dx^i \otimes dx^j. \tag{6.44}$$

By making a coordinate transformation

$$\widehat{x}^j = (4a/K_0)\left(x^j + a^{-1}b^{(j)}\right)$$

and dropping hats, we can prove the validity of (6.35). ■

Remark: In example 5.3.12, problem 8 of exercises 5.3, and problem 1 of exercises 5.4, spaces of constant curvature are involved.

Now consider the following linear coordinate transformation:

$$\begin{aligned} \widehat{x}^i &= c^i + l^i{}_j x^j, \\ l^i{}_j d_{ik} l^k{}_m &= d_{jm}, \\ \sum_{k=1}^{N} \varepsilon_{(k)} l^k{}_j l^k{}_m &= \varepsilon_{(j)}\delta_{jm}. \end{aligned} \tag{6.45}$$

The set of all such coordinate transformations under the usual composition rule forms a group called the **generalized Poincaré group** $\mathcal{IO}(p,n;\mathbb{R})$. (See (5.17) and (6.5) and problem 2 in exercises 6.1.)

Under the $N(N+1)/2$ parameter group of transformations (6.44), the metric form (6.35) remains intact. (Or, in other words, the metric (6.35) admits $N(N+1)/2$ Killing vectors.) See Appendix 2.

Now, we shall state a deep theorem.

Theorem 6.2.7 *A differentiable manifold M_N with a metric admits a group of motions involving $N(N+1)/2$ parameters if and only if it is a space of constant curvature.*

(For the proof of the theorem above, see the book by Eisenhart [10].)

Exercises 6.2

1. (i) Express the sectional curvature as

$$K(x, \vec{\mathbf{t}}, \vec{\mathbf{v}}) \equiv \tilde{K}(x, \tilde{\mathbf{t}}, \tilde{\mathbf{v}}) := \frac{R^{likj}(x) t_l v_i t_j v_k}{(g^{hm} g^{in} - g^{hk} g^{im}) t_h v_i t_m v_n}.$$

Prove that

$$4\tilde{K}(x, \tilde{\mathbf{t}}, \tilde{\mathbf{v}}) = \frac{\mathbf{R}^{\cdots\cdot}(x)\left(\tilde{\mathbf{t}}, \tilde{\mathbf{v}}, \tilde{\mathbf{t}}, \tilde{\mathbf{v}}\right)}{\wedge\mathbf{g}^{\cdots\cdot}(x)\left(\tilde{\mathbf{t}} \wedge \tilde{\mathbf{v}}, \tilde{\mathbf{t}} \wedge \tilde{\mathbf{v}}\right)}.$$

(ii) Let $\{\vec{\mathbf{e}}_a(x)\}_1^N$ be an orthonormal basis set. *Suspending* the summation convention for this problem, prove that the Ricci tensor components satisfy

$$\mathbf{R}^{\cdot\cdot}(x)\left(\vec{\mathbf{e}}_b(x), \vec{\mathbf{e}}_b(x)\right) = -\varepsilon_{(b)} \sum_{a \neq b}{}' K\left(x, \vec{\mathbf{e}}_a, \vec{\mathbf{e}}_b\right).$$

2. (i) Prove that in a regular domain of a space of constant curvature, the orthonormal components of the curvature tensor satisfy the criteria of a **symmetric space**, namely

$$\nabla_a R_{bcde}(x) \equiv 0.$$

(ii) Prove that the **Kretschmann invariant** for a space of constant curvature satisfies

$$R_{ijkl}(x) R^{ijkl}(x) = R_{abcd}(x) R^{abcd}(x) = 2N(N-1)(K_0)^2.$$

6.3 Einstein Spaces

We shall define a space *more general* than a space of constant curvature. We assume that in a regular domain of such a space, the metric field is of class C^3. In a coordinate chart, let the equations

$$\begin{aligned}
\mathbf{R}_{..}(x) &= [R(x)/N]\,\mathbf{g}_{..}(x), \\
R_{ij}(x) &= [R(x)/N]\,g_{ij}(x), \\
R_{ab}(x) &= [R(x)/N]\,d_{ab}(x), \\
x &\in D \subset \mathbb{R}^N,
\end{aligned} \tag{6.46}$$

hold. Such a domain D is called a domain of an **Einstein space**. (See the book by Petrov [32].)

Consider an *arbitrary non-zero* vector field $\vec{\mathbf{v}}(x) \in T_x(\mathbb{R}^N)$. By (6.46), we derive that

$$R_{ij}(x)v^j(x) = [R(x)/N]\,v_i(x). \tag{6.47}$$

Therefore, the Ricci tensor in such a space possesses an invariant eigenvalue $R(x)/N$. The corresponding subset of "eigenvectors" is given by $T_x(\mathbb{R}^N) - \{\vec{\mathbf{o}}_x\}$.

Now we shall state and prove a theorem about Einstein spaces and spaces of constant curvature.

Theorem 6.3.1 *A regular domain of a space of constant curvature must be a regular domain of an Einstein space.*

Proof. By (6.31) for $N > 1$, we obtain

$$R_{ij}(x) = -(N-1)K_0 g_{ij}(x) = (1/N)R(x)g_{ij}(x).$$

Thus, by (6.46) the domain is that of an Einstein space. For $N = 1$, the proof is trivial. ■

Remark: The converse of the theorem above is *not* true.

The next theorem deals with the scalar invariant of an Einstein space.

Theorem 6.3.2 *In the case where $N > 2$, the scalar invariant $R(x)$ in a regular domain $D \subset \mathbb{R}^N$ corresponding to that of an Einstein space is constant-valued.*

Proof. By (5.119) and (5.123), it follows that

$$\nabla_j R^j{}_i - (1/2)\partial_i R \equiv 0$$

in $D \subset \mathbb{R}^N$. Using (6.46), the identity above yields

$$[(N-2)/N]\,\partial_i R \equiv 0.$$

Therefore, for $N > 2$, $\partial_i R \equiv 0$ and $R(x) = k =$ constant. ■

Remark: A regular domain of any two-dimensional manifold M_2 must be a domain of an Einstein space. (See problem 3(ii) of exercises 5.2.)

Example 6.3.3 Consider the four-dimensional space-time manifold and the following metric field in a chosen chart:

$$\begin{aligned} \mathbf{g}_{..}(x) &= \left[\sin\left((\sqrt{3\lambda}/2)\right)x^3\right]^{4/3}(dx^1 \otimes dx^1 + dx^2 \otimes dx^2) + dx^3 \otimes dx^3 \\ &\quad - \left[\cos\left((\sqrt{3\lambda}/2)x^3\right)\right]^2 \left[\sin\left((\sqrt{3\lambda}/2)x^3\right)\right]^{-2/3} dx^4 \otimes dx^4, \\ D &:= \left\{x \in \mathbb{R}^4, x^1 \in \mathbb{R}, 0 < x^3 < \left(\pi/\sqrt{3\lambda}\right), x^4 \in \mathbb{R}\right\}. \end{aligned}$$

Here, $\lambda > 0$ is the cosmological constant. By direct computation by (5.114), we obtain

$$R_{ij}(x) = -\lambda g_{ij}(x) = [R(x)/4]\, g_{ij}(x).$$

Thus, $D \subset \mathbb{R}^N$ is a domain of an Einstein space. (This example is from Novotny and Horsky [29].) □

The next theorem is due to Thomas discussed by Petrov [32].

Theorem 6.3.4 *Let $D \subset \mathbb{R}^N$ be a regular domain corresponding to that of a positive-definite differentiable manifold M_N. The domain corresponds to that of an Einstein space if and only if*

$$N R_{ij}(x) R^{ij}(x) - [R(x)]^2 = 0. \tag{6.48}$$

For the proof, see Petrov [32].

A special class of Einstein spaces with vanishing curvature invariant are called **Ricci-flat spaces**. (For this class, $R_{ij}(x) \equiv 0$.) In the cases where $N = 2$ and $N = 3$, the Ricci flatness is *equivalent* to the usual flatness. For $N \geq 4$, Ricci flatness is a *weaker* condition than flatness. In Einstein's theory of gravitation, a Ricci-flat domain indicates the *absence of sources generating gravitation.*

(See the references [18], [25], [27] and [38].)

Example 6.3.5 A space-time metric field

$$\begin{aligned} \mathbf{g}_{..}(x) &= (x^1)^{2c_1} dx^1 \otimes dx^1 + (x^1)^{2c_2} dx^2 \otimes dx^2 \\ &\quad + (x^1)^{2c_3} dx^3 \otimes dx^3 - (x^1)^{2c_4} dx^4 \otimes dx^4, \\ D &:= \left\{x \in \mathbb{R}^4 : x^1 > 0, (x^2, x^3, x^4) \in \mathbb{R}^3\right\}, \\ c_2 + c_3 + c_4 &= 1 + c_1, \;\; (c_2)^2 + (c_3)^2 + (c_4)^2 = (c_1 + 1)^2 \end{aligned}$$

is an example of a Ricci-flat domain. (This metric was first derived by Kasner [23].) □

Exercises 6.3

1. Show that a regular domain of a three-dimensional Einstein space must be that of a space of constant curvature.

2: Consider the following metric for $N \geq 2$:

$$\mathbf{g}_{..}(x) := \sum_{\alpha=1}^{N-1} \left[h_{(\alpha)}(x^N)\right]^2 dx^\alpha \otimes dx^\alpha - dx^N \otimes dx^N .$$

Prove that for

$$h_{(\alpha)}(x^N) = (\sin \beta x^N)^{1/N-1} \left[\tan\left(\beta x^N/2\right)\right]^{\beta_\alpha/\beta},$$

$\beta := \beta_1 + \cdots + \beta_{N-1} \neq 0,\ \ \alpha \in \{1, \ldots, N-1\}$, the resulting metric field yields an Einstein space (Petrov [31]).

3. Consider the following metric field in a regular domain of a space-time manifold:

$$\mathbf{g}_{..} := W(x^1, x^2, x^3) dx^1 \otimes dx^1 + 2dx^1 \otimes dx^4 + dx^2 \otimes dx^2 + dx^3 \otimes dx^3,$$
$$W(x^1, x^2, x^3) := \left[(x^2)^2 - (x^3)^2\right] f(x^1) - 2x^2 x^3 h(x^1).$$

(Here, f and h are arbitrary C^3-functions.) Show that the metric represents a Ricci-flat domain.

Remark: The metric above represents the plane gravitational-wave solution of Bondi, Pirani, and Robinson [4].

6.4 Conformally Flat Spaces

We shall now investigate **conformal mappings**. This term is well known in complex analysis. A mapping from a domain of a complex plane into another complex plane such that the angle (and the sense) between each pair of intersecting curves is preserved is called a **conformal mapping**. We generalize this concept to the conformal mapping of a regular domain of M_N into $\bar{M}_N$. (For a global definition, it is assumed that M_N and $\bar{M}_N$ have *identical atlases.*) The **conformal mapping** can be defined (in a specific chart for both spaces) as

$$\begin{aligned}
\bar{\mathbf{g}}_{..}(x) &:= \exp\left[2\mu(x)\right] \mathbf{g}_{..}(x), \\
\bar{\mathbf{g}}^{..}(x) &= \exp\left[-2\mu(x)\right] \mathbf{g}^{..}(x), \\
\bar{\tilde{\mathbf{e}}}^a(x) &= \exp\left[\mu(x)\right] \tilde{\mathbf{e}}^a(x), \\
\bar{\vec{\mathbf{e}}}_a(x) &= \exp\left[-\mu(x)\right] \vec{\mathbf{e}}_a(x), \\
x &\in D \subset \mathbb{R}^N.
\end{aligned} \tag{6.49}$$

Here, μ is assumed to be a thrice-differentiable function into $\mathbb{R}$. We note that by (6.49) we can obtain

$$\begin{aligned} \bar{\mathbf{g}}_{..}(x)(\vec{\mathbf{v}}(x), \vec{\mathbf{v}}(x)) &= \exp[2\mu(x)]\, \mathbf{g}_{..}(x)(\vec{\mathbf{v}}(x), \vec{\mathbf{v}}(x)), \\ \bar{\mathbf{g}}_{..}(x)(\vec{\mathbf{v}}(x), \vec{\mathbf{w}}(x)) &= \exp[2\mu(x)]\, \mathbf{g}_{..}(x)(\vec{\mathbf{v}}(x), \vec{\mathbf{w}}(x)). \end{aligned} \tag{6.50}$$

The equation above shows that spacelike, timelike, and null vectors are mapped into spacelike, timelike, and null vectors, respectively. Moreover, for a positive-definite metric $\mathbf{g}_{..}(x)$, the *angle* between two non-zero vectors is *exactly preserved.*

By (6.49) and (5.41), we obtain the transformation between Christoffel symbols as

$$\overline{\left\{ {k \atop ij} \right\}} = \left\{ {k \atop ij} \right\} + \delta^k{}_j \partial_i \mu + \delta^k{}_i \partial_j \mu - g_{ij} g^{kl} \partial_l \mu. \tag{6.51}$$

By (6.49), (5.93), and (5.96), (after a long calculation) we obtain for the transformed curvature tensor components

$$\begin{aligned} \bar{R}^i{}_{jkl}(x) &= R^i{}_{jkl}(x) + \delta^i{}_l \left[\nabla_k \nabla_j \mu - (\partial_k \mu)(\partial_j \mu)\right] \\ &- \delta^i{}_k \left[\nabla_l \nabla_j \mu - (\partial_l \mu)(\partial_j \mu)\right] \\ &+ \left(\delta^i{}_l g_{jk} - \delta^i{}_k g_{jl}\right) \left[g^{mn} (\partial_m \mu)(\partial_n \mu)\right] \\ &+ g^{im} \left\{ g_{kj} \left[\nabla_l \nabla_m \mu - (\partial_l \mu)(\partial_m \mu)\right] - g_{jl} \left[\nabla_k \nabla_m \mu - (\partial_k \mu)(\partial_m \mu)\right]\right\}, \end{aligned} \tag{6.52}$$

$$\begin{aligned} e^{2\mu} \bar{R}^a{}_{bcd}(x) &= R^a{}_{bcd}(x) + \delta^a{}_d \left[\nabla_c \nabla_b \mu - (\partial_c \mu)(\partial_b \mu)\right] \\ &- \delta^a{}_c \left[\nabla_d \nabla_b \mu - (\partial_d \mu)(\partial_b \mu)\right] \\ &+ \left(\delta^a{}_d d_{bc} - \delta^a{}_c d_{bd}\right) \left[d^{ef} (\partial_e \mu)(\partial_f \mu)\right] \\ &+ d^{ae} \left\{ d_{bc} \left[\nabla_d \nabla_e \mu - (\partial_d \mu)(\partial_e \mu)\right] - d_{bd} \left[\nabla_c \nabla_e \mu - (\partial_c \mu)(\partial_e \mu)\right]\right\}. \end{aligned} \tag{6.53}$$

Here, the covariant derivatives are performed in the chart for M_N. By (6.49), (6.52), (6.53), (5.114), (5.115), (5.116), (5.117), (5.119), and (5.120), we get

$$\begin{aligned} \bar{R}_{jk}(x) &= R_{jk}(x) + (N-2)\left[\nabla_k \nabla_j \mu - (\partial_k \mu)(\partial_j \mu)\right] \\ &\quad + g_{jk} \left[\nabla^i \nabla_i \mu + (N-2) g^{mn} (\partial_m \mu)(\partial_n \mu)\right], \\ e^{2\mu} \bar{R}_{ab}(x) &= R_{ab}(x) + (N-2)\left[\nabla_a \nabla_b \mu - (\partial_a \mu)(\partial_b \mu)\right] \\ &\quad + d_{ab} \left[\nabla^c \nabla_c \mu + (N-2) d^{cd} (\partial_c \mu)(\partial_d \mu)\right], \\ \bar{R}(x) &= \exp[-2\mu(x)] \left\{ R(x) + 2(N-1)(\nabla^i \nabla_i \mu) \right. \\ &\quad \left. + (N-1)(N-2)\left[g^{ij} (\partial_i \mu)(\partial_j \mu)\right]\right\}, \\ \bar{G}_{jk}(x) &= G_{jk}(x) + (N-2)\left[\nabla_k \nabla_j \mu - (\partial_k \mu)(\partial_j \mu)\right] \\ &\quad - (1/2)(N-2) g_{jk} \left[2(\nabla^i \nabla_i \mu) + (N-3) g^{il} (\partial_i \mu)(\partial_l \mu)\right]. \end{aligned} \tag{6.54}$$

Now we shall define Weyl's **conformal tensor**. (See problem 4 of exercises 5.2.) For $N \geq 3$, Weyl's conformal tensor is defined by

$$
\begin{aligned}
C^l{}_{ijk}(x) &:= R^l{}_{ijk}(x) + \frac{1}{N-2}\left[\delta^l{}_j R_{ik} - \delta^l{}_k R_{ij} + g_{ik} R^l{}_j - g_{ij} R^l{}_k\right] \\
&\quad + \frac{R(x)}{(N-1)(N-2)}\left[\delta^l{}_k g_{ij} - \delta^l{}_j g_{ik}\right],
\end{aligned}
\tag{6.55}
$$

$$
\begin{aligned}
C^d_{abc}(x) &:= R^d{}_{abc}(x) + \frac{1}{N-2}\left[\delta^d{}_b R_{ac} - \delta^d{}_c R_{ab} + d_{ac} R^d{}_b - d_{ab} R^d{}_c\right] \\
&\quad + \frac{R(x)}{(N-1)(N-2)}\left[\delta^d{}_c d_{ab} - \delta^d{}_b d_{ac}\right].
\end{aligned}
\tag{6.56}
$$

An important theorem about the conformal tensor will be stated and proved now.

Theorem 6.4.1 *Consider a conformal mapping* $\bar{\mathbf{g}}_{..}(x) = \exp[2\mu(x)]\,\mathbf{g}_{..}(x)$ *of class* C^3 *in a domain* $D \subset \mathbb{R}^N (N \geq 3)$. *Then the components of the conformal tensor remain invariant; i.e.,*

$$
\bar{C}^l{}_{ijk}(x) \equiv C^l{}_{ijk}(x). \tag{6.57}
$$

Proof. By equations (6.49), (6.54), and (6.52), equation (6.57) follows. ■

Recall the algebraic identities of the Riemann-Christoffel tensor in (5.100)–(5.107). Similar and other algebraic identities of the conformal tensor are

$$
\begin{aligned}
C_{ijkl}(x) &\equiv -C_{jikl}(x) \equiv -C_{ijkl} \equiv C_{klij}(x), \\
C^l{}_{ijk}(x) &+ C^l{}_{jki}(x) + C^l{}_{kij}(x) \equiv 0, \\
C^j_{jkl}(x) &\equiv 0, \;\; C^k{}_{jkl}(x) \equiv 0, \;\; C^l{}_{jkl}(x) \equiv 0.
\end{aligned}
\tag{6.58}
$$

By Bianchi identities (5.108) for $N \geq 3$, the conformal tensor satisfies the following differential identities:

$$
\begin{aligned}
&\nabla_l C^h{}_{ijk} + \nabla_j C^h{}_{ikl} + \nabla_k C^h{}_{ilj} \equiv (N-2)^{-1}\left(\delta^h{}_j R_{ikl} + \delta^h{}_k R_{ilj} + \delta^h{}_l R_{ijk}\right. \\
&\qquad \left. + g_{ik} R^h{}_{jl} + g_{il} R^h{}_{kj} + g_{ij} R^h{}_{lk}\right), \\
&R_{ijk}(x) := \nabla_k R_{ij} - \nabla_j R_{ik} + [2(N-1)]^{-1}\left(g_{ik}\partial_j R - g_{ij}\partial_k R\right) \equiv -R_{ikj}(x), \\
&R^i{}_{ik}(x) \equiv 0.
\end{aligned}
\tag{6.59}
$$

The third-order tensor field $R_{ijk}(x)$ can be called the **Schouten-Cotton tensor**.

The contraction of the identities for $N \geq 3$ in (6.57) leads to

$$\nabla_h C^h{}_{ilj} = \left(\frac{N-3}{N-2}\right) R_{ilj}(x). \tag{6.60}$$

A regular domain of a **conformally flat** space M_N is characterized by the conditions

$$\begin{aligned} \mathbf{g}_{..}(x) &= \exp[-2\mu(x)]\,\overline{\mathbf{g}}_{..}(x), \quad g_{ij}(x) = \exp[-2\mu(x)]\,\overline{g}_{ij}(x), \\ \overline{R}^i{}_{jkl}(x) &\equiv 0. \end{aligned} \tag{6.61}$$

We shall now state and prove the main theorem regarding the conformal flatness.

Theorem 6.4.2 *A regular domain of M_N for $N \geq 4$ is conformally flat if and only if $C^l{}_{ijk}(x) \equiv 0$ in the corresponding domain $D \subset \mathbb{R}^N$.*

Proof. (i) Assume that the domain is conformally flat. Then, by (6.61), (5.112), (5.116), and (6.55),

$$\begin{aligned} &\overline{R}^i{}_{jkl}(x) \equiv 0, \;\; \overline{R}_{jk}(x) \equiv 0, \;\; \overline{R}(x) \equiv 0, \\ &\overline{C}^l{}_{ijk}(x) \equiv 0. \end{aligned}$$

By theorem 6.4.1, we must have

$$C^l{}_{ijk}(x) \equiv 0.$$

(ii) Assume that $C^l{}_{ijk}(x) \equiv 0$ in $D \subset \mathbb{R}^N$. By (6.60), we get

$$R_{ijk}(x) \equiv 0.$$

Now, consider an overdetermined system of partial differential equations

$$\begin{aligned} \nabla_j \nabla_i w = \;& (\partial_i w)(\partial_j w) - (1/2) g_{ij} g^{kl} (\partial_k w)(\partial_l w) \\ & + (N-2)^{-1} \left\{ [2(N-1)]^{-1} R g_{ij} - R_{ij} \right\}. \end{aligned} \tag{6.62}$$

To investigate the *integrability*, we derive from (6.62), (6.59), and (5.110) that

$$\begin{aligned} (\partial_l w) R^l{}_{ijk} &= \nabla_k \nabla_j \nabla_i w - \nabla_j \nabla_k \nabla_i w, \\ (\partial_l w) C^l{}_{ijk}(x) &+ R_{ijk}(x) = 0. \end{aligned} \tag{6.63}$$

By our previous assumptions, the equation above (which is the integrability condition for (6.62)) is *identically* satisfied. Thus, a solution $w(x)$ of (6.62) locally exists. We identify $\mu(x) := w(x)$. Thus, $\overline{g}_{ij}(x) = \exp[2w(x)] g_{ij}(x)$. By theorem 6.4.1, we have

$$\overline{C}^l{}_{ijk}(x) \equiv 0.$$

Equations (6.54) yield

$$\begin{aligned}&(N-2)^{-1}\left\{[2(N-1)]^{-1}\overline{R}\,\overline{g}_{ij}-\overline{R}_{ij}\right\}=\\&\quad-\nabla_i\nabla_j w+(\partial_i w)(\partial_j w)-(1/2)g_{ij}g^{kl}(\partial_k w)(\partial_l w)\\&\quad+(N-2)^{-1}\left\{[2(N-1)]^{-1}R\,g_{ij}-R_{ij}\right\}.\end{aligned}$$

Since $w(x)$ satisfies (6.62), we obtain

$$[2(N-1)]^{-1}\overline{R}\,\overline{g}_{ij}-\overline{R}_{ij}\equiv 0,\;\;\overline{R}_{ij}(x)\equiv 0,\;\;\overline{R}(x)\equiv 0.$$

Thus, by $\overline{C}^l{}_{ijk}(x)\equiv 0$, we get $\overline{R}^l{}_{ijk}(x)\equiv 0$. Therefore, the metric $\mathbf{g}_{..}(x)$ is conformally flat. ■

Remarks: (i) Any regular domain of M_2 is conformally flat.

(ii) A regular domain of M_3 is conformally flat if and only if $R_{ijk}(x)\equiv 0$.

(iii) Any regular domain of a space of constant curvature M_N is conformally flat.

(iv) In a regular Ricci-flat domain, $C^l{}_{ijk}(x)\equiv R^l_{ijk}(x)$.

Remark: The last equation has been utilized to investigate gravitational waves. (See the reference [25].)

Consider a conformally flat space M_N. It has a coordinate chart such that

$$g_{ij}(x)=\exp[2\nu(x)]d_{ij}\,. \tag{6.64}$$

We consider another chart intersecting the preceding one such that

$$\widehat{g}_{ij}(\widehat{x})=\exp[2\widehat{\mu}(\widehat{x})]d_{ij}\,. \tag{6.65}$$

It is natural to investigate the possible coordinate transformations that take (6.64) into (6.65). The answer was given by Liouville a long time ago (See Bianchi [1].). To understand these transformations, consider the following coordinate transformations:

$$\widehat{x}^i=\lambda x^i\neq 0; \tag{6.66}$$

$$\widehat{x}^i=x^i+c^i; \tag{6.67}$$

$$\widehat{x}^i=l^i{}_j x^j,\;\; d_{ik}l^i{}_j l^k{}_m=d_{jm}; \tag{6.68}$$

$$\widehat{x}^i=\left[x^i/(d_{jk}x^j x^k)\right]; \tag{6.69}$$

$$\begin{aligned}\widehat{x}^i&=\left[x^i-b^i(x_l x^l)\right]/\left[1-2b_j x^j+(b_k b^k)(x_m x^m)\right],\\x_l&:=d_{lk}x^k,\;\;x_l x^l\neq 0.\end{aligned} \tag{6.70}$$

The first of the transformations in (6.66) is called a **dilation** (or a scale transformation). The second transformation (6.68) is the **generalized Lorentz group** $O(p, n, \mathbb{R})$ (with $p+n = N$). The next transformation (6.69) is the **inversion**. The last transformation (6.70) is called a **special conformal transformation**. It can be obtained by the composition (inversion) $\circ$ (translation) $\circ$ (inversion). The compositions of the transformations (6.66) to (6.68) and (6.70) constitute the $(1/2)(N+1)(N+2)$ parameter **conformal group** $C(p, n; \mathbb{R})$.

Theorem 6.4.3 *A conformally flat metric field* $g_{ij}(x) = \exp[2\nu(x)]d_{ij}$ *for* $N \geq 3$ *goes over to a conformally flat metric* $\widehat{g}_{ij}(\widehat{x}) = \exp[2\widehat{\mu}(\widehat{x})]d_{ij}$ *if and only if the coordinate transformation belongs to the conformal group* $C(p, n; \mathbb{R})$.

Proof. We shall provide only a *partial* proof. Assume that the transformation belongs to the conformal group. It is not difficult to prove that the conformally flat metric $\exp[2\nu(x)]d_{ij}$ goes over to another conformally flat metric under each of the transformations in (6.66), (6.67), and (6.68). Let us consider the inversion in (6.69). We derive the differential forms by (3.46):

$$\tilde{\mathbf{X}}'(d\widehat{x}^i) = (x^l x_l)^{-2} dx^i - 2(x^l x_l)^{-4} d_{jk} x^i x^j dx^k.$$

Therefore, the flat metric

$$\begin{aligned} {}_2\mathbf{X}'(d_{ij} d\widehat{x}^i \otimes d\widehat{x}^j) &= (x^l x_l)^{-8} d_{ij} \left[(x^h x_h)^2 dx^i - 2d_{nk} x^i x^n dx^k\right] \\ &\quad \otimes \left[(x^p x_p)^2 dx^j - 2d_{ml} x^j x^m dx^l\right] \\ &= (x^l x_l)^{-4} d_{ij} dx^i \otimes dx^j + \mathbf{O}_{..}(x). \end{aligned}$$

Thus, we deduce that

$$\begin{aligned} {}_2\widehat{\mathbf{X}}'(\exp[2\nu(x)] d_{ij} dx^i \otimes dx^j) &= (x^l x_l)^4 \exp[2\nu(x)] d_{ij} d\widehat{x}^i \otimes d\widehat{x}^j \\ &= (\widehat{x}^l \widehat{x}_l)^{-4} \exp[2\widehat{\nu}(\widehat{x})] d_{ij} d\widehat{x}^i \otimes d\widehat{x}^j \\ &=: \exp[2\widehat{\mu}(\widehat{x})] d_{ij} d\widehat{x}^i \otimes d\widehat{x}^j. \end{aligned}$$

Therefore, we conclude that any transformation in $C(p, n; \mathbb{R})$ takes a conformally flat metric into another conformally flat metric. ■

Example 6.4.4 Consider a manifold M_2 with signature zero. Moreover, let us choose a conformally flat chart

$$\mathbf{g}(x) = \exp[2\mu(x)](dx^1 \otimes dx^1 - dx^2 \otimes dx^2); \quad x \in D \subset \mathbb{R}^2.$$

We now make a coordinate transformation

$$\widehat{x}^1 = x^1 + x^2, \quad \widehat{x}^2 = x^1 - x^2.$$

The transformed metric is given by

$$\widehat{\mathbf{g}}_{..}(\widehat{x}) = (1/2)\exp\left[2\widehat{\mu}(\widehat{x})\right]\left(d\widehat{x}^1 \otimes d\widehat{x}^2 + d\widehat{x}^2 \otimes d\widehat{x}^1\right).$$

(The new chart is called the **doubly null coordinate chart**.) We make another coordinate transformation to null coordinates so that the conformal flatness is preserved. Therefore, we have

$$\bar{x}^1 = \bar{X}^1(\widehat{x}), \quad \bar{x}^2 = \bar{X}^2(\widehat{x}), \quad \frac{\partial(\bar{x}^1, \bar{x}^2)}{\partial(\widehat{x}^1, \widehat{x}^2)} \neq 0;$$

$$0 = \bar{g}^{11}(\widehat{x}) = \exp\left[2\widehat{\mu}(\widehat{x})\right] \frac{\partial \bar{X}^1(\widehat{x})}{\partial \widehat{x}^1} \frac{\partial \bar{X}^1(\widehat{x})}{\partial \widehat{x}^2},$$

$$0 = \bar{g}^{22}(\widehat{x}) = \exp\left[2\widehat{\mu}(\widehat{x})\right] \frac{\partial \bar{X}^2(\widehat{x})}{\partial \widehat{x}^1} \frac{\partial \bar{X}^2(\widehat{x})}{\partial \widehat{x}^2}.$$

The general solutions of the partial differential equations above are provided by

$$\bar{x}^1 = f^1(\widehat{x}^1), \quad \bar{x}^2 = f^2(\widehat{x}^2);$$

$$\text{or else} \quad \bar{x}^1 = g^1(\widehat{x}^2), \quad \bar{x}^2 = g^2(\widehat{x}^1).$$

Here, functions f^1, f^2, g^1, and g^2 are arbitrary C^1. As a proper subset of these functions, we can allow arbitrary real-analytic functions that involve a *denumerably infinite number* of parameters! This conclusion shows why the case of $N = 2$ was excluded in theorem 6.4.3. □

Example 6.4.5 Consider the space-time manifold (of signature $+2$). Let a metric field be furnished by

$$\begin{aligned}
\mathbf{g}_{..} &= \left[A(x^4)\right]^2 \Big\{ \left[1 - \varepsilon(x^1)^2\right]^{-1} dx^1 \otimes dx^1 \\
&\quad + (x^1)^2 \left[dx^2 \otimes dx^2 + (\sin x^2)^2 dx^3 \otimes dx^3\right] \Big\} - dx^4 \otimes dx^4; \\
\varepsilon &:= 0, \pm 1; \\
D &:= \left\{x \in \mathbb{R}^4 : x^1 > 0,\ 0 < x^2 < \pi,\ -\pi < x^3 < \pi,\ x^4 > 0\right\}.
\end{aligned}$$

The metric above is expressed in a geodesic normal chart or comoving, synchronized coordinate system. (See example 5.4.5.) If the function A is twice differentiable, the domain $D \subset \mathbb{R}^4$ corresponds to a *conformally flat* one in M_4. This metric is the well-known Friedmann-Robertson-Walker (FRW) metric of cosmology. □

Now, we shall state and prove the last theorem of this section.

Theorem 6.4.6 *Let the curvature tensor components in a regular domain for* $N > 2$ *satisfy*

$$\begin{aligned} R_{lijk}(x) &= \kappa(N-2)^{-1}\big[g_{lj}T_{ik} + g_{ik}T_{lj} - g_{lk}T_{ij} - g_{ij}T_{lk} \\ &\quad +2(N-1)^{-1}T(x)\,(g_{lk}g_{ij} - g_{lj}g_{ik})\big], \qquad (6.71) \\ T(x) &:= T^i{}_i(x). \end{aligned}$$

Here, κ *is a positive constant and* $\mathbf{T}_{..}(x)$ *is a differentiable tensor field. Then,*

$$(i) \qquad G_{ij}(x) = -\kappa T_{ij}(x); \qquad (6.72)$$

$$(ii) \qquad T_{ij}(x) \equiv T_{ji}(x); \quad \nabla_i T^{ij} = 0; \qquad (6.73)$$

$$(iii) \qquad \nabla_k T_{ij} - \nabla_j T_{ik} + (N-1)^{-1}(g_{ik}\partial_j T - g_{ij}\partial_k T) = 0; \qquad (6.74)$$

$$(iv) \qquad C^l{}_{ijk}(x) \equiv 0; \quad R_{ijk}(x) \equiv 0. \qquad (6.75)$$

Proof. (i) By (6.71), it follows that

$$\begin{aligned} &R_{ij}(x) = \kappa\left[(N-2)^{-1}T(x)g_{ij}(x) - T_{ij}(x)\right], \\ &R(x) = 2\kappa(N-2)^{-1}T(x), \\ &R_{ij}(x) - (1/2)g_{ij}(x)R(x) = -\kappa T_{ij}(x). \end{aligned}$$

(ii) By the symmetry of Einstein's tensor $G_{ij}(x)$ and the second contracted Bianchi identity (5.123), the proof follows.

(iii) By the first contracted Bianchi identity (5.121), the proof emerges.

(iv) Substituting $R_{lijk}(x)$ from (6.71) and consequently $R_{ij}(x)$ and $R(x)$ in (6.55) and (6.59), the proof is established. ■

Remark: In the case where $N = 4$, (6.72) represents Einstein's gravitational equation in the presence of matter. Therefore, (6.71) yields the class of conformally flat gravitational fields. (See the paper of Das [8] and also Willmore's book [39].)

Exercises 6.4

1. Prove that in a regular domain of a three-dimensional manifold, the conformal tensor $\mathbf{C}^{\cdot}{}_{...}(x) \equiv \mathbf{O}^{\cdot}{}_{...}(x)$.

2. Suppose that a *non-null*, four-dimensional vector field $\vec{\mathbf{v}}(x) = v^i(x)\frac{\partial}{\partial x^i}$ satisfies $v^i(x)C_{ijkl}(x) \equiv 0$ in a domain $D \subset \mathbb{R}^4$. Prove that the domain is conformally flat (Lovelock and Rund [26].)

3. Consider the conformal mapping (6.49) in a domain of a four-dimensional manifold. Let $\mathbf{F}_{..}(x) = (1/2)F_{ij}dx^i \wedge dx^j$ be a differentiable 2-form (electromagnetic) field in the same domain. Show the invariance of the electromagnetic field equations

$$\begin{aligned} 0 = \bar{\nabla}_i \bar{F}^{ij} \quad &\Leftrightarrow \quad \nabla_i F^{ij} = 0, \\ 0 = \bar{\nabla}_i F_{jk} + \bar{\nabla}_j F_{ki} + \bar{\nabla}_k F_{ik} \quad &= \quad \nabla_i F_{jk} + \nabla_j F_{ki} + \nabla_k F_{ij}. \end{aligned}$$

Chapter 7

Hypersurfaces, Submanifolds, and Extrinsic Curvature

7.1 Two-Dimensional Surfaces Embedded in a Three-Dimensional Space

Let us consider the Frenet-Serret formula (5.188) for a parametrized curve in a three-dimensional Euclidean space. The first curvature $\kappa(s)$ of the curve appears in both of the equations

$$\kappa(s) = \delta_{ij} N^i(s) \frac{dT^j(s)}{ds} = \delta_{ij} N^i(s) \frac{d^2 \mathcal{X}^j(s)}{ds^2}, \tag{7.1}$$

$$\begin{aligned} \kappa(s) &= -\delta_{ij} T^i(s) \frac{dN^j(s)}{ds} = -\delta_{ij} \frac{d\mathcal{X}^i(s)}{ds} \frac{dN^j(s)}{ds}, \\ s &\in (s_1, s_2) \subset \mathbb{R}. \end{aligned} \tag{7.2}$$

We shall generalize this concept of curvature to the **extrinsic curvature** of a two-dimensional surface $\sum_2$ embedded in a three-dimensional Euclidean space. (The intrinsic curvature of such a surface is controlled by the Riemann-Christoffel tensor of $\sum_2$.) Consider a non-degenerate, parametrized surface ξ of class C^3 (with the image $\sum_2$). Let it be given in a Cartesian coordinate chart by equations

$$\begin{aligned} x &= \xi(u) \in \mathbb{R}^3, \\ x^i &= \xi^i(u) \equiv \xi^i(u^1, u^2), \ \ (u^1, u^2) \in \mathcal{D}_2 \subset \mathbb{R}^2; \\ \partial_\mu \xi^i &:= \frac{\partial \xi^i(u^1, u^2)}{\partial u^\mu}, \\ \text{Rank } \left[\partial_\mu \xi^i\right] &= 2; \\ \mu &\in \{1, 2\}, \ \ i \in \{1, 2, 3\}. \end{aligned} \tag{7.3}$$

The three-dimensional vectors

$$\vec{\mathbf{t}}_{(\mu)}(\xi(u)) := (\partial_\mu \xi^i) \frac{\partial}{\partial x^i}_{|\xi(u)} \tag{7.4}$$

are *tangential* to the coordinate curves on $\sum_2$. (See fig. 7.1.)

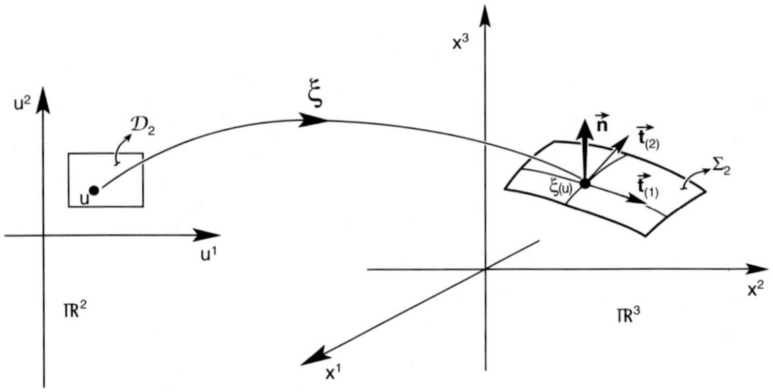

Figure 7.1: A two-dimensional surface $\sum_2$ embedded in $\mathbb{R}^3$.

These two vectors span a two-dimensional (tangential) vector *subspace* of $T_{\xi(u)}(\mathbb{R}^3)$. It is isomorphic to the intrinsic tangent planes $T_{\xi(u)}(\sum_2)$ and $T_u(\mathbb{R}^2)$. For any two vectors $\vec{\mathbf{V}}(\xi(u))$, $\vec{\mathbf{W}}(\xi(u))$, in the two-dimensional vector subspace, we can express

$$\begin{aligned} \vec{\mathbf{V}}(\xi(u)) &= [v^\mu(u) \partial_\mu \xi^i] \frac{\partial}{\partial x^i}_{|\xi(u)}, \\ \vec{\mathbf{W}}(\xi(u)) &= [w^\mu(u) \partial_\mu \xi^i] \frac{\partial}{\partial x^i}_{|\xi(u)}. \end{aligned} \tag{7.5}$$

Now, the inner products for three-dimensional vectors are provided by

$$\mathbf{g}..(x)\left[\frac{\partial}{\partial x^i}, \frac{\partial}{\partial x^j}\right] \equiv \mathbf{I}..(x)\left[\frac{\partial}{\partial x^i}, \frac{\partial}{\partial x^j}\right] := \delta_{ij}. \tag{7.6}$$

By the equation above, the previous equations (7.4) and (7.5) yield

$$\mathbf{I}..(\xi(u))\left(\vec{\mathbf{t}}_{(\mu)}(\xi(u)), \vec{\mathbf{t}}_{(\nu)}(\xi(u))\right) = \delta_{ij}(\partial_\mu \xi^i)(\partial_\nu \xi^j) =: \bar{g}_{\mu\nu}(u), \tag{7.7}$$

$$\begin{aligned} \mathbf{I}..(\xi(u))\left(\vec{\mathbf{V}}(\xi(u)), \vec{\mathbf{W}}(\xi(u))\right) &= \delta_{ij}\left[v^\mu(u)\partial_\mu \xi^i\right]\left[w^\nu(u)\partial_\nu \xi^j\right] \\ &= \bar{g}_{\mu\nu}(u) v^\mu(u) w^\nu(u). \end{aligned} \tag{7.8}$$

From the equations above, we identify

$$\bar{\mathbf{g}}..(u) = \bar{g}_{\mu\nu}(u) du^\mu \otimes du^\nu \tag{7.9}$$

as the *intrinsic* metric for $\sum_2$. Moreover, the vectors

$$\begin{aligned} \vec{\mathbf{v}}(u) &:= v^\mu(u)\frac{\partial}{\partial u^\mu}, \\ \vec{\mathbf{w}}(u) &:= w^\nu(u)\frac{\partial}{\partial u^\nu}, \end{aligned}$$

belong to the *intrinsic* tangent plane $T_u\left(\mathbb{R}^2\right)$. We note that by (3.46) and (3.49)

$$\begin{aligned} (\boldsymbol{\xi}^{-1})'(dx^i{}_{|\xi(u)}) &= \partial_\mu \xi^i du^\mu, \\ {}_2(\boldsymbol{\xi}^{-1})'(\mathbf{I}..(\xi(u))) &= \delta_{ij}\left(\partial_\mu \xi^i\right)\left(\partial_\nu \xi^j\right) du^\mu \otimes du^\nu \\ &= \bar{g}_{\mu\nu}(u) du^\mu \otimes du^\nu = \bar{\mathbf{g}}..(u). \end{aligned} \tag{7.10}$$

(See (5.155).)

We can work out the unique outer normal vector from (5.157) and (5.159). It turns out to be

$$\begin{aligned} \nu^l(\xi(u)) &= (1/2)\delta^{li}\varepsilon_{ijk}\left[(\partial_1 \xi^j)(\partial_2 \xi^k) - (\partial_1 \xi^k)(\partial_2 \xi^j)\right], \\ n^l(\xi(u)) &= \nu^l(\xi(u))/\sqrt{\delta_{ij}\nu^i \nu^j}, \\ \vec{\mathbf{n}}(\xi(u)) &= n^i(\xi(u))\frac{\partial}{\partial x^i}{}_{|\xi(u)}. \end{aligned} \tag{7.11}$$

In the simpler language of vector calculus, which uses the *cross product*, we can simplify (7.11) into

$$\begin{aligned} \vec{\mathbf{n}}(\xi(u)) &= \left[\vec{\mathbf{t}}_{(1)}(\xi(u)) \times \vec{\mathbf{t}}_{(2)}(\xi(u))\right] / \|\vec{\mathbf{t}}_{(1)}(\xi(u)) \times \vec{\mathbf{t}}_{(2)}(\xi(u))\| \\ &= \left[\vec{\mathbf{t}}_{(1)}(\xi(u)) \times \vec{\mathbf{t}}_{(2)}(\xi(u))\right] / \sqrt{\bar{g}(u)}; \\ \bar{g}(u) &:= \det\left[\bar{g}_{\mu\nu}(u)\right] > 0. \end{aligned} \tag{7.12}$$

It can be proved that

$$\delta_{ij} n^i(\xi(u)) n^j(\xi(u)) \equiv 1, \tag{7.13}$$

$$\delta_{ij} n^i(\xi(u)) \partial_\mu \xi^j(\xi(u)) \equiv 0. \tag{7.14}$$

Consider now the rate of change of tangent vectors $\vec{\mathbf{t}}_{(\mu)}(\xi(u))$ along coordinate curves on $\sum_2$. For that purpose, we define vectors

$$\vec{\mathbf{t}}_{(\mu\nu)}(\xi(u)) := \left(\partial_\mu \partial_\nu \xi^i\right) \frac{\partial}{\partial x^i}_{|\xi(u)} \equiv \vec{\mathbf{t}}_{(\nu\mu)}(\xi(u)). \tag{7.15}$$

Since $\{\vec{\mathbf{t}}_{(1)}(\xi(u)), \vec{\mathbf{t}}_{(2)}(\xi(u)), \vec{\mathbf{n}}(\xi(u))\}$ is a basis set for $T_{\xi(u)}(\mathbb{R}^3)$, we can express the vector field

$$\vec{\mathbf{t}}_{(\mu\nu)}(\xi(u)) = C^\lambda_{\mu\nu}(u) \vec{\mathbf{t}}_{(\lambda)}(\xi(u)) + K_{\mu\nu}(u) \vec{\mathbf{n}}(\xi(u)). \tag{7.16}$$

Here, $C^\lambda_{\mu\nu}(u)$ and $K_{\mu\nu}(u)$ are suitable coefficients occurring in the linear combination (7.16). We define the tangential and normal parts as

$$\begin{aligned}
\text{tangent}\left(\vec{\mathbf{t}}_{(\mu\nu)}(\xi(u))\right) &:= C^\lambda_{\mu\nu}(u) \vec{\mathbf{t}}_{(\lambda)}(\xi(u)), \\
\text{normal}\left(\vec{\mathbf{t}}_{(\mu\nu)}(\xi(u))\right) &:= K_{\mu\nu}(u) \vec{\mathbf{n}}(\xi(u)), \\
\vec{\mathbf{t}}_{(\mu\nu)}(\xi(u)) &\equiv \text{tangent}\,(\vec{\mathbf{t}}_{(\mu\nu)}(\xi(u))) + \text{normal}\,(\vec{\mathbf{t}}_{(\mu\nu)}(\xi(u))).
\end{aligned} \tag{7.17}$$

(See the book by Faber [12].)
By (7.15) and (7.16), we can prove the symmetries

$$C^\lambda_{\nu\mu}(u) \equiv C^\lambda_{\mu\nu}(u), \tag{7.18}$$

$$K_{\nu\mu}(u) \equiv K_{\mu\nu}(u). \tag{7.19}$$

Let us try to determine the coefficients $C^\lambda{}_{\mu\nu}(u)$ in terms of known quantities of $\sum_2$. Using (7.7), (7.8), (7.13), (7.14), and (7.16), we get

$$\begin{aligned}
\delta_{ij}(\partial_\lambda \xi^i)(\partial_\mu \partial_\nu \xi^j) = \mathbf{I}_{..}(\xi(u)) \left(\vec{\mathbf{t}}_{(\lambda)}, \vec{\mathbf{t}}_{(\mu\nu)}\right) &= C^\rho{}_{\mu\nu}(u) \bar{g}_{\rho\lambda}(u) \\
&=: \bar{C}_{\mu\nu\lambda}(u) \equiv \bar{C}_{\nu\mu\lambda}(u).
\end{aligned} \tag{7.20}$$

Differentiating the equation $\bar{g}_{\mu\lambda}(u) = \delta_{ij}(\partial_\mu \xi^i)(\partial_\lambda \xi^j)$, we derive that

$$\begin{aligned}
\partial_\nu \bar{g}_{\mu\lambda} &= \delta_{ij} \left[(\partial_\nu \partial_\mu \xi^i)(\partial_\lambda \xi^j) + (\partial_\mu \xi^i)(\partial_\nu \partial_\lambda \xi^j)\right] \\
&= \bar{C}_{\nu\mu\lambda}(u) + \bar{C}_{\nu\lambda\mu}(u).
\end{aligned} \tag{7.21}$$

Therefore, we deduce by (5.40) and (7.21) that

$$\begin{aligned}
&\overline{[\mu\nu, \lambda]} = (1/2)\left[\partial_\mu \bar{g}_{\nu\lambda} + \partial_\nu \bar{g}_{\lambda\mu} - \partial_\lambda \bar{g}_{\mu\nu}\right] = \bar{C}_{\mu\nu\lambda}(u), \\
&\overline{\left\{ \begin{matrix} \lambda \\ \mu\nu \end{matrix} \right\}} = C^\lambda{}_{\mu\nu}(u).
\end{aligned} \tag{7.22}$$

Thus, the coefficients $C^{\lambda}{}_{\mu\nu}(u)$ are components of the *intrinsic* Christoffel symbols of $\sum_2$. We can rewrite (7.16) as

$$\vec{\mathbf{t}}_{(\mu\nu)}(\xi(u)) = \overline{\left\{ \begin{array}{c} \lambda \\ \mu\nu \end{array} \right\}} \vec{\mathbf{t}}_{(\lambda)}(\xi(u)) + K_{\mu\nu}(u)\vec{\mathbf{n}}(\xi(u)). \tag{7.23}$$

By the equation above, we can express

$$K_{\mu\nu}(u) = \mathbf{I}_{..}\,(\xi(u))\left(\vec{\mathbf{t}}_{(\mu\nu)}(\xi(u)), \vec{\mathbf{n}}(\xi(u))\right) = \delta_{ij}\left(\partial_\mu\partial_\nu\xi^i\right)n^j(\xi(u)). \tag{7.24}$$

The components $K_{\mu\nu}(u)$ are generalizations of the first curvature $\kappa(s)$ of a curve in (7.1).

The symmetric tensor

$$\mathbf{K}_{..}\,(u) := K_{\mu\nu}(u)du^\mu \otimes du^\nu \tag{7.25}$$

is called the **extrinsic curvature** tensor. The form $\Phi_{\mathrm{II}} := K_{\alpha\beta}(u)du^\alpha du^\beta$ is called the **second fundamental form**. Whereas the first fundamental form $\Phi_{\mathrm{I}} \equiv ds^2 = \bar{g}_{\mu\nu}(u)du^\mu du^\nu$ determines the *intrinsic geometry* of the surface, the second fundamental form Φ_{II} reveals how it *curves* in the external three-dimensional space. Besides (7.24), there is an *alternate way* to express $K_{\mu\nu}(u)$. For that purpose, we derive from (7.13) that

$$\delta_{ij}n^i(\xi(u))\,\partial_\mu\left[n^j(\xi(u))\right] \equiv 0. \tag{7.26}$$

Therefore, the vectors $\partial_\mu\vec{\mathbf{n}} := \partial_\mu\left[n^j(\xi(u))\right]\frac{\partial}{\partial x^j}\big|_{\xi(u)}$ must be tangential to the surface. Thus, we can express

$$\begin{aligned} \partial_\mu\vec{\mathbf{n}} &= A^\nu{}_\mu(u)\mathbf{t}_{(\nu)}(\xi(u)), \\ \partial_\mu\left[n^j(\xi(u))\right] &= A^\nu{}_\mu(u)\partial_\nu\xi^j, \end{aligned} \tag{7.27}$$

for some coefficients $A^\nu{}_\mu(u)$. Differentiating (7.14) and using (7.27) and (7.24), we obtain that

$$\begin{aligned} &\delta_{ij}\left[(\partial_\nu\partial_\mu\xi^i)n^j + (\partial_\mu\xi^i)(\partial_\nu n^j)\right] \equiv 0, \\ &K_{\nu\mu}(u) = -\delta_{ij}(\partial_\mu\xi^i)[A^\lambda{}_\nu(u)\partial_\lambda\xi^j] = -\bar{g}_{\mu\lambda}(u)A^\lambda{}_\nu(u), \\ &\bar{K}^\rho{}_\nu(u) := \bar{g}^{\mu\rho}K_{\nu\mu}(u) = -A^\rho{}_\nu(u). \end{aligned} \tag{7.28}$$

Substituting the above into (7.27), we get

$$\partial_\mu\vec{\mathbf{n}} = -\bar{K}^\nu{}_\mu(u)\vec{\mathbf{t}}_{(\nu)}(\xi(u)), \tag{7.29}$$

$$\partial_\mu\left[n^j(\xi(u))\right] = -\bar{K}^\nu{}_\mu(u)\partial_\nu\xi^j, \tag{7.30}$$

$$\delta_{ij}\left(\partial_\lambda\xi^i\right)\left(\partial_\mu n^j\right) = -K_{\mu\lambda}(u). \tag{7.31}$$

Equation (7.29), which was derived by Weingarten, shows that the rate of change of the surface normal in space is governed by the extrinsic curvature. Equation (7.31) resembles the equation (7.2) for the principal curvature of a curve.

Now, we shall consider the "**invariant eigenvalue**" problem posed in the following equations:

$$[K_{\mu\nu}(u) - k(u)\bar{g}_{\mu\nu}(u)]\, v^{\nu}(u) = 0, \tag{7.32}$$

$$\det\left[K_{\mu\nu}(u) - k(u)\bar{g}_{\mu\nu}(u)\right] = 0, \tag{7.33}$$

$$(\bar{g})k^2 - (\bar{g}_{11}K_{11} + \bar{g}_{22}K_{22} - 2\bar{g}_{12}K_{12})\, k + \det\left[K_{\mu\nu}\right] = 0. \tag{7.34}$$

Since $\mathbf{\bar{g}}_{..}(u)$ is positive-definite, (7.33) or (7.34) always yields two real roots. These roots, denoted by $k_{(1)}(u)$ and $k_{(2)}(u)$, are the **principal roots** discussed in (6.22) and (6.23). By (7.34), we can prove that the Gaussian curvature and the mean curvatures of the surface are furnished by

$$K(u) = k_{(1)}(u)k_{(2)}(u) = \det\left[K_{\mu\nu}(u)\right]/\bar{g}(u), \tag{7.35}$$

$$\begin{aligned} 2\mu(u) &= k_{(1)}(u) + k_{(2)}(u) \\ &= (\bar{g}_{11}K_{22} + \bar{g}_{22}K_{11} - 2\bar{g}_{12}K_{12})\,/\bar{g}(u) = \bar{g}^{\mu\nu}K_{\mu\nu}. \end{aligned} \tag{7.36}$$

We shall mention the following classification for a point on $\sum_2$.

(i) If $k_{(1)}(u) = k_{(2)}(u)$, the point is called **umbilic**.

(ii) If $K(u) = k_{(1)}(u)k_{(2)}(u) > 0$, the point is called **elliptic**.

(iii) If $K(u) = k_{(1)}(u)k_{(2)}(u) < 0$, the point is called **hyperbolic**.

(iv) If $K(u) = 0$ but $\mu(u) \neq 0$, the point is called **parabolic**.

(v) Finally, If $k_{(1)}(u) = k_{(2)}(u) = 0$, the point is called **planar**.

(See fig. 6.2 again.)

Example 7.1.1 We shall investigate a smooth surface of revolution in this example. Consider a non-degenerate, parametrized curve of class C^3 given by

$$\begin{aligned} &x = \mathcal{X}(u^1) := \left(f(u^1), 0, h(u^1)\right), u^1 \in (a,b) \subset \mathbb{R}; \\ &f(u^1) > 0, \left[f'(u^1)\right]^2 + \left[h'(u^1)\right]^2 > 0. \end{aligned} \tag{7.37}$$

The image is called the **profile curve** and is totally contained in the $x^1 - x^3$ plane of $\mathbb{R}^3$. In the case where the profile curve is revolved around the x^3-axis, the resulting surface of revolution is furnished by

$$x = \xi(u^1, u^2) = \left(f(u^1)\cos u^2, f(u^1)\sin u^2, h(u^1)\right), \qquad (7.38)$$
$$\mathcal{D}_2 := \left\{u \in \mathbb{R}^2 : a < u^1 < b, -\pi < u^2 < \pi\right\}.$$

(See fig. 7.2.)

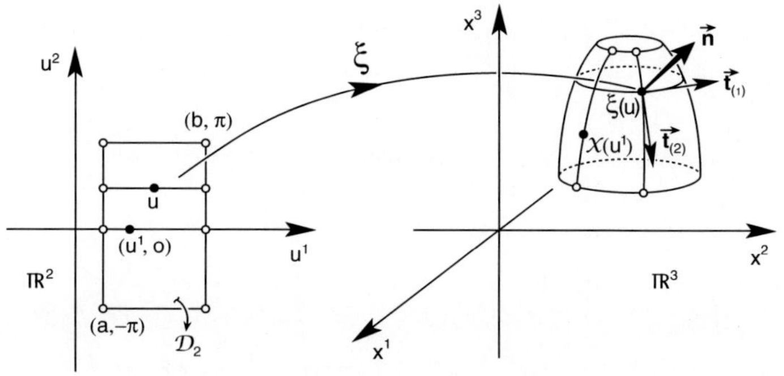

Figure 7.2: A smooth surface of revolution.

By (7.38), (7.4), and (7.8), we obtain

$$\vec{\mathbf{t}}_{(1)}(\xi(u)) = \left[f'(u^1)\left(\cos u^2 \frac{\partial}{\partial x^1} + \sin u^2 \frac{\partial}{\partial x^2}\right) + h'(u^1)\frac{\partial}{\partial x^3}\right]_{|\xi(u)},$$

$$\vec{\mathbf{t}}_{(2)}(\xi(u)) = \left[f(u^1)\left(-\sin u^2 \frac{\partial}{\partial x^1} + \cos u^2 \frac{\partial}{\partial x^2}\right)\right]_{|\xi(u)},$$

$$\bar{g}_{11}(u) = \left[f'(u^1)\right]^2 + \left[h'(u^1)\right]^2 > 0, \qquad (7.39)$$

$$\bar{g}_{22}(u) = \left[f(u^1)\right]^2 > 0,$$

$$\bar{g}_{12}(u) = \bar{g}_{21}(u) \equiv 0,$$

$$\bar{g}(u) := \det[\bar{g}_{\mu\nu}(u)] = [f(u^1)]^2 \left\{\left[f'(u^1)\right]^2 + \left[h'(u^1)\right]^2\right\} > 0.$$

The unit outer normal, by (7.11) and (7.12), is given by

$$\vec{\mathbf{n}}(\xi(u)) = \left[(f')^2 + (h')^2\right]^{-1/2} \cdot \qquad (7.40)$$
$$\left\{-h'(u^1)\left[\cos u^2 \frac{\partial}{\partial x^1} + \sin u^2 \frac{\partial}{\partial x^2}\right] + f'(u^1)\frac{\partial}{\partial x^3}\right\}_{|\xi(u)}.$$

Using (7.39) and (7.15), we get

$$\vec{\mathbf{t}}_{(11)}(\xi(u)) = \left\{ f''(u^1) \left[\cos u^2 \frac{\partial}{\partial x^1} + \sin u^2 \frac{\partial}{\partial x^2} \right] + h''(u^1) \frac{\partial}{\partial x^3} \right\}_{|\xi(u)},$$
$$\vec{\mathbf{t}}_{(22)}(\xi(u)) = -f(u^1) \left[\cos u^2 \frac{\partial}{\partial x^1} + \sin u^2 \frac{\partial}{\partial x^2} \right]_{|\xi(u)}, \tag{7.41}$$
$$\vec{\mathbf{t}}_{(12)}(\xi(u)) \equiv \vec{\mathbf{t}}_{(21)}(\xi(u)) = f'(u^1) \left[-\sin u^2 \frac{\partial}{\partial x^1} + \cos u^2 \frac{\partial}{\partial x^2} \right]_{|\xi(u)}.$$

(It is instructive to note that the vectors $\vec{\mathbf{t}}_{(2)}$, $\vec{\mathbf{t}}_{(22)}$, and $\vec{\mathbf{t}}_{(12)}$ are all "horizontal".) Now, by (7.41), (7.40), and (7.24), the extrinsic curvature components are

$$K_{11}(u) = \left[(f')^2 + (h')^2 \right]^{-1/2} \left[f'(u^1)h''(u^1) - f''(u^1)h'(u^1) \right],$$
$$K_{22}(u) = \left[(f')^2 + (h')^2 \right]^{-1/2} f(u^1)h'(u^1), \tag{7.42}$$
$$K_{12}(u) \equiv K_{21}(u) \equiv 0.$$

From (7.42), (7.34), (7.35), and (7.36), we deduce that

$$k_{(1)}(u) = \left[(f')^2 + (h')^2 \right]^{-3/2} \left[f'(u^1)h''(u^1) - f''(u^1)h'(u^1) \right],$$
$$k_{(2)}(u) = \left[f(u^1) \right]^{-1} h'(u^1) \left[(f')^2 + (h')^2 \right]^{-1/2}, \tag{7.43}$$
$$K(u) = \left[f(u^1) \right]^{-1} h'(u^1) \left[(f')^2 + (h')^2 \right]^{-2} \left[f'(u^1)h''(u^1) - f''(u^1)h'(u^1) \right],$$
$$2\mu(u) = \left[f(u^1) \right]^{-1} \left[(f')^2 + (h')^2 \right]^{-3/2} \left[f(f'h'' - f''h') + h'((f')^2 + (h')^2) \right].$$

None of the geometric entities above depend on the rotational parameter u^2. Moreover, the principal directions of curvature are along the directions of coordinate curves. In many engineering problems, one deals with metallic or ceramic surfaces of revolution. As an esoteric application, we cite the example of spherically symmetric wormholes in general relativity. In that arena, the surfaces of revolution are pertinent. □

Exercises 7.1

1. Consider a non-degenerate Monge surface embedded in a three-dimensional Euclidean space. It is parametrically specified by

$$x = \xi(u) := (u^1, u^2, f(u^1, u^2)), u \equiv (u^1, u^2) \in \mathcal{D}_2 \subset \mathbb{R}^2.$$

(Assume that f is of class C^4.)

(i) Prove that the extrinsic curvature is furnished by

$$K_{\mu\nu}(u) = \left[1 + \delta^{\gamma\rho}(\partial_\gamma f)(\partial_\rho f)\right]^{-1/2}(\partial_\mu \partial_\nu f).$$

(ii) Show that the vanishing of Gaussian curvature $K(u)$ in a domain implies the existence of a non-constant differentiable function F satisfying $F(\partial_1 f, \partial_2 f) = 0$.

2. Suppose that a non-degenerate, twice-differentiable parametrized surface is given by (7.3). Prove that the extrinsic curvature is provided by

$$\sqrt{\bar{g}(u)}K_{\mu\nu}(u) = \det\begin{bmatrix} \partial_\mu\partial_\nu\xi^1 & \partial_\mu\partial_\nu\xi^2 & \partial_\mu\partial_\nu\xi^3 \\ \partial_1\xi^1 & \partial_1\xi^2 & \partial_1\xi^3 \\ \partial_2\xi^1 & \partial_2\xi^2 & \partial_2\xi^3 \end{bmatrix}.$$

7.2 $(N-1)$-Dimensional Hypersurfaces

Now we shall discuss the **embedding** of an $(N-1)$-dimensional differentiable manifold M_{N-1} into a differentiable manifold M_N. The manifold M_{N-1} is said to be embedded in M_N provided there exists a *one-to-one* mapping $F: M_{N-1} \to M_N$ such that F is of class C^r $(r \in \mathbb{Z}^+)$. When the mapping F is *not* demanded to be globally one-to-one, it is called an **immersion**. In the sequel, we shall require F to be an embedding of class $C^r, r \geq 3$. The image $F(M_{N-1})$ is called the **embedded manifold**. Under the restrictions of coordinate charts in both M_{N-1} and M_N, we obtain a parametrized hypersurface ξ with the image $\sum_{N-1}$ in $D \subset \mathbb{R}^N$. (See fig. 7.3.)

(See the book by Eisenhart [10].)

We explain various mappings in fig. 7.3 by the following equations:

$$\begin{aligned}
&q = F(p), \\
&\chi_{|..} := \chi_{|U \cap F(M_{N-1})}, \\
&x = \chi_{|..}(q) = \left[\chi_{|..} \circ F \circ \phi^{-1}\right](u) =: \xi(u), \\
&x^i = \xi^i(u) \equiv \xi^i(u^1, \ldots, u^{N-1}), \\
&\text{Rank}[\partial_\mu \xi^i] = N - 1 > 0, \\
&\xi \in C^r(\mathcal{D}_{N-1} \subset \mathbb{R}^{N-1}; \mathbb{R}^N);\ r \geq 3.
\end{aligned} \tag{7.44}$$

Here, Roman indices are taken from $\{1, \ldots, N\}$, whereas the Greek indices are taken from $\{1, \ldots, N-1\}$. The summation convention is followed for *both* sets of indices.

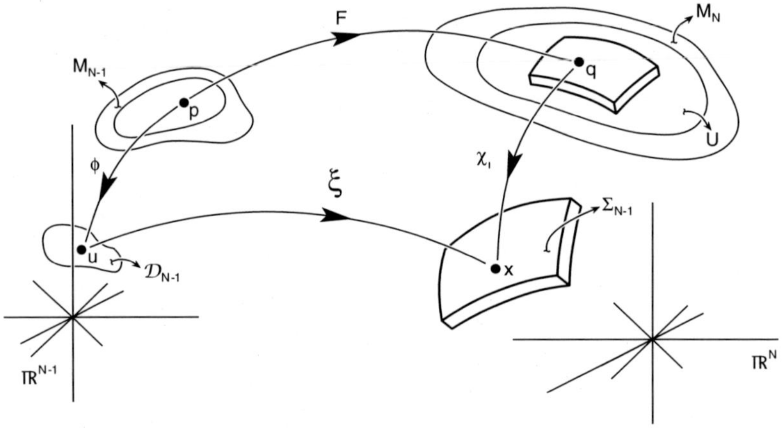

Figure 7.3: The image $\sum_{N-1}$ of a parametrized hypersurface ξ.

The metric $\mathbf{g}..(x)$ in ${}^0_2\mathcal{T}_x(\mathbb{R}^N)$ and the induced metric $\bar{\mathbf{g}}..(u)$ in ${}^0_2\mathcal{T}_u(\mathbb{R}^{N-1})$ are furnished (as in (7.10)) respectively by

$$\mathbf{g}..(x) = g_{ij}(x)\, dx^i \otimes dx^j, \tag{7.45}$$

$$\begin{aligned}\bar{g}_{\mu\nu}(u)\, du^\mu \otimes du^\nu &= \bar{\mathbf{g}}..(u) := {}_2\left(\boldsymbol{\xi}^{-1}\right)'\left(\mathbf{g}..(x)_{|\xi(u)}\right)\\ &= \left[g_{ij}\left(\xi(u)\right)(\partial_\mu \xi^i)(\partial_\nu \xi^j)\right] du^\mu \otimes du^\nu.\end{aligned} \tag{7.46}$$

An outer normal to $\sum_{N-1}$ is provided by (5.157) as

$$\nu^k\left(\xi(u)\right) := \frac{1}{(N-1)!} g^{kj}\left(\xi(u)\right) \varepsilon_{jj_2\cdots j_N} \frac{\partial(x^{j_2}, \ldots, x^{j_N})}{\partial(u^1, \ldots, u^{N-1})}. \tag{7.47}$$

From the assumption about the rank of $[\partial_\mu \xi^i]$, it follows that $g_{kj}(..)\nu^k \nu^j > 0$ for a positive-definite metric $\mathbf{g}..(x)$. However, for a pseudo-Riemannian manifold M_N, there exist *three* possibilities. (i) The hypersurface $\sum_{N-1}$ is called a **spacelike hypersurface** if $g_{kj}\nu^k\nu^j < 0$. (ii) It is called a **timelike hypersurface** if $g_{kj}\nu^k\nu^j > 0$. (iii) Moreover, $\sum_{N-1}$ is called a **null hypersurface** provided $g_{kj}\nu^k\nu^j \equiv 0$. In the sequel of this section, we shall *exclude all null hypersurfaces.*

Recall the coordinate transformation given in (3.2) for M_N. We may have a similar coordinate transformation, called a **reparametrization**, in M_{N-1}.

The mapping symbols occurring in fig. 7.4 are elaborated below:

$$u^{\#} = U^{\#}(u), u^{\#_\alpha} = U^{\#_\alpha}(u), \text{Rank}\left[\frac{\partial u^{\#_\alpha}}{\partial u^\beta}\right] = N-1, U := [U^{\#}]^{-1}, \tag{7.48}$$

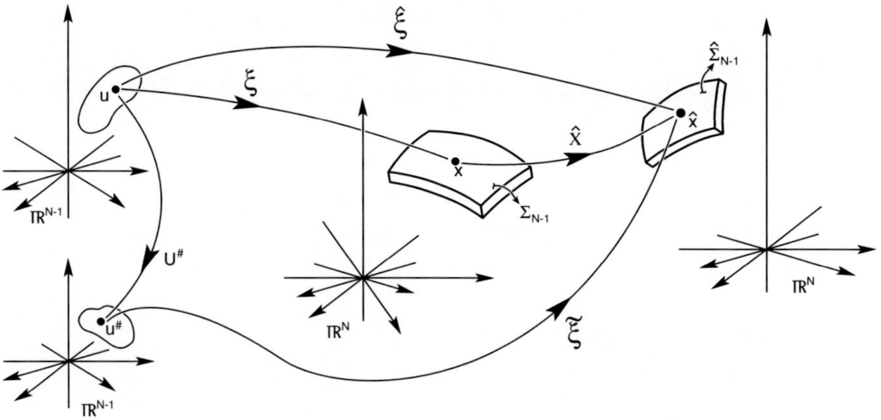

Figure 7.4: Coordinate transformation and reparametrization of hypersurface ξ.

$$x = \xi(u) = \xi^{\#}(u^{\#}) := \left[\xi \circ (U^{\#})^{-1}\right](u^{\#}), \tag{7.49}$$

$$\widehat{x} = \widehat{X}_{|\sum_{N-1}}(x) = \left[\widehat{X} \circ \xi\right](u) =: \widehat{\xi}(u), \tag{7.50}$$

$$\widehat{x} = \left[\widehat{X} \circ \xi \circ (U^{\#})^{-1}\right](u^{\#}) =: \tilde{\xi}(u^{\#}). \tag{7.51}$$

Let us consider transformation properties of $\partial_\mu \xi^i$ under each of the transformations (7.49), (7.50), and (7.51). By the chain rules, we have

$$\frac{\partial \xi^{\#_i}(u^{\#})}{\partial u^{\#_\mu}} = \frac{\partial U^\nu(u^{\#})}{\partial u^{\#_\mu}} \cdot \frac{\partial \xi^i(u)}{\partial u^\nu}, \tag{7.52}$$

$$\frac{\partial \widehat{\xi^i}(u)}{\partial u^\mu} = \frac{\partial \widehat{X}^i(x)}{\partial x^k}_{|\xi(u)} \cdot \frac{\partial \xi^k(u)}{\partial u^\mu}, \tag{7.53}$$

$$\begin{aligned} \frac{\partial \tilde{\xi}^i(u^{\#})}{\partial u^{\#_\mu}} &= \frac{\partial}{\partial u^{\#_\mu}} \left\{ \widehat{X}^i \left[\xi \left(U(u^{\#}) \right) \right] \right\} \\ &= \frac{\partial \widehat{X}^i(x)}{\partial x^k}_{|\xi^{\#}(u^{\#})} \cdot \frac{\partial \xi^k(u)}{\partial u^\nu}_{|U(u^{\#})} \cdot \frac{\partial U^\nu(u^{\#})}{\partial u^{\#_\mu}}. \end{aligned} \tag{7.54}$$

Therefore, we conclude that (i) under a reparametrization, $\partial_\mu \xi^i$ behave like components of a covariant vector; (ii) under a coordinate transformation, $\partial_\mu \xi^i$ behave like components of a contravariant vector, and (iii) under a combined transformation, $\partial_\mu \xi^i$ behave like components of a *mixed* second-order "hybrid" tensor! These transformation rules prompt us to define *hybrid* components

$T^{\mu_1\cdots\mu_p}{}_{\nu_1\cdots\nu_q}{}^{i_1\cdots i_r}{}_{j_1\cdots j_s}(u,\xi(u))$ by the following transformation rules (under the *combined* transformations):

$$\begin{aligned}\tilde{T}^{\mu_1\cdots\mu_p}{}_{\nu_1\cdots\nu_q}{}^{i_1\cdots i_r}{}_{j_1\cdots j_s}(u^{\#},\tilde{\xi}(u^{\#})) = \\ \frac{\partial u^{\#_{\mu_1}}}{\partial u^{\lambda_1}}\cdots\frac{\partial u^{\#_{\mu_p}}}{\partial u^{\lambda_p}}\cdot\frac{\partial u^{\rho_1}}{\partial u^{\#_{\nu_1}}}\cdots\frac{\partial u^{\rho_q}}{\partial u^{\#_{\nu_q}}}\cdot\frac{\partial \widehat{x}^{i_1}}{\partial x^{k_1}}\cdots\frac{\partial \widehat{x}^{i_r}}{\partial x^{k_r}}\cdot\frac{\partial x^{l_1}}{\partial \widehat{x}^{j_1}}\cdots\frac{\partial x^{l_s}}{\partial \widehat{x}^{j_s}}\cdot \\ T^{\lambda_1\cdots\lambda_p}{}_{\rho_1\cdots\rho_q}{}^{k_1\cdots k_r}{}_{l_1\cdots l_s}(u,\xi(u)).\end{aligned} \tag{7.55}$$

Example 7.2.1 Consider $N=3$ and $M_N=\mathbb{E}_3$, the Euclidean space. A smooth, non-degenerate parametrized surface is given by

$$\begin{aligned}\xi(u) &:= \left(\alpha^1(u^1)+u^2\mathcal{X}^1(u^1),\alpha^2(u^1)+u^2\mathcal{X}^2(u^1),\alpha^3(u^1)+u^2\mathcal{X}^3(u^1)\right),\\ &\quad (u^1,u^2)\in(a,b)\times(c,d)\subset\mathbb{R}^2.\end{aligned}$$

Such a surface is called a **ruled surface**. At each point $\xi(u_0)$ on $\sum_2$, a line segment (or ruling) passes through wholly lying in $\sum_2$. These rulings are given by the line segments

$$\mathcal{X}_{(0)}(u^2) := \left(\alpha^1(u_0^1)+u^2\mathcal{X}^1(u_0^1),\alpha^2(u_0^1)+u^2\mathcal{X}^2(u_0^1),\alpha^3(u_0^1)+u^2\mathcal{X}^3(u_0^1)\right), \quad u^2\in(c,d).$$

Let us consider a reparametrization

$$u^{\#_1}=u^1, u^{\#_2}=2u^2, \left[\frac{\partial u^{\#_\mu}}{\partial u^\nu}\right]=\begin{bmatrix}1&0\\0&2\end{bmatrix}.$$

Moreover, we introduce a translation of Cartesian axes given by

$$\widehat{x}^1=x^1+1,\ \widehat{x}^2=x^2+2,\ \widehat{x}^3=x^3+3,\ \left[\frac{\partial\widehat{x}^i}{\partial x^j}\right]=\left[\delta^i_j\right].$$

In this problem, we have

$$\left[\partial_\mu\xi^i\right]=\begin{bmatrix}\alpha^{1\prime}(u^1)+u^2\mathcal{X}^{1\prime}(u^1), & \mathcal{X}^1(u^1)\\ \alpha^{2\prime}(u^1)+u^2\mathcal{X}^{2\prime}(u^1), & \mathcal{X}^2(u^1)\\ \alpha^{3\prime}(u^1)+u^2\mathcal{X}^{3\prime}(u^1), & \mathcal{X}^3(u^1)\end{bmatrix}.$$

The rank of the matrix above is assumed to be 2. By the combined transformations, we have

$$\begin{aligned}\tilde{\xi}(u^{\#}) &= \left(1+\alpha^1(u^{\#_1})+(1/2)u^{\#_2}\mathcal{X}^1(u^{\#_1}), 2+\alpha^2(u^{\#_1})+(1/2)u^{\#_2}\mathcal{X}^2(u^{\#_1}),\right.\\ &\quad \left.3+\alpha^3(u^{\#_1})+(1/2)u^{\#_2}\mathcal{X}^3(u^{\#_1})\right).\end{aligned}$$

The image of $\tilde{\xi}$ is again a ruled surface.

The transformation (7.54) can be obtained in this example by the matrix multiplication

$$\begin{aligned}\left[\frac{\partial\tilde{\xi}^i(u^\#)}{\partial u^{\#_\mu}}\right] &= \left[\frac{\partial\widehat{x}^i}{\partial x^k}\right]\left[\frac{\partial\xi^k}{\partial u^\nu}\right]\left[\frac{\partial u^\nu}{\partial u^{\#_\mu}}\right] \\ &= \begin{bmatrix} \alpha^{1\prime}(u^{\#_1}) + (1/2)u^{\#_2}\mathcal{X}^{1\prime}(u^{\#_1}), & (1/2)\mathcal{X}^1(u^{\#_1}) \\ \alpha^{2\prime}(u^{\#_1}) + (1/2)u^{\#_2}\mathcal{X}^{2\prime}(u^{\#_1}), & (1/2)\mathcal{X}^2(u^{\#_1}) \\ \alpha^{3\prime}(u^{\#_1}) + (1/2)u^{\#_2}\mathcal{X}^{3\prime}(u^{\#_1}), & (1/2)\mathcal{X}^3(u^{\#_1}) \end{bmatrix} \qquad \square\end{aligned}$$

Now, we shall define a suitable covariant derivative for a *hybrid* tensor. For that purpose, we need to introduce the concept of a **total partial derivative**. When f is a differentiable mapping from $\mathcal{D}_{N-1}\times\mathbb{R}^N$ into $\mathbb{R}$, the *total partial derivative* is given by

$$\frac{d}{du^\mu}\left[f(u,x)_{|x=\xi(u)}\right] := \frac{\partial f(u,x)}{\partial u^\mu}_{|x=\xi(u)} + \partial_\mu\xi^k\left[\frac{\partial f(u,x)}{\partial x^k}\right]_{|x=\xi(u)}. \tag{7.56}$$

The first term on the right-hand side is the *explicit* differentiation, whereas the second term is due to the *implicit* differentiation.

The general definition of a covariant derivative of a hybrid tensor is a generalization of (5.83) together with (7.56). This is furnished by

$$\begin{aligned}&\tilde{\nabla}_\lambda T^{\mu_1\cdots\mu_p}{}_{\nu_1\cdots\nu_q}{}^{i_1\cdots i_r}{}_{j_1\cdots j_s}(u,\xi(u)) := \frac{d}{du^\lambda}T^{\mu_1\cdots\mu_p}{}_{\nu_1\cdots\nu_q}{}^{i_1\cdots i_r}{}_{j_1\cdots j_s}(u,\xi(u)) \\ &+\sum_{h=1}^{p}\overline{\left\{\begin{matrix}\mu_h\\ \lambda\rho\end{matrix}\right\}}T^{\mu_1\cdots\mu_{h-1}\rho\mu_{h+1}\cdots\mu_p}{}_{\nu_1\cdots\nu_q}{}^{i_1\cdots i_r}{}_{j_1\cdots j_s} \\ &-\sum_{w=1}^{q}\overline{\left\{\begin{matrix}\rho\\ \lambda\nu_w\end{matrix}\right\}}T^{\mu_1\cdots\mu_p}{}_{\nu_1\cdots\nu_{w-1}\rho\nu_{w+1}\cdots\nu_q}{}^{i_1\cdots i_r}{}_{j_1\cdots j_s} \\ &+\sum_{\alpha=1}^{r}\partial_\lambda\xi^k\left\{\begin{matrix}i_\alpha\\ k\ell\end{matrix}\right\}T^{\mu_1\cdots\mu_p}{}_{\nu_1\cdots\nu_q}{}^{i_1\cdots i_{\alpha-1}\ell i_{\alpha+1}\cdots i_r}{}_{j_1\cdots j_s} \qquad (7.57) \\ &-\sum_{\beta=1}^{s}\partial_\lambda\xi^k\left\{\begin{matrix}\ell\\ kj_\beta\end{matrix}\right\}T^{\mu_1\cdots\mu_p}{}_{\nu_1\cdots\nu_q}{}^{i_1\cdots i_r}{}_{j_1\cdots j_{\beta-1}\ell j_{\beta+1}\cdots j_s} \\ &=\left[\overline{\nabla}_\lambda T^{\mu_1\cdots\mu_p}{}_{\nu_1\cdots\nu_q}{}^{i_1\cdots i_r}{}_{j_1\cdots j_s}(u,x)+\partial_\lambda\xi^k\nabla_k T^{\mu_1\cdots\mu_p}{}_{\nu_1\cdots\nu_q}{}^{i_1\cdots i_r}{}_{j_1\cdots j_s}(u,x)\right]_{|x=\xi(u)}.\end{aligned}$$

In the last line, *the covariant derivative acts only on the Greek indices in the first term and only on the Roman indices in the second term*! Now we shall provide some simple examples of these differentiations.

Example 7.2.2

$$\begin{aligned}\tilde{\nabla}_\lambda f &= \frac{d}{du^\lambda} f(u,\xi(u)) \\ &= \left[\overline{\nabla}_\lambda f(u,x) + \partial_\lambda \xi^k \nabla_k f(u,x)\right]_{|x=\xi(u)};\end{aligned} \tag{7.58}$$

$$\tilde{\nabla}_\lambda V_\mu(u) = \overline{\nabla}_\lambda V_\mu(u); \tag{7.59}$$

$$\tilde{\nabla}_\lambda T_j(\xi(u)) = \partial_\lambda \xi^k \left[\nabla_k T_j(x)\right]_{|\xi(u)}; \tag{7.60}$$

$$\begin{aligned}&\left[\tilde{\nabla}_\lambda W_j(u,\xi(u))\right] du^\lambda \otimes dx^j{}_{|\xi(u)} \\ &= \left[\partial_\lambda W_j(u,x)\, du^\lambda \otimes dx^j + \nabla_k\, W_j(u,x)\, dx^k \otimes dx^j\right]_{|x=\xi(u)};\end{aligned} \tag{7.61}$$

$$\tilde{\nabla}_\lambda \partial_\mu \xi^i(u) = \overline{\nabla}_\lambda(\partial_\mu \xi^i) + \left\{ \begin{matrix} i \\ k\ell \end{matrix} \right\}_{|\xi(u)} \partial_\lambda \xi^k \cdot \partial_\mu \xi^\ell; \tag{7.62}$$

$$\tilde{\nabla}_\lambda T^\mu{}_j(u,\xi(u)) = \overline{\nabla}_\lambda T^\mu{}_j + \partial_\lambda \xi^k \left[\nabla_k T^\mu{}_j\right]_{|\xi(u)}; \tag{7.63}$$

$$\tilde{\nabla}_\lambda \bar{g}_{\mu\nu}(u) = \overline{\nabla}_\lambda \bar{g}_{\mu\nu}(u) \equiv 0; \tag{7.64}$$

$$\tilde{\nabla}_\lambda g_{ij}(\xi(u)) = \left[\nabla_k g_{ij}\right]_{|\xi(u)} \cdot \partial_\lambda \xi^k \equiv 0; \tag{7.65}$$

$$\begin{aligned}\tilde{\nabla}_\nu \tilde{\nabla}_\mu f(u,\xi(u)) &= \overline{\nabla}_\nu \overline{\nabla}_\mu f + \left[\overline{\nabla}_\nu \overline{\nabla}_\mu \xi^k(u) + \left\{ \begin{matrix} k \\ \ell j \end{matrix} \right\}_{|..} \partial_\nu \xi^\ell \cdot \partial_\mu \xi^j\right] \cdot \nabla_k f \\ &\quad + \partial_\mu \xi^\ell \cdot \partial_\nu \partial_\ell f + \partial_\nu \xi^\ell \cdot \partial_\mu \partial_\ell f + \partial_\nu \xi^\ell \cdot \partial_\mu \xi^k \left[\nabla_\ell \nabla_k f\right]_{|..} \\ &= \tilde{\nabla}_\mu \tilde{\nabla}_\nu f.\end{aligned} \tag{7.66}$$

□

The connection $\tilde{\nabla}_\mu$ generates a *new* hybrid tensor from a given differentiable hybrid tensor. Under the combined transformation (7.48) – (7.51), the following transformation rule holds:

$$\begin{aligned}&\tilde{\nabla}^{\#}_\sigma \tilde{T}^{\mu_1\cdots\mu_p}{}_{\nu_1\cdots\nu_q}{}^{i_1\cdots i_r}{}_{j_1\cdots j_s}(u^{\#}, \tilde{\xi}(u^{\#})) \\ &= \frac{\partial u^\gamma}{\partial u^{\#\sigma}} \cdot \frac{\partial u^{\#\mu_1}}{\partial u^{\lambda_1}} \cdots \frac{\partial u^{\#\mu_p}}{\partial u^{\lambda_p}} \cdot \frac{\partial u^{\rho_1}}{\partial u^{\#\nu_1}} \cdots \frac{\partial u^{\rho_q}}{\partial u^{\#\nu_q}} \cdot \frac{\partial \widehat{x}^{i_1}}{\partial x^{k_1}} \cdots \frac{\partial \widehat{x}^{i_r}}{\partial x^{k_r}} \cdot \frac{\partial x^{\ell_1}}{\partial \widehat{x}^{j_1}} \cdots \frac{\partial x^{\ell_s}}{\partial \widehat{x}^{j_s}} \cdot \\ &\quad \tilde{\nabla}_\gamma T^{\lambda_1\cdots\lambda_p}{}_{\rho_1\cdots\rho_q}{}^{k_1\cdots k_r}{}_{\ell_1\cdots\ell_s}(u,\xi(u)).\end{aligned} \tag{7.67}$$

Another important property of the hybrid covariant derivative is that the *Leibnitz property* (4.25) holds. Therefore, we have

$$\begin{aligned}\tilde{\nabla}_\sigma &\left[A^{\mu_1\cdots\mu_p}{}_{\nu_1\cdots\nu_q}{}^{i_1\cdots i_r}{}_{j_1\cdots j_s}\cdot B^{\lambda_1\cdots\lambda_m}{}_{\rho_1\cdots\rho_n}{}^{k_1\cdots k_w}{}_{\ell_1\cdots\ell_v}\right]\\ =\ &\left[\tilde{\nabla}_\sigma A^{\mu_1\cdots\mu_p}{}_{\nu_1\cdots\nu_q}{}^{i_1\cdots i_r}{}_{j_1\cdots j_s}\right]\cdot B^{\lambda_1\cdots\lambda_m}{}_{\rho_1\cdots\rho_n}{}^{k_1\cdots k_w}{}_{\ell_1\cdots\ell_v}\\ &+A^{\mu_1\cdots\mu_p}{}_{\nu_1\cdots\nu_q}{}^{i_1\cdots i_r}{}_{j_1\cdots j_s}\cdot\left[\tilde{\nabla}_\sigma B^{\lambda_1\cdots\lambda_m}{}_{\rho_1\cdots\rho_n}{}^{k_1\cdots k_w}{}_{\ell_1\cdots\ell_v}\right].\end{aligned}\tag{7.68}$$

Now we shall derive the commutator $[\tilde{\nabla}_\nu\tilde{\nabla}_\lambda-\tilde{\nabla}_\lambda\tilde{\nabla}_\nu]\partial_\mu\xi^i(u)$. By (7.57) and (7.58)–(7.66), we obtain that

$$\begin{aligned}&\tilde{\nabla}_\nu\tilde{\nabla}_\lambda\partial_\mu\xi^i(u)-\tilde{\nabla}_\lambda\tilde{\nabla}_\nu\partial_\mu\xi^i(u)\\ &=\overline{\nabla}_\nu\left[\tilde{\nabla}_\lambda\partial_\mu\xi^i\right]+\partial_\nu\xi^k\cdot\left\{\nabla_k\left[\tilde{\nabla}_\lambda\partial_\mu\xi^i(u)\right]_{|..}\right\}\\ &\quad-\overline{\nabla}_\lambda\left[\tilde{\nabla}_\nu\partial_\mu\xi^i\right]-\partial_\lambda\xi^k\cdot\left\{\nabla_k\left[\tilde{\nabla}_\nu\partial_\mu\xi^i(u)\right]_{|..}\right\}\\ &=\overline{\nabla}_\nu\left[\overline{\nabla}_\lambda\overline{\nabla}_\mu\xi^i+\left\{{i\atop k\ell}\right\}_{|..}\partial_\lambda\xi^k\cdot\partial_\mu\xi^\ell\right]+\partial_\nu\xi^k\left[0+\left\{{i\atop kj}\right\}_{|..}\tilde{\nabla}_\lambda\partial_\mu\xi^j\right]\\ &\quad-\overline{\nabla}_\lambda\left[\overline{\nabla}_\nu\overline{\nabla}_\mu\xi^i+\left\{{i\atop k\ell}\right\}_{|..}\partial_\nu\xi^k\cdot\partial_\mu\xi^\ell\right]-\partial_\lambda\xi^k\left[0+\left\{{i\atop kj}\right\}_{|..}\tilde{\nabla}_\nu\partial_\mu\xi^j\right]\\ &=\left(\overline{\nabla}_\nu\overline{\nabla}_\lambda-\overline{\nabla}_\lambda\overline{\nabla}_\nu\right)\overline{\nabla}_\mu\xi^i+\left[\partial_j\left\{{i\atop k\ell}\right\}\right]_{|\xi(u)}\left[\partial_\nu\xi^j\cdot\partial_\lambda\xi^k-\partial_\lambda\xi^j\cdot\partial_\nu\xi^k\right]\partial_\mu\xi^\ell\\ &\quad+\left\{{i\atop jk}\right\}\left\{{j\atop \ell h}\right\}\left[\partial_\lambda\xi^\ell\cdot\partial_\nu\xi^k-\partial_\nu\xi^\ell\cdot\partial_\lambda\xi^k\right]\partial_\mu\xi^h.\end{aligned}$$

Using (5.110) and (5.93), we get

$$\begin{aligned}&\tilde{\nabla}_\nu\tilde{\nabla}_\lambda\partial_\mu\xi^i(u)-\tilde{\nabla}_\lambda\tilde{\nabla}_\nu\partial_\mu\xi^i(u)\\ &\quad=\overline{R}^\sigma{}_{\mu\nu\lambda}(u)\cdot\partial_\sigma\xi^i(u)+R^i{}_{jk\ell}(x)_{|\xi(u)}\cdot\partial_\mu\xi^j\cdot\partial_\nu\xi^k\cdot\partial_\lambda\xi^\ell.\end{aligned}\tag{7.69}$$

The unit normal vector $\vec{\mathbf{n}}(\xi(u))$ obtained from (7.47) satisfies by (7.64), (7.65), and (7.68) the following equations:

$$g_{ij}(\xi(u))(\partial_\mu\xi^i)n^j(\xi(u))\equiv 0\,,\tag{7.70i}$$

$$g_{ij}(\xi(u))n^i(\xi(u))n^j(\xi(u))=\pm 1=:\varepsilon_{(n)}\,,\tag{7.70ii}$$

$$g_{ij}(\cdot\cdot)(\tilde{\nabla}_\lambda n^i)n^j(\cdot\cdot)\equiv 0\,,\tag{7.70iii}$$

$$g_{ij}(\cdot\cdot)[(\tilde{\nabla}_\lambda\partial_\mu\xi^i)n^j+\partial_\mu\xi^i(\tilde{\nabla}_\lambda n^j)]\equiv 0\,.\tag{7.70iv}$$

We generalize (7.23) and (7.24) to define the extrinsic curvature as

$$K_{\mu\nu}(u) \equiv k_{\mu\nu}(\xi(u)) := g_{ij}(\xi(u))n^i(\xi(u))\tilde{\nabla}_\mu(\partial_\nu \xi^j)\,, \tag{7.71a}$$

$$\tilde{\nabla}_\mu(\partial_\nu \xi^j) = \varepsilon_{(n)} n^j(\xi(u))K_{\mu\nu}(u)\,, \tag{7.71b}$$

$$\partial_\mu\partial_\nu\xi^j = \overline{\left\{ {\lambda \atop \mu\nu} \right\}}\partial_\lambda\xi^j - \left\{ {j \atop ik} \right\}_{|\xi(u)}(\partial_\mu\xi^i)(\partial_\nu\xi^k) + \varepsilon_{(n)}n^j(\xi(u))K_{\mu\nu}(u)\,, \tag{7.71c}$$

$$K_{\nu\mu}(u) \equiv K_{\mu\nu}(u)\,. \tag{7.71d}$$

By (7.70iv) and (7.71a), we can express

$$\begin{aligned} K_{\mu\nu}(u) &= -g_{ij}(x)_{|\xi(u)}(\tilde{\nabla}_\mu n^i)(\partial_\nu\xi^j) \\ &= -(\nabla_i n_j)_{|\xi(u)}(\partial_\mu\xi^i)(\partial_\nu\xi^j)\,. \end{aligned} \tag{7.72}$$

By (7.70iii), it is clear that the vectors $\tilde{\nabla}_\mu n^i$ are *tangential* to the hypersurface $\sum_{N-1}$. Therefore, these are expressible as linear combinations

$$\tilde{\nabla}_\mu n^i = A^\sigma_\mu(u)\partial_\sigma\xi^i\,. \tag{7.73}$$

(Compare the equation above with (7.27).) Substituting (7.73) into (7.72), we deduce that

$$K_{\mu\nu}(u) = -A^\sigma_\mu(u)g_{ij}(\cdot\cdot)(\partial_\sigma\xi^i)(\partial_\nu\xi^j) = -\overline{g}_{\sigma\nu}(u)A^\sigma_\mu(u)\,, \tag{7.74a}$$

$$\overline{K}^\rho_\mu(u) := \overline{g}^{\rho\nu}(u)K_{\mu\nu}(u) = -A^\rho_\mu(u)\,. \tag{7.74b}$$

Putting (7.74b) into (7.73), we finally derive that

$$\tilde{\nabla}_\mu n^i = \partial_\mu\xi^k\left[\nabla_k n^i\right]_{|\xi(u)} = -\overline{K}^\rho_\mu(u)\partial_\rho\xi^i\,, \tag{7.75a}$$

$$\partial_\mu\xi^k\left[\partial_k n^i + \left\{ {i \atop kj} \right\} n^j\right]_{|\xi(u)} = -\overline{g}^{\rho\nu}(u)K_{\mu\nu}(u)\partial_\rho\xi^i\,. \tag{7.75b}$$

The equations above are generalizations of the *Weingarten equations* (7.30) and (7.31). Figure 7.5 depicts the generalized Weingarten equations (7.75a,b) geometrically.

In fig. 7.5, the dotted vector at $\xi(U(t+\Delta t))$ is the parallelly transported normal vector. The difference vector $\Delta\vec{\mathbf{n}}$ indicates the change due to the extrinsic curvature.

Equation (7.33),

$$\det[K_{\mu\nu}(u) - k(u)\overline{g}_{\mu\nu}(u)] = 0,$$

in the present context may *not* admit *real* invariant eigenvalues! However, *when* $\overline{\mathbf{g}}_{\cdot\cdot}(u)$ is *positive-definite*, there exist $N-1$ real, invariant eigenvalues

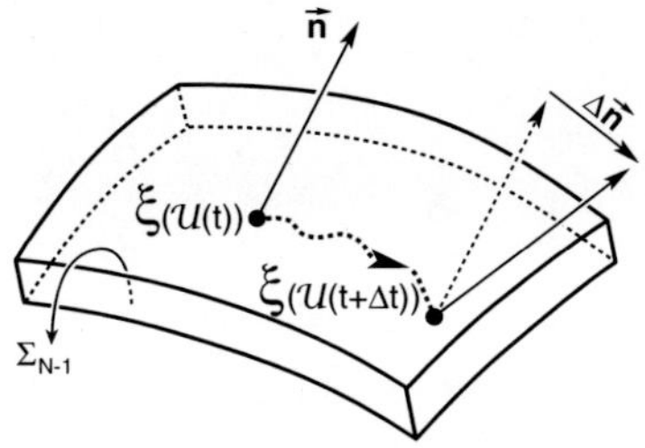

Figure 7.5: Change of normal vector due to extrinsic curvature.

$k_{(1)}(u), \ldots, k_{(N-1)}(u)$. These are also called **principal**, **normal curvatures** of $\sum_{N-1}$.

Suppose that we are confronted with the question of the existence of a hypersurface ξ of class C^r $(r \geq 3)$ for some prescribed extrinsic curvature $K_{\mu\nu}(x)$. Then, we have to investigate the *integrability conditions* of the partial differential equations (7.71b) or (7.71c). These conditions will be summarized in the following lemma.

Lemma 7.2.3 *Suppose that ξ is an $(N-1)$-dimensional parametrized hypersurface embedded in $\mathbb{R}^N$ according to* (7.44)*. Then, the integrability conditions of the partial differential equations* (7.71b) *or* (7.71c) *are furnished by the following hybrid tensor equation:*

$$\begin{aligned}\Theta^i_{\mu\nu\lambda}(u,\xi(u)) &:= \left[\overline{R}^{\rho}_{\mu\nu\lambda}(u) + \varepsilon_{(n)}(K_{\mu\nu}\overline{K}^{\rho}_{\lambda} - K_{\mu\lambda}\overline{K}^{\rho}_{\nu})\right]\partial_\rho \xi^i \\ &+ R^i_{j\ell k}(\xi(u))\partial_\mu \xi^j \partial_\lambda \xi^\ell \partial_\nu \xi^k + \varepsilon_{(n)}\left[\tilde{\nabla}_\nu K_{\mu\lambda} - \tilde{\nabla}_\lambda K_{\mu\nu}\right] n^i(\xi(u)) \\ &= 0 .\end{aligned} \tag{7.76}$$

Proof. The integrability conditions of (7.71b) or (7.71c) are

$$\tilde{\nabla}_\nu \tilde{\nabla}_\lambda(\partial_\mu \xi^i) - \tilde{\nabla}_\lambda \tilde{\nabla}_\nu(\partial_\mu \xi^i) = \varepsilon_{(n)}\left[\tilde{\nabla}_\nu(n^i K_{\mu\lambda}) - \tilde{\nabla}_\lambda(n^i K_{\mu\nu})\right]. \tag{7.77}$$

By the Leibnitz rule (7.68) and the Weingarten equations (7.75a,b), we derive that

$$\tilde{\nabla}_\nu(K_{\mu\lambda} n^i) - \tilde{\nabla}_\lambda(n^i K_{\mu\nu}) = (\tilde{\nabla}_\nu K_{\mu\lambda} - \tilde{\nabla}_\lambda K_{\mu\nu}) n^i + (K_{\mu\nu}\overline{K}^{\rho}_{\lambda} - K_{\mu\lambda}\overline{K}^{\rho}_{\nu})\partial_\rho \xi^i .$$

With the equation above, together with (7.69), integrability conditions (7.77) yield

$$\begin{aligned}\overline{R}^{\rho}_{\mu\lambda\nu}(u)\cdot\partial_\rho\xi^i \;+\; & R^i_{jk\ell}(\xi(u))\cdot\partial_\mu\xi^i\partial_\nu\xi^k\partial_\lambda\xi^\ell \\ &= \varepsilon_{(n)}(\tilde{\nabla}_\nu K_{\mu\lambda} - \tilde{\nabla}_\lambda K_{\mu\nu}) n^i \\ &+ \varepsilon_{(n)}(K_{\mu\nu}\overline{K}^{\rho}_{\lambda} - K_{\mu\lambda}\overline{K}^{\rho}_{\nu})\partial_\rho\xi^i .\end{aligned}$$

Therefore, (7.76) is established. ■

The lemma above leads to the following generalizations of the **Gauss** and **Codazzi-Mainardi** equations.

Theorem 7.2.4 *Let ξ be a non-degenerate, non-null $(N-1)$-dimensional parametric hypersurface of class C^r $(r \geq 3)$ into $\mathbb{R}^N$. Then, the integrability conditions* (7.71b) *imply that*

$$
\begin{aligned}
(i) \qquad \overline{R}_{\sigma\mu\nu\lambda}(u) &= \varepsilon_{(n)}(K_{\mu\lambda}K_{\sigma\nu} - K_{\mu\nu}K_{\sigma\lambda}) \\
&+ R_{ijk\ell}(\xi(u))\partial_\sigma\xi^i \cdot \partial_\mu\xi^j \cdot \partial_\nu\xi^k \cdot \partial_\lambda\xi^\ell\,;
\end{aligned} \tag{7.78i}
$$

$$
(ii) \qquad \tilde{\nabla}_\nu K_{\mu\lambda} - \tilde{\nabla}_\lambda K_{\mu\nu} = R_{ijk\ell}(\xi(u))\partial_\mu\xi^j \cdot \partial_\nu\xi^k \cdot \partial_\lambda\xi^\ell \cdot n^i(\xi(u))\,. \tag{7.78ii}
$$

Proof. By (7.76), (7.46), and (7.70i), we deduce that the "tangential component" implies

$$
\begin{aligned}
0 = (g_{ij}\partial_\sigma\xi^j)\theta^i_{\mu\nu\lambda} &= \overline{g}_{\sigma\rho}\overline{g}^{\rho\gamma}\left[\overline{R}_{\gamma\mu\nu\lambda} + \varepsilon_{(n)}(K_{\mu\nu}K_{\lambda\gamma} - K_{\mu\lambda}K_{\nu\gamma})\right] \\
&\quad -(g_{ih}\partial_\sigma\xi^h)R^i_{jk\ell}\partial_\mu\xi^\ell \cdot \partial_\lambda\xi^\ell \cdot \partial_\nu\xi^k + 0\,.
\end{aligned}
$$

Thus, (7.78i) follows.

By (7.76) and (7.70i, ii), we derive that the "normal component" implies

$$
0 = (g_{ih}n^h)\theta^i_{\mu\nu\lambda} = 0 - (g_{ih}n^h)R^i_{jk\ell}\partial_\mu\xi^j \cdot \partial_\nu\xi^k \cdot \partial_\lambda\xi^\ell + \left[\tilde{\nabla}_\nu K_{\mu\lambda} - \tilde{\nabla}_\lambda K_{\mu\nu}\right].
$$

Therefore, (7.78ii) follows. ■

Now, we shall provide some examples.

Example 7.2.5 Consider the three-dimensional Euclidean space, a Cartesian chart, and an embedded surface $\sum_2$. In this case, $R_{ijk\ell}(x)_{|\xi(u)} \equiv 0$. The equations (7.78i, ii) reduce to

$$
\overline{R}_{1212}(u) = K_{11}K_{22} - (K_{12})^2 = \det[K_{\mu\nu}(u)], \tag{7.79i}
$$

$$
\tilde{\nabla}_\lambda K_{\mu\nu} - \tilde{\nabla}_\nu K_{\mu\lambda} = \overline{\nabla}_\lambda K_{\mu\nu} - \overline{\nabla}_\nu K_{\mu\lambda} = 0\,. \tag{7.79ii}
$$

(The equations (7.79i, ii) were first discovered by Gauss and Codazzi-Mainardi, respectively.) Using (7.79i) and (7.35), it follows that the Gaussian curvature

$$
K(u) = \det[K_{\mu\nu}(u)]\,/\,\overline{g}(u) = R_{1212}(u)\,/\,\overline{g}(u)\,. \tag{7.80}
$$

(This equation appeared in (6.25).) The left-hand side $K(u)$ is related to the *extrinsic curvature*, whereas the right-hand side is given by the *intrinsic properties* of the surface. The surprising equality of these two was called by Gauss "**Theorema Egregium**." □

Example 7.2.6 Consider an N-dimensional manifold with a coordinate chart such that

$$\begin{aligned} \mathbf{g}_{\cdot\cdot}(x) &= g_{ij}(x)dx^i \otimes dx^j\,, \\ g^{NN}(x) &\neq 0\,. \end{aligned} \tag{7.81}$$

Let an $(N-1)$-dimensional hypersurface be furnished by

$$\begin{aligned} x^\mu &= \xi^\mu(u) := u^\mu\,, \\ x^N &= \xi^N(u) := c = \text{const.}\,, \\ u &= (\mathbf{x}) \equiv (x^1, \ldots, x^{N-1})\,, \\ \partial_\mu \xi^\nu &\equiv \delta^\nu_\mu,\ \partial_\mu \xi^N \equiv 0\,. \end{aligned} \tag{7.82}$$

The intrinsic metric is provided by

$$\overline{g}_{\mu\nu}(u) \equiv \overline{g}_{\mu\nu}(\mathbf{x}) := g_{\mu\nu}(x)_{|_{x^N=c}}\,. \tag{7.83}$$

The restricted coordinate transformations in (5.263) reduce to the parametric transformations in (7.48). Moreover, the subtensor fields in (5.264) are intrinsic tensors of the $(N-1)$-dimensional hypersurface.

The unit normal to the hypersurface is given by

$$n^i(x)_{|_{x^N=c}} = \left[g^{iN}(x)/\sqrt{|g^{NN}(x)|}\right]_{|_{x^N=c}}\,. \tag{7.84}$$

The equation (7.71a) by (7.82), (7.83), and (7.84) yields

$$K_{\mu\nu}(u) \equiv K_{\mu\nu}(\mathbf{x}) = \left[\frac{1}{\sqrt{|g^{NN}(x)|}} \left\{ {N \atop \mu\nu} \right\}\right]_{|_{x^N=c}}\,. \tag{7.85}$$

The Weingarten equations (7.75a,b) *confirm* the equation above. In the case of a geodesic normal coordinate chart in (5.271),

$$\begin{aligned} & g^{NN}(x) \equiv \varepsilon_N, \quad g^{N\mu}(x) \equiv 0\,, \\ & K_{\mu\nu}(\mathbf{x}) = -\frac{\varepsilon_N}{2} \frac{\partial g_{\mu\nu}(x)}{\partial x^N}_{|_{x^N=c}} =: -(\varepsilon_N/2)\dot{g}_{\mu\nu}(\mathbf{x})\,. \end{aligned}$$

The Gauss equations (7.78i) reduce to (5.274), and the Codazzi-Mainardi equations (7.78ii) reduce to (5.275). □

The hypersurface in (7.82) can alternately be characterized by the equation

$$f(x) := x^N = c\,. \tag{7.86}$$

In the general case of a smooth hypersurface, there exist coordinate charts in M_N such that the embedded hypersurface can *always* be expressed by (7.86).

Exercises 7.2

1. Consider the metric tensor

$$\begin{aligned} \mathbf{g}_{..}(x) &:= g_{\mu\nu}(\mathbf{x})dx^{\mu} \otimes dx^{\nu} + g_{NN}(\mathbf{x})dx^{N} \otimes dx^{N}\,, \\ \mathbf{x} &:= (x^1, \ldots, x^{N-1}), \quad \mu, \nu \in \{1, \ldots, N-1\}\,. \end{aligned}$$

Prove that the geodesics on the hypersurface $x^N = \text{const.}$ are also geodesics in the enveloping manifold M_N. (*Remark*: Such a hypersurface is called a **totally geodesic hypersurface**.)

2. Consider a spherically symmetric metric in the space-time manifold furnished by

$$\begin{aligned} \mathbf{g}_{..}(x) = \exp[2\alpha(x^1, x^4)]dx^1 \otimes dx^1 + [Y(x^1, x^4)]^2[dx^2 \otimes dx^2 + \sin^2 x^2\, dx^3 \otimes dx^3] \\ - \exp[2\gamma(x^1, x^4)]dx^4 \otimes dx^4\,. \end{aligned}$$

(Compare this with (5.125).) Let a three-dimensional non-null parametrized (boundary) hypersurface be specified by

$$\begin{aligned} x^1 &= \xi^1(u) := B(u^3)\,, \\ x^2 &= \xi^2(u) := u^1\,, \\ x^3 &= \xi^3(u) := u^2\,, \\ x^4 &= \xi^4(u) := u^3\,. \end{aligned}$$

Prove that the non-zero extrinsic curvature components are given by

$$\begin{aligned} K_{11}(u) &= -\{Y(x^1, x^4)|e^{-2\alpha} - e^{-2\gamma}[B'(x^4)]^2|^{-1/2}[e^{-2\alpha}\partial_1 y + e^{-2\gamma}\cdot\partial_4 y\cdot B']\}_{|\xi(u)}, \\ K_{22}(u) &= (\sin^2 u^1)K_{11}(u)\,, \\ K_{33}(u) &= \Big\{|e^{-2\alpha} - e^{-2\gamma}(B')^2|^{-1/2}\big[B'' + e^{2(\gamma-\alpha)}\partial_1\gamma + (\partial_4(2\alpha - \gamma))B' \\ &\quad +(\partial_1(\alpha - 2\gamma))(B')^2 - e^{2(\alpha-\gamma)}\cdot(\partial_4\alpha)\cdot(B')^3\big]\Big\}_{|\xi(u)}\,; \\ B' &:= \frac{d}{du^3}B(u^3)\,. \end{aligned}$$

3. Suppose that M_N is a space of constant curvature. (See (6.31).) Prove that the Gauss equation (7.78i) and the Codazzi-Mainardi equation (7.78ii) for a hypersurface reduce respectively to

$$\overline{R}_{\sigma\mu\nu\lambda}(u) = \varepsilon_{(n)}(K_{\mu\lambda}K_{\sigma\nu} - K_{\mu\nu}K_{\sigma\lambda}) + K_0(\overline{g}_{\mu\lambda}\overline{g}_{\sigma\nu} - \overline{g}_{\mu\nu}\overline{g}_{\sigma\lambda})$$

and

$$\overline{\nabla}_{\nu}K_{\mu\lambda} - \overline{\nabla}_{\lambda}K_{\mu\nu} = 0\,.$$

4. A **stationary space-time** manifold M_4 is locally characterized by the metric tensor

$$\mathbf{g}_{..}(x) = e^{-\omega(\mathbf{x})} g'_{\mu\nu}(\mathbf{x}) dx^\mu \otimes dx^\nu - e^{\omega(\mathbf{x})}[dx^4 + a_\mu(\mathbf{x})dx^\mu] \otimes [dx^4 + a_\nu(\mathbf{x})dx^\nu],$$
$$\mathbf{x} := (x^1, x^2, x^3)\,.$$

A $x^4 = \text{const.}$ spatial hypersurface M_3 must be given by the metric

$$\overline{\mathbf{g}}_{..}(\mathbf{x}) = [e^{-\omega(\mathbf{x})} g'_{\mu\nu}(\mathbf{x}) - e^{\omega(\mathbf{x})} a_\mu(\mathbf{x}) a_\nu(\mathbf{x})] dx^\mu \otimes dx^\nu\,.$$

Using the notation $f_{\mu\nu}(x) := \partial_\nu a_\mu - \partial_\mu a_\nu$, show that

(i) $\overline{e}^{\omega/2} g'_{\mu\nu} R + R_{\mu\nu} - a_\mu R_{\nu 4} - a_\nu R_{\mu 4} + a_\mu a_\nu R_{44}$
$= G'_{\mu\nu}(\mathbf{x}) + (1/2)(\partial_\mu \omega \cdot \partial_\nu \omega - (1/2) g'_{\mu\nu} g'^{\lambda\rho} \partial_\lambda \omega \cdot \partial_\rho \omega)$;

(ii) $-2e^{-2\omega} R_{44} = g'^{\mu\nu} \nabla'_\mu \nabla'_\nu \omega + (1/2) e^{2\omega} f'^{\mu\nu} f_{\mu\nu}$;

(iii) $2a_\mu R_{44} - 2R_{\mu 4} = \nabla'_\nu [e^{2\omega} f'^\nu{}_\mu]$;

(iv) $R = e^\omega [R' + (1/2) g'^{\mu\nu} \partial_\mu \omega \cdot \partial_\nu \omega - g'^{\mu\nu} \nabla'_\mu \nabla'_\nu \omega - (1/4) e^{2\omega} g'^{\mu\nu} g'^{\rho\sigma} f_{\mu\rho} f_{\nu\sigma}]$;
$f'^{\mu\nu} := g'^{\mu\lambda} g'^{\nu\rho} f_{\lambda\rho}$ (Gegenberg [14] & Kloster, Som, and Das [24]).

7.3 D-Dimensional Submanifolds

Let us consider now the more general case of embedding a **submanifold** M_D into M_N, where $1 \leq D < N$. The lowercase Roman indices as usual take $\{1, \ldots, N\}$. The Greek indices now take $\{1, \ldots, D\}$. Moreover, the capital Roman indices assume values $\{1, \ldots, N - D\}$. The summation convention operates on *three* distinct sets of indices. Choosing appropriate coordinate charts in M_D and M_N, we have a **parametrized submanifold** $\xi \in C^r(\mathcal{D}_D \subset \mathbb{R}^D; \mathbb{R}^N)$, $r \geq 3$. (Compare this with fig. 7.3.) The corresponding image $\sum_D$

is in an appropriate domain of $\mathbb{R}^N$. We can generalize (7.44) by the following:

$$\begin{aligned}
&x = \xi(u) \in \mathbb{R}^N\,,\\
&x^i = \xi^i(u) \equiv \xi^i(u^1, \ldots, u^D)\,,\\
&\text{Rank}\,[\partial_\mu \xi^i] = D \geq 1\,,\\
&\overline{g}_{\mu\nu}(u) du^\mu \otimes du^\nu := [g_{ij}(\xi(u)) \partial_\mu \xi^i \partial_\nu \xi^j] du^\mu \otimes du^\nu\,.
\end{aligned} \tag{7.87}$$

A normal vector $\nu^i(\xi(u))$ to $\sum_D$ satisfies

$$[\partial_\mu \xi^i] \nu_i = 0\,. \tag{7.88}$$

The above is a system of D linear homogeneous equations in N unknown components ν_i. Since the coefficient matrix $[\partial_\mu \xi^i]$ is of rank D, there exist $N - D$ linearly independent solution covectors $\tilde{\boldsymbol{\nu}}_A(\xi(u))$ $(A \in \{1, \ldots, N - D\})$. The image $\sum_D$ is a **non-null submanifold**, provided that there exist D *non-null* solution covectors $\tilde{\boldsymbol{\nu}}_A(\xi(u))$. We can normalize these covectors to provide $N - D$ orthonormal vectors satisfying

$$g_{ij}(\xi(u)) n^i_A(\xi(u)) n^j_B(\xi(u)) = d_{AB}\,, \quad A,\, B \in \{1, \ldots, N - D\}\,. \tag{7.89}$$

Moreover, the normal vectors $n^i_A(\cdot\cdot)$ by (7.88) satisfy

$$g_{ij}(\xi(u))(\partial_\mu \xi^i) n^j_A(\xi(u)) \equiv 0\,. \tag{7.90}$$

We define the *hybrid* covariant derivatives as in (7.57), treating the capital indices as "invariant labels." Thus, we obtain from (7.87), (7.89), and (7.90) the equations

$$g_{ij}(\tilde{\nabla}_\lambda \partial_\mu \xi^i) \partial_\nu \xi^j \equiv 0\,, \tag{7.91i}$$

$$g_{ij}(\tilde{\nabla}_\lambda n^i_A) n^j_B \equiv 0\,, \tag{7.91ii}$$

$$g_{ij}[(\tilde{\nabla}_\lambda \partial_\mu \xi^i) n^j_A + \partial_\mu \xi^i \tilde{\nabla}_\lambda n^j_A] \equiv 0\,. \tag{7.91iii}$$

We generalize the definitions (7.71a, b, c, d) and the consequences to obtain

$$K_{A\mu\nu}(u) \equiv k_{A\mu\nu}(\xi(u)) := g_{ij}(\xi(u)) n^i_A(\xi(u)) \tilde{\nabla}_\mu(\partial_\nu \xi^j)\,, \tag{7.92a}$$

$$\tilde{\nabla}_\mu(\partial_\nu \xi^j) = d^{AB} n^j_A(\xi(u)) K_{B\mu\nu}(u)\,, \tag{7.92b}$$

$$\partial_\mu \partial_\nu \xi^j = \overline{\left\{ \begin{matrix} \lambda \\ \mu\nu \end{matrix} \right\}} \partial_\lambda \xi^j - \left\{ \begin{matrix} j \\ ik \end{matrix} \right\}_{|\xi(u)} (\partial_\mu \xi^i)(\partial_\nu \xi^k) + d^{AB} n^j_A K_{B\mu\nu}\,, \tag{7.92c}$$

$$K_{A\nu\mu}(u) \equiv K_{A\mu\nu}(u)\,. \tag{7.92d}$$

By (7.70iv) and (7.92a), we deduce that

$$K_{A\mu\nu}(u) = -g_{ij}(\tilde{\nabla}_\mu n^i_{\ A})(\partial_\nu \xi^j) = -(\nabla_i n_{Aj})(\partial_\mu \xi^i)(\partial_\nu \xi^j) \,. \tag{7.93}$$

Let us define the *hybrid tensor*

$$\mu_{AB\nu}(\xi(u)) := n_{Ai}(\xi(u))(\tilde{\nabla}_\nu n^i_{\ B}) \,. \tag{7.94}$$

By (7.91ii), we derive that

$$\mu_{AB\nu}(\xi(u)) \equiv -\mu_{BA\nu}(\xi(u)) \,. \tag{7.95}$$

Now, $\partial_\mu \xi^i$ and $n^i_{\ A}(\xi(u))$ are components of basis vectors for $T_{\xi(u)}(\mathbb{R}^N)$, which is endowed with the metric $\mathbf{g}_{..}(\xi(u))$. For any vector $\vec{\mathbf{V}}(\xi(u)) \in T_{\xi(u)}(\mathbb{R}^N)$, we can express as the linear combination

$$\begin{aligned} \vec{\mathbf{V}}(\xi(u)) &= \text{tangent}\,(\vec{\mathbf{V}}(\xi(u))) + \text{normal}\,(\vec{\mathbf{V}}(\xi(u)))\,, \\ V^i(\xi(u)) &= (g_{jk} V^j \partial_\nu \xi^k)\overline{g}^{\nu\mu} \partial_\mu \xi^i + (g_{jk} V^j n^k_{\ B}) d^{BA} n^i_{\ A} \,. \end{aligned} \tag{7.96}$$

Substituting (7.94) and (7.92a) into (7.96), we get

$$\tilde{\nabla}_\mu n^i_{\ C} = -\overline{K}^{\nu}_{\ C\mu}(u)\partial_\nu \xi^i + d^{AB} \mu_{AC\mu}(\xi(u)) n^i_{\ B}(\xi(u)) \,, \tag{7.97i}$$

$$\partial_\mu \xi^k (\nabla_k n^i_{\ C})_{|..} = -\overline{g}^{\nu\rho} K_{C\mu\rho} \partial_\nu \xi^i + d^{AB} \mu_{AC\mu} n^i_{\ B} \,. \tag{7.97ii}$$

These are the generalizations of *Weingarten equations* (7.75a, b).

Now, we shall obtain another useful equation. By (7.68) and (7.97i, ii), we deduce that

$$\begin{aligned} &\tilde{\nabla}_\nu (n^i_{\ A} K_{B\mu\lambda}) - \tilde{\nabla}_\lambda (n^i_{\ A} K_{B\mu\nu}) \\ &= (K_{B\mu\nu} \overline{K}^{\rho}_{\ A\lambda} - K_{B\mu\lambda} \overline{K}^{\rho}_{\ A\nu}) \partial_\rho \xi^i + (\tilde{\nabla}_\nu K_{B\mu\lambda} - \tilde{\nabla}_\lambda K_{B\mu\nu}) n^i_{\ A} \\ &+ d^{DE} (\mu_{DA\nu} K_{B\mu\lambda} - \mu_{DA\lambda} K_{B\mu\nu}) n^i_{\ E} \,. \end{aligned} \tag{7.98}$$

The integrability conditions for the partial differential equations (7.92b) are given by

$$\tilde{\nabla}_\nu \tilde{\nabla}_\lambda (\partial_\mu \xi^i) - \tilde{\nabla}_\lambda \tilde{\nabla}_\nu (\partial_\mu \xi^i) = d^{AB} [\tilde{\nabla}_\nu (n^i_{\ A} K_{B\mu\lambda}) - \tilde{\nabla}_\lambda (n^i_{\ A} K_{B\mu\nu})] \,. \tag{7.99}$$

Using (7.69) and (7.98), the integrability conditions yield

$$\begin{aligned}
&\overline{R}^{\rho}_{\mu\lambda\nu}(u)\partial_\rho\xi^i + R^i_{jk\ell}(\xi(u))\partial_\mu\xi^j\partial_\nu\xi^k\partial_\lambda\xi^\ell \\
&\quad = d^{AB}\Big[(K_{B\mu\nu}\overline{K}^{\rho}_{A\lambda} - K_{B\mu\lambda}\overline{K}^{\rho}_{A\nu})\partial_\rho\xi^i \\
&\quad +(\tilde{\nabla}_\nu K_{B\mu\lambda} - \tilde{\nabla}_\lambda K_{B\mu\nu})n^i_A \\
&\quad +d^{DE}(\mu_{DA\nu}K_{B\mu\lambda} - \mu_{DA\lambda}K_{B\mu\nu})n^i_E\Big],
\end{aligned} \tag{7.100i}$$

$$\begin{aligned}
\theta^i_{\mu\nu\lambda}(u,\xi(u)) &:= \big[\overline{R}^{\rho}_{\mu\nu\lambda}(u) + d^{AB}(K_{B\mu\nu}\overline{K}^{\rho}_{A\lambda} - K_{B\mu\lambda}\overline{K}^{\rho}_{A\nu})\big]\partial_\rho\xi^i \\
&+R^i_{j\ell k}\partial_\mu\xi^j\cdot\partial_\lambda\xi^\ell\cdot\partial_\nu\xi^k \\
&+d^{AB}\Big[\tilde{\nabla}_\nu K_{B\mu\lambda} - \tilde{\nabla}_\lambda K_{B\mu\nu} \\
&+d^{DE}(\mu_{EB\lambda}K_{D\mu\nu} - \mu_{EB\nu}K_{D\mu\lambda})\Big]n^i_A(\xi(u)) \\
&=0\,.
\end{aligned} \tag{7.100ii}$$

This is the generalization of (7.76). By (7.87), (7.90), and (7.100ii), we derive that the tangential component satisfies

$$\begin{aligned}
(g_{ih}\partial_\sigma\xi^h)\theta^i_{\mu\nu\lambda} &= 0\,, \\
\overline{R}_{\sigma\mu\nu\lambda}(u) &= d^{AB}(K_{B\mu\lambda}K_{A\sigma\nu} - K_{B\mu\nu}K_{A\sigma\lambda}) \\
&+R_{ijk\ell}(\xi(u))\partial_\sigma\xi^i\cdot\partial_\mu\xi^j\cdot\partial_\nu\xi^k\cdot\partial_\lambda\xi^\ell\,.
\end{aligned} \tag{7.101}$$

This is the generalization of equation (7.78i) of Gauss. It shows that the intrinsic curvature of the embedded submanifold is partly due to the extrinsic curvature and partly due to the curvature of the larger enveloping space itself.

Using (7.87), (7.89), and (7.90), we deduce that the normal component satisfies

$$\begin{aligned}
(g_{ih}n^h_C)\theta^i_{\mu\nu\lambda} &= 0\,, \\
\tilde{\nabla}_\nu K_{C\mu\lambda} - \tilde{\nabla}_\lambda K_{C\mu\nu} &+ d^{BE}(\mu_{EC\lambda}K_{B\mu\nu} - \mu_{EC\nu}K_{B\mu\lambda}) \\
&= R_{ijk\ell}(\xi(u))\partial_\mu\xi^j\cdot\partial_\nu\xi^k\cdot\partial_\lambda\xi^\ell\cdot n^i_C(\xi(u))\,.
\end{aligned} \tag{7.102}$$

The equation above is the generalization of the Codazzi-Mainardi equation (7.78ii).

Now, we shall derive some more useful equations. By (7.57), (7.60), and (5.110), we have

$$\tilde{\nabla}_\mu n^i_A = \frac{d}{du^\mu} n^i_A(\xi(u)) + \left\{ \begin{matrix} i \\ jk \end{matrix} \right\}_{|..} \partial_\mu \xi^j \cdot n^k_A \,,$$

$$\tilde{\nabla}_\nu \tilde{\nabla}_\mu n^i_A - \tilde{\nabla}_\mu \tilde{\nabla}_\nu n^i_A = \frac{d}{du^\nu}(\tilde{\nabla}_\mu n^i_A) - \overline{\left\{ \begin{matrix} \lambda \\ \mu\nu \end{matrix} \right\}} \tilde{\nabla}_\lambda n^i_A + \partial_\nu \xi^j \left\{ \begin{matrix} i \\ jk \end{matrix} \right\}_{|..} \tilde{\nabla}_\mu n^k_A$$
$$-(\nu \leftrightarrow \mu)$$
$$= \frac{d}{du^\nu}\left[\frac{dn^i_A}{du^\mu} + \left\{ \begin{matrix} i \\ jk \end{matrix} \right\}_{|..} \partial_\mu \xi^j \cdot n^k_A \right] - 0 + \partial_\nu \xi^j \left\{ \begin{matrix} i \\ jk \end{matrix} \right\}_{|..} \times$$
$$\left[\frac{dn^k_A}{du^\mu} + \left\{ \begin{matrix} k \\ \ell h \end{matrix} \right\}_{|..} \partial_\mu \xi^\ell \cdot n^h_A \right] - (\nu \leftrightarrow \mu)$$
$$= 0 + \left\{ \left[\partial_\ell \left\{ \begin{matrix} i \\ jk \end{matrix} \right\} \right]_{|..} \partial_\nu \xi^\ell \cdot \partial_\mu \xi^j + \left\{ \begin{matrix} i \\ jk \end{matrix} \right\}_{|..} \partial_\nu \partial_\mu \xi^j \right\} n^k_A$$
$$+ \left\{ \begin{matrix} i \\ jk \end{matrix} \right\}_{|..} \left[\partial_\mu \xi^j \cdot \frac{dn^k_A}{du^\nu} + \partial_\nu \xi^j \cdot \frac{dn^k_A}{du^\mu} \right]$$
$$+ \left\{ \begin{matrix} i \\ jk \end{matrix} \right\}_{|..} \left\{ \begin{matrix} k \\ \ell h \end{matrix} \right\}_{|..} \partial_\nu \xi^j \cdot \partial_\mu \xi^\ell \cdot n^h_A - (\nu \leftrightarrow \mu)$$
$$= \left\{ \left[\partial_\ell \left\{ \begin{matrix} i \\ jk \end{matrix} \right\} \right]_{|..} (\partial_\nu \xi^\ell \cdot \partial_\mu \xi^j - \partial_\mu \xi^\ell \cdot \partial_\nu \xi^j) + 0 \right\} n^k_A + 0$$
$$+ \left\{ \begin{matrix} i \\ jk \end{matrix} \right\}_{|..} \left\{ \begin{matrix} k \\ \ell h \end{matrix} \right\}_{|..} (\partial_\nu \xi^j \cdot \partial_\mu \xi^\ell - \partial_\mu \xi^j \cdot \partial_\nu \xi^\ell) n^h_A$$
$$= R^i_{\ell kj}(\xi(u)) \partial_\mu \xi^j \partial_\nu \xi^k n^\ell_A(\xi(u)) \,. \tag{7.103}$$

Using (7.68), (7.92b), and (7.102), we deduce that

$$\tilde{\nabla}_\mu(\overline{K}^\rho_{A\nu} \cdot \partial_\rho \xi^i) - (\mu \leftrightarrow \nu)$$
$$= (\tilde{\nabla}_\mu \overline{K}^\rho_{A\nu}) \cdot \partial_\rho \xi^i + \overline{K}^\rho_{A\nu}(d^{BC} n^i_C K_{B\mu\rho}) - (\mu \leftrightarrow \nu)$$
$$= \overline{g}^{\rho\lambda}(\tilde{\nabla}_\mu K_{A\nu\lambda} - \tilde{\nabla}_\nu K_{A\mu\lambda}) \partial_\rho \xi^i + d^{BC}(\overline{K}^\rho_{A\nu} K_{B\mu\rho} - \overline{K}^\rho_{A\mu} K_{B\nu\rho}) n^i_C$$
$$= \overline{g}^{\rho\lambda} \partial_\rho \xi^i \left[R_{hj\ell k} \partial_\lambda \xi^j \cdot \partial_\mu \xi^\ell \cdot \partial_\nu \xi^k \cdot n^h_A + d^{DE}(\mu_{EA\mu} K_{D\lambda\nu} - \mu_{EA\nu} K_{D\lambda\mu}) \right]$$
$$+ d^{BC}(\overline{K}^\rho_{A\nu} K_{B\mu\rho} - \overline{K}^\rho_{A\mu} K_{B\nu\rho}) n^i_C \,. \tag{7.104}$$

Now, we use (7.97i) to obtain the following expression:

$$\begin{aligned}\tilde{\nabla}_\mu(\mu_{AD\nu}n^i_E)-\tilde{\nabla}_\nu(\mu_{AD\mu}n^i_E) &= (\tilde{\nabla}_\mu\mu_{AD\nu}-\tilde{\nabla}_\nu\mu_{AD\mu})n^i_E\\ &\quad+(\mu_{AD\mu}\overline{K}^\rho_{E\nu}-\mu_{AD\nu}\overline{K}^\rho_{E\mu})\partial_\rho\xi^i\\ &\quad+d^{BC}(\mu_{AD\mu}\mu_{EB\nu}-\mu_{AD\nu}\mu_{EB\mu})n^i_C.\end{aligned} \tag{7.105}$$

The integrability conditions of Weingarten equation (7.97i) are furnished by

$$\begin{aligned}(\tilde{\nabla}_\nu\tilde{\nabla}_\mu-\tilde{\nabla}_\mu\tilde{\nabla}_\nu)n^i_A &= \tilde{\nabla}_\mu(\overline{K}^\rho_{A\nu}\cdot\partial_\rho\xi^i)-\tilde{\nabla}_\nu(\overline{K}^\rho_{A\mu}\cdot\partial_\rho\xi^i)\\ &\quad+d^{DE}[\tilde{\nabla}_\mu(\mu_{AD\nu}n^i_E)-\tilde{\nabla}_\nu(\mu_{AD\mu}n^i_E)]\,.\end{aligned} \tag{7.106}$$

Using (7.103), (7.104), and (7.105), the integrability conditions (7.106) yield

$$\begin{aligned}R^i_{\ell kj}(\xi(u))\partial_\mu\xi^i\cdot\partial_\nu\xi^k\cdot n^\ell_A &= \overline{g}^{\rho\lambda}\partial_\rho\xi^i\cdot R_{hj\ell k}\partial_\lambda\xi^j\cdot\partial_\mu\xi^\ell\cdot\partial_\nu\xi^k\cdot n^h_A\\ &\quad+d^{BC}(\overline{K}^\rho_{A\nu}K_{B\mu\rho}-\overline{K}^\rho_{A\mu}K_{B\nu\rho})n^i_A\\ &\quad+d^{DE}\Big[(\tilde{\nabla}_\mu\mu_{AD\nu}-\tilde{\nabla}_\nu\mu_{AD\mu})n^i_E\\ &\qquad+d^{BC}(\mu_{AD\mu}\mu_{EB\nu}-\mu_{AD\nu}\mu_{EB\mu})n^i_C\Big],\end{aligned} \tag{7.107i}$$

or

$$\begin{aligned}\Phi^i{}_{A\mu\nu}(u,\xi(u)) &:= d^{DE}\rho_{AD\mu\nu}(\xi(u))n^i_E\\ &\quad+d^{BC}(\overline{K}^\rho_{A\nu}K_{B\mu\rho}-\overline{K}^\rho_{A\mu}K_{B\nu\rho})n^i_C\\ &\quad+(g^{ih}-\overline{g}^{\lambda\rho}\partial_\rho\xi^i\cdot\partial_\lambda\xi^h)R_{h\ell jk}\partial_\mu\xi^j\cdot\partial_\nu\xi^k\cdot n^\ell_A\\ &= 0,\end{aligned} \tag{7.107ii}$$

$$\rho_{AB\mu\nu}(\xi(u)) := \tilde{\nabla}_\mu\mu_{AB\nu}-\tilde{\nabla}_\nu\mu_{AB\mu}+d^{FC}(\mu_{AF\mu}\mu_{CB\nu}-\mu_{AF\nu}\mu_{CB\mu}), \tag{7.107iii}$$

$$\rho_{AB\nu\mu}\equiv-\rho_{AB\mu\nu}\,. \tag{7.107iv}$$

Now, we shall state and prove a theorem due to Voss (for a positive-definite metric) and Ricci (for an indefinite metric).

Theorem 7.3.1 *Let ξ be a non-degenerate, non-null, D-dimensional parametrized submanifold of class C^r, $(r\geq 3)$ in $\mathbb{R}^N$. Then, the integrability conditions* (7.107ii) *imply that*

$$\begin{aligned}\rho_{AF\mu\nu}(\xi(u)) &+(\overline{K}^\rho_{A\nu}\cdot K_{F\mu\rho}-\overline{K}^\rho_{A\mu}\cdot K_{F\nu\rho})\\ &+R_{i\ell jk}\partial_\mu\xi^j\cdot\partial_\nu\xi^k\cdot n^\ell_A\cdot n^i_F=0\,.\end{aligned} \tag{7.108}$$

Proof. The equations (7.108) follow from (7.107ii, iii) and $n_{Fi}\Phi^i{}_{A\mu\nu}=0$.
We shall discuss some examples now. ■

Example 7.3.2 Consider as the embedding space the three-dimensional Euclidean manifold $\mathbb{E}_3$. We choose a Cartesian coordinate chart so that

$$\mathbf{g}_{..}(x) \equiv \mathbf{I}_{..}(x) = \delta_{ij} dx^i \otimes dx^j .$$

Let a non-degenerate parametrized curve ξ of class C^3 be embedded in $\mathbb{R}^3$. (See (5.170).) In this example, $D = 1$ and the image of the curve ξ is denoted by $\sum_1$. Roman indices take from $\{1, 2, 3\}$, the Greek indices take from $\{1\}$, and the capital Roman indices take from $\{1, 2\}$. The curve is given by

$$x^i = \xi^i(u) \equiv \xi^i(u^1), \quad u^1 \in (a, b) \subset \mathbb{R} .$$

Without loss of generality, we choose u to be the arc length parameter. Thus, we must have

$$\mathbf{I}_{..}(\vec{\boldsymbol{\xi}}'(u), \vec{\boldsymbol{\xi}}'(u)) = \delta_{ij} \frac{d\xi^i(u)}{du^1} \frac{d\xi^j(u)}{du^1} \equiv 1 = \overline{g}_{11}(u) ,$$

$$\mathbf{I}_{..}(\vec{\boldsymbol{\xi}}'(u), \vec{\boldsymbol{\xi}}''(u)) = \delta_{ij} \frac{d\xi^i(u)}{du^1} \frac{d^2\xi^j(u)}{(du^1)^2} \equiv 0 .$$

Assuming that $\vec{\boldsymbol{\xi}}''(u) \not\equiv \vec{\mathbf{O}}_{\xi(u)}$, we define one normal vector as

$$\begin{aligned} \vec{\mathbf{n}}_{(1)}(\xi(u)) &:= \vec{\boldsymbol{\xi}}''(u)/\|\vec{\boldsymbol{\xi}}''(u)\| , \\ \frac{d^2\xi^i(u)}{(du^1)^2} &= \|\vec{\boldsymbol{\xi}}''(u)\| n^i_{(1)}(\xi(u)) =: \kappa(u) n^i_{(1)}(\xi(u)) . \end{aligned} \tag{7.109}$$

Recall from (5.188) that $\kappa(u)$ is the *principal curvature.*

The second normal is defined by

$$n^i_{(2)}(\xi(u)) := \delta^{ij} \varepsilon_{jk\ell} \frac{d\xi^k(u)}{du^1} n^\ell_{(1)}(\xi(u)) ,$$

$$\delta_{ij} n^i_{(A)} n^j_{(B)} = \delta_{AB} .$$

Now, (7.92b) yields

$$\frac{d^2\xi^i(u)}{(du^1)^2} = \delta^{(11)} n^i_{(1)} K_{(1)11} + \delta^{(22)} n^i_{(2)} K_{(2)11} . \tag{7.110i}$$

Comparing with (7.109), we obtain that

$$\kappa(u) = K_{(1)11}(u),\; K_{(2)11}(u) \equiv 0 . \tag{7.110ii}$$

The second and third of the equations (5.188) and the Weingarten equations (7.97i) imply that

$$\begin{aligned}
\frac{dn^i_{(1)}(\cdot\cdot)}{du^1} &= -\kappa(u)\frac{d\xi^i(u)}{du^1} + \tau(u)n^i_{(2)}(\cdot\cdot) = -K_{(1)11}\frac{d\xi^i(u)}{du^1} + \delta^{(22)}\mu_{(21)1}n^i_{(2)}\,, \\
\frac{dn^i_{(2)}(\cdot\cdot)}{du^1} &= -\tau(u)n^i_{(1)}(\cdot\cdot) = -K_{(2)11}\frac{d\xi^i(u)}{du^1} + \delta^{(11)}\mu_{(12)1}n^i_{(1)}\,.
\end{aligned} \tag{7.111}$$

By comparison, we infer that

$$K_{(2)11}(u) \equiv 0, \quad \mu_{(21)1}(u) \equiv -\mu_{(12)1}(u) = \tau(u)\,.$$

In this example, we have established the equivalence of equations (7.92b) and (7.97i) with Frenet-Serret formulas (5.188). □

Example 7.3.3 Every regular subspace allows local parametrization:

$$\begin{aligned}
x^\alpha &= \xi^\alpha(u) := u^\alpha\,, \\
x^{D+A} &= \xi^{D+A}(u) := c^A = \text{const.}\,, \\
\partial_\nu \xi^\mu &= \delta^\mu_\nu,\ \partial_\nu \xi^{D+A} \equiv 0\,, \\
\overline{g}_{\mu\nu}(u) &:= g_{ij}(\xi(u))\partial_\mu\xi^i \cdot \partial_\nu\xi^j = g_{\mu\nu}(u,c)\,.
\end{aligned} \tag{7.112}$$

We could have represented the image $\sum_D$ by the linear constraints

$$\begin{aligned}
f_{(D+A)}(x) &:= x^{D+A} - c^A = 0\,, \\
\partial_i f_{(D+A)} &= \delta_i^{D+A},\ g^{ij}(\cdot\cdot)\partial_i f_{(D+A)} \cdot \partial_j f_{D+A} \equiv g^{D+A,D+A}(\cdot\cdot)\,.
\end{aligned} \tag{7.113}$$

(In this example, we suspend *the summation convention* on capital indices.) The non-null subspace must satisfy

$$g^{D+A,D+A}(\xi(u)) \neq 0 \ \text{ for every } \ A \in \{1,\ldots,N-D\}.$$

The unit normal vectors are provided by

$$\begin{aligned}
n_{Ai}(\xi(u)) &= \delta_i^{D+A}/\sqrt{|g^{D+A,D+A}|}\,, \\
n^i_A(\xi(u)) &= g^{i,D+A}(\xi(u))/\sqrt{|g^{D+A,D+A}|}\,.
\end{aligned} \tag{7.114}$$

By (7.93) and (7.114), the components of the exterior curvature tensor are given by

$$\begin{aligned}
K_{A\mu\nu}(u) &= \left[\left\{ {j \atop ik} \right\} \partial_\mu\xi^i \cdot \partial_\nu\xi^k \cdot n_{Aj}\right]_{|\xi(u)} \\
&= \left[|g^{D+A,D+A}(x)|^{-1/2}\left\{ {D+A \atop \mu\nu} \right\}\right]_{|x=(u,c)}\,.
\end{aligned} \tag{7.115}$$

Using (7.94) and (7.114), we can derive that

$$\mu_{AB\nu}(u,c) = \delta_i^{D+A}|g^{D+A,D+A}(\cdot\cdot)|^{-1/2} \cdot \partial_\nu \xi^j [\nabla_j n_B^i]_{|\cdot\cdot}$$

$$= \left\{ |g^{D+A,D+A}(x)|^{-1/2} \left[\frac{\partial}{\partial x^\nu}(g^{D+A,D+B} \cdot |g^{D+B,D+B}|^{-1/2}) \right.\right. \tag{7.116}$$

$$\left.\left. + g^{k,D+B}(x)|g^{D+B,D+B}|^{-1/2} \left\{ \begin{matrix} D+A \\ \nu k \end{matrix} \right\} \right] \right\}_{|(u,c)} . \quad \square$$

Example 7.3.4 In this example, we have the usual summation conventions *except* for two cases. The capital Roman index D is not summed. Moreover, if two repeated capital Roman indices *both* occur as superscripts (or else subscripts), the summation is suspended. Let the metric tensor $\mathbf{g}_{\cdot\cdot}(x)$ be provided by

$$\begin{aligned} g_{ij}(x)dx^i \otimes dx^j &:= g'_{\mu\nu}(\mathbf{x})dx^\mu \otimes dx^\nu + \delta_{BC}\tilde{e}^B(\mathbf{x}) \otimes \tilde{e}^C(\mathbf{x}) , \\ \tilde{e}^B(\mathbf{x}) &:= dx^{D+B} + A_\mu^B(\mathbf{x})dx^\mu , \\ \mathbf{x} &:= (x^1, \ldots, x^D) . \end{aligned} \tag{7.117}$$

It follows from (7.117) that

$$\begin{aligned} [g_{ij}(\mathbf{x})] &= \left[\begin{array}{c|c} g'_{\mu\nu} + \delta_{EC}A_\mu^E A_\nu^C & A_\mu^C \delta_{CB} \\ \hline \delta_{AC}A_\nu^C & \delta_{AB} \end{array} \right], \\ [g^{ij}(\mathbf{x})] &= \left[\begin{array}{c|c} g'^{\mu\nu} & -g'^{\mu\nu}A_\nu^B \\ \hline -A_\mu^A g'^{\mu\nu} & \delta^{AB} + g'^{\lambda\rho}A_\lambda^A A_\rho^B \end{array} \right], \\ \det[g_{ij}(\mathbf{x})] &= \det[g'_{\mu\nu}(\mathbf{x})] . \end{aligned} \tag{7.118}$$

Some of the Christoffel symbols, as computed from (7.118), are furnished by

$$\left\{ \begin{matrix} D+E \\ \mu\nu \end{matrix} \right\} = (1/2)[\nabla'_\mu A_\nu^E + \nabla'_\nu A_\mu^E] - (1/2)g'^{\lambda\rho}\delta_{AB}A_\lambda^E(A_\mu^A F_{\nu\rho}^B + A_\nu^A F_{\mu\rho}^B) , \tag{7.119i}$$

$$F_{\mu\nu}^B(\mathbf{x}) := \partial_\mu A_\nu^B - \partial_\nu A_\mu^B \equiv -F_{\nu\mu}^B(\mathbf{x}) . \tag{7.119ii}$$

Comparing (7.119ii) with (5.88) and (3.77), we conclude that A_μ^B and $F_{\mu\nu}^B$ are generalizations of electromagnetic potentials and fields.

For a D-dimensional subspace, we choose the parametrization (7.112) ($x^\mu \equiv u^\mu$) again. Therefore, we get for the intrinsic metric

$$\overline{g}_{\mu\nu}(u) = g'_{\mu\nu}(u) + \delta_{EC} A^E_\mu(u) A^C_\nu(u)\,. \tag{7.120}$$

The unit normal vectors are given by (7.114) and (7.118) as

$$\begin{aligned} n_{Bi}(u,c) &= \delta_i^{D+B}|1 + g'^{\lambda\rho} A^B_\lambda A^B_\rho|^{-1/2}\,, \\ n^i_B(u,c) &= g^{i,D+B}(u)|1 + g'^{\lambda\rho} A^B_\lambda A^B_\rho|^{-1/2}\,, \\ n^\mu_B(u,c) &= -g'^{\mu\nu}(u) A^B_\nu(u)|1 + g'^{\lambda\rho} A^B_\lambda A^B_\rho|^{-1/2}\,, \\ n^{D+A}_B(u,c) &= [\delta^{AB} + g'^{\mu\nu} A^A_\mu A^B_\nu]\,|1 + g'^{\lambda\rho} A^B_\lambda A^B_\rho|^{-1/2}\,. \end{aligned} \tag{7.121}$$

The extrinsic curvature components are furnished by (7.115), (7.118), (7.119i), and (7.121) as

$$\begin{aligned} K_{C\mu\nu}(u) &= (1/2)|1 + g'^{\lambda\rho} A^C_\lambda A^C_\rho|^{-1/2} \\ &\quad \times [\nabla'_\mu A^C_\nu + \nabla'_\nu A^C_\mu - g'^{\alpha\beta}\delta_{AB} A^C_\alpha (A^A_\mu F^B_{\nu\beta} + A^A_\nu F^B_{\mu\beta})]. \end{aligned} \tag{7.122}$$

Using (7.116), (7.118), and (7.121), we obtain

$$\begin{aligned} \mu_{AB\nu}(u) = \Bigg\{ &|1 + g'^{\sigma\tau} A^A_\sigma A^A_\tau|^{-1/2} \Bigg[\frac{\partial}{\partial x^\nu} \Big((\delta^{AB} + g'^{\alpha\beta} A^A_\alpha A^B_\beta)|1 + g'^{\lambda\rho} A^B_\lambda A^B_\rho|^{-1/2} \Big) \\ &- g'^{\mu\gamma} A^B_\gamma |1 + g'^{\lambda\rho} A^B_\lambda A^B_\rho|^{-1/2} \begin{Bmatrix} D+A \\ \mu\nu \end{Bmatrix} \\ &- (1/2) g'^{\gamma\delta} A^A_\gamma F_{C\nu\delta} (\delta^{CB} + g'^{\alpha\beta} A^C_\alpha A^B_\beta)|1 + g'^{\lambda\rho} A^B_\lambda A^B_\rho|^{-1/2} \Bigg] \Bigg\}_{|x=u}. \end{aligned} \tag{7.123}$$

The equation (7.101) of Gauss yields

$$\begin{aligned} R_{\sigma\mu\nu\lambda}(\mathbf{x}, c)_{|\mathbf{x}=u} = \overline{R}_{\sigma\mu\nu\lambda}(u) + d^{AB} \Bigg\{ &|(1 + g'^{\lambda\rho} A^A_\lambda A^A_\rho)(1 + g'^{\alpha\beta} A^B_\alpha A^B_\beta)|^{-1/2} \\ &\times \Bigg[\begin{Bmatrix} D+B \\ \mu\nu \end{Bmatrix} \begin{Bmatrix} D+A \\ \sigma\lambda \end{Bmatrix} - \begin{Bmatrix} D+B \\ \mu\lambda \end{Bmatrix} \begin{Bmatrix} D+A \\ \sigma\nu \end{Bmatrix} \Bigg] \Bigg\}_{|\mathbf{x}=u}. \end{aligned} \tag{7.124}$$

This example is relevant in the unification of gravitation with an **Abelian gauge field**. □

There exists another approach to extrinsic curvature of a *non-null* subspace based *solely* on tensors in the enveloping space M_N. In this approach, the

projection tensor h$^{\cdot}_{\cdot}$ is defined by the components

$$\begin{aligned} h^i_j(\xi(u)) &:= \delta^i_j - d^{AB} n^i_A(\xi(u)) n_{Bj}(\xi(u)) \,, \\ n_{Ci} h^i_j &\equiv 0 \,, \\ \partial_\mu \xi^j h^i_j &\equiv \partial_\mu \xi^i \,, \\ h^i_j h^j_k &\equiv h^i_k \,, \\ \delta^i_j &= h^i_j + d^{AB} n^i_A n_{Bj} \,. \end{aligned} \tag{7.125}$$

(The equations above can be proved by using (7.89) and (7.90).) The equation (7.96) can be expressed as

$$\begin{aligned} \vec{\mathbf{V}}(\xi(u)) &\equiv \text{tangent}\,(\vec{\mathbf{V}}(\xi(u))) + \text{normal}\,(\vec{\mathbf{V}}(\xi(u))) \,, \\ V^i(\xi(u)) &\equiv (h^i_j V^j) + (d^{AB} n_{Bj} V^j) n^i_A \,. \end{aligned} \tag{7.126}$$

The components $h^i_j V^j$ yield vectors tangential to the subspace $\sum_D$. In a similar fashion, we can decompose a general tensor into unique tangential and normal components. Especially, the tangential components are provided by

$$\text{tangential}\,[T^{i_1 \cdots i_r}_{j_1 \cdots j_s}(\xi(u))] := h^{i_1}_{k_1} \cdots h^{i_r}_{k_r} h^{\ell_1}_{j_1} \cdots h^{\ell_s}_{j_s} T^{k_1 \cdots k_r}_{\ell_1 \cdots \ell_s}(\xi(u)) \,. \tag{7.127}$$

The set of all tangential tensors of order $r+s$ spans a D^{r+s}-dimensional subspace of ${}^r_s\mathcal{T}_{\xi(u)}(\mathbb{R}^N)$. This subspace is *isomorphic to the intrinsic tensor space* ${}^r_s\mathcal{T}_u(\mathbb{R}^D)$. We shall show this correspondence in the following simple cases:

$$\begin{aligned} W_\mu(u) &= (\partial_\mu \xi^j)[h^\ell_j W_\ell(\xi(u))] \,, \\ V^\mu(u) \partial_\mu \xi^i &= h^i_k V^k(\xi(u)) \,, \\ \partial_\mu \xi^i \cdot \partial_\nu \xi^j [h^k_i h^\ell_j g_{k\ell}] &= \partial_\mu \xi^i \cdot \partial_\nu \xi^j g_{ij}(\xi(u)) \;=\; \overline{g}_{\mu\nu}(u) \,, \\ T^\nu_\mu(u) \partial_\nu \xi^i &= \partial_\mu \xi^j [h^i_k h^\ell_j T^k_\ell(\xi(u))] \,. \end{aligned} \tag{7.128}$$

The *tangential* covariant derivative is defined by

$$T^{i_1 \cdots i_r}_{j_1 \cdots j_s \,\|q} := [\nabla_m T^{k_1 \cdots k_r}_{\ell_1 \cdots \ell_s}] h^{i_1}_{k_1} \cdots h^{i_r}_{k_r} h^{\ell_1}_{j_1} \cdots h^{\ell_s}_{j_s} h^m_q \,. \tag{7.129}$$

Now, we shall provide some simple examples.

Example 7.3.5

$$\begin{aligned} h_{ij\,\|k} &= [\nabla_\ell (g_{mn} - d^{AB} n_{Am} n_{Bn})] h^m_i h^n_j h^\ell_k \\ &= -d^{AB} [(\nabla_\ell n_{Am}) n_{Bn} + n_{Am} \nabla_\ell n_{Bn}] h^m_i h^n_j h^\ell_k \;\equiv\; 0 \,. \end{aligned}$$

$$\begin{aligned} n^i_{A\,\|j} &= (\nabla_k n^\ell_A) h^i_\ell h^k_j \,, \\ \partial_\mu \xi^j n^i_{A\,\|j} &= (\partial_\mu \xi^k)(\nabla_k n^\ell_A) h^i_\ell \;=\; \tilde{\nabla}_\mu n^i_A - d^{CB} n^i_C n_{B\ell} \tilde{\nabla}_\mu n^\ell_A \;=\; \tilde{\nabla}_\mu n^i_A \,. \end{aligned} \qquad \square$$

The extrinsic curvature in this approach is defined by the components

$$\begin{aligned}\chi_{Aij}(\xi(u)) &:= -h_i^k h_j^\ell \nabla_k n_{A\ell_{|\xi(u)}} \\ &= -(1/2) h_i^k h_j^\ell (\nabla_k n_{A\ell} + \nabla_\ell n_{Ak})_{|\xi(u)} \,.\end{aligned} \tag{7.130}$$

By (7.93) and (7.130), we deduce that

$$\begin{aligned}K_{A\mu\nu}(u) &= -\partial_\mu \xi^k \cdot \partial_\nu \xi^\ell \nabla_k n_{A\ell_{|\xi(u)}} \\ &= \partial_\mu \xi^i \cdot \partial_\nu \xi^j \chi_{Aij}(\xi(u)) \,.\end{aligned} \tag{7.131}$$

Exercises 7.3

1. Consider a four-dimensional *Euclidean* space and a Cartesian coordinate chart. Let a two-dimensional submanifold of real-analytic class C^ω be given by

$$x^1 = u^1, \quad x^2 = u^2, \quad x^3 = V(u), \quad x^4 = W(u), \quad u \in \mathcal{D}_2 \,.$$

Moreover, let the functions V and W satisfy the Cauchy-Riemann equations

$$\partial_1 V = \partial_2 W, \quad \partial_2 V = -\partial_1 W \,.$$

Prove that $\delta^{AB} K_{A\mu\nu} \overline{g}^{\mu\nu} n^i_B \equiv 0$.

(*Remark*: Such a submanifold is called a **minimal submanifold**.)

2. Consider a non-null submanifold of class C^r $(r \geq 3)$ and the normal vector subspace spanned by $\left\{ n^i_A \frac{\partial}{\partial x^i}_{|\xi(u)} \right\}_1^{N-D}$. Let the orthonormal basis vectors $\vec{\mathbf{n}}_A(\xi(u))$ undergo a generalized (variable) Lorentz transformation

$$\widehat{n}^i_A(\xi(u)) = L^B_A(\xi(u)) n^i_B(\xi(u)) \,,$$

$$L^B_A d_{BC} L^C_E = d_{AE} \,.$$

Assume the induced transformations

$$\widehat{K}_{A\mu\nu} = L^B_A K_{B\mu\nu} \,,$$

$$\widehat{\mu}_{AB\nu} = L^C_A L^E_B \mu_{CE\nu} + d_{CE} L^C_A \tilde{\nabla}_\nu (L^E_B) \,.$$

Prove that under such transformations, (7.101), (7.102), and (7.108) remain covariantly intact.

3. Consider the hybrid tensor components $\rho_{AB\mu\nu}(\xi(u))$ in (7.107iii). Assuming differentiability, prove the Bianchi-type differential identities

$$\tilde{\nabla}_\lambda \rho_{AB\mu\nu} + \tilde{\nabla}_\mu \rho_{AB\nu\lambda} + \tilde{\nabla}_\nu \rho_{AB\lambda\mu} \equiv 0 \,.$$

Appendix 1

Fiber Bundles

Let E and B be two sets endowed with topologies. (These are also called topological spaces.) Let E be "larger" than B in some sense. Moreover, let there exist a continuous **projection** mapping $\Pi : E \to B$. The triple (E, B, Π) is called a **bundle**. The topological space B is called the **base space**.

(See the reference [5].)

Example (A1.1). Let M and $M^{\#}$ be two C^0-manifolds. The Cartesian product manifold $M \times M^{\#}$ can be defined. The bundle $(M \times M^{\#}, M, \Pi)$ is a **product bundle** or **trivial bundle** with the projection mapping

$$\Pi(p, p^{\#}) := p$$

for all p in M and all $p^{\#}$ in $M^{\#}$. (See fig. A1.1). □

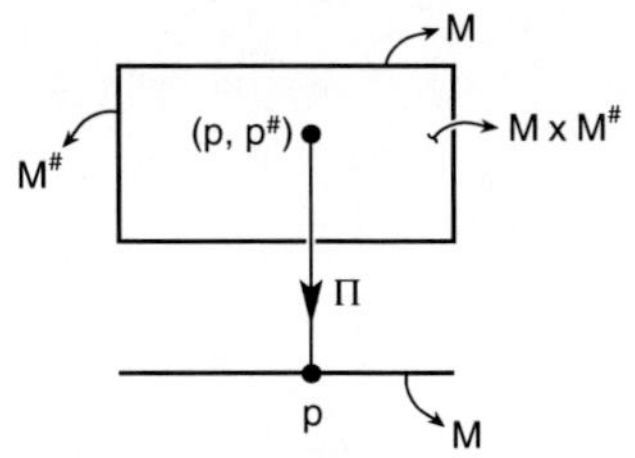

Figure A1.1: A product or trivial bundle.

Example (A1.2). A circular cylinder can be constructed by glueing the opposite edges of a rectangular sheet of paper. Topologically speaking, the cylinder is the product manifold $S^1 \times \mathcal{I}$, where S^1 is the closed circle and $\mathcal{I}$ is an open linear segment. The product bundle $(S^1 \times \mathcal{I}, S^1, \Pi)$ is obtained by defining the projection mapping $\Pi(p,q) := p$ for $p \in S^1$ and $q \in \mathcal{I}$. The usual chart on the circle is furnished by $\theta \equiv x = \chi(p)$, $\theta \in [-\pi, \pi]$. The points $\theta = -\pi$ and $\theta = \pi$ *must be identified.* (See fig. A1.2) (*Distinguish* between the projection Π and the angle π!)

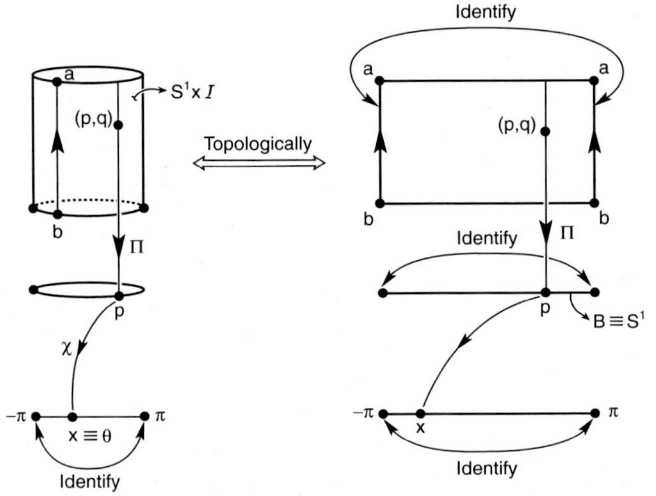

Figure A1.2: The product bundle $(S^1 \times \mathcal{I}, S^1, \Pi)$.

Consider the special bundles such that the subset $F_p := \Pi^{-1}(p)$ (for some $p \in B$) can be mapped in a continuous and one-to-one manner into a typical set F. (Such a mapping is called a **homeomorphism**.) In that case, F_p is called the **fiber** over p, and F is the **typical fiber**. It may happen that there exists a group G of homeomorphisms taking F into itself. Moreover, let the base space allow a covering of countable open subset U_j's. (See fig. A1.3).

A **fiber bundle** (E, B, Π, G) is a bundle (E, B, Π) with a **typical fiber** F and the group of homeomorphisms G of F into itself and a countable open covering of B by U_j's. Moreover, the following assumptions must hold.

I. Locally, the bundle is a product manifold.

II. There exist homeomorphisms H_j and h_j such that

$$H_j: \quad \Pi^{-1}(U_j) \longrightarrow U_j \times F \quad \text{for every } j.$$

$$h_j: \quad \Pi^{-1}(U_j) \longrightarrow F \quad \text{for every } j.$$

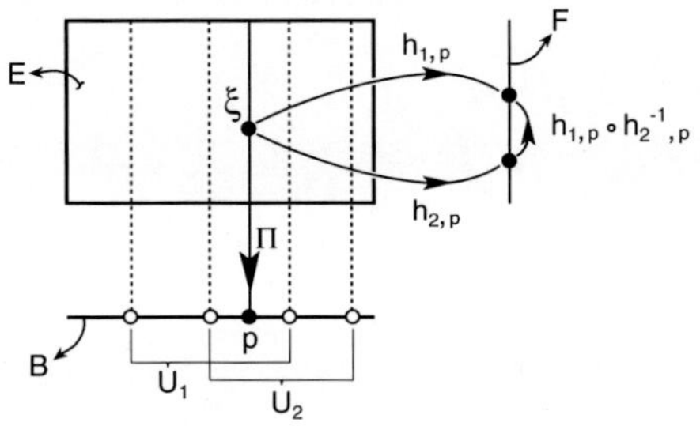

Figure A1.3: A fiber bundle (E, B, Π, G).

Thus, $H_j(\xi) := (\Pi(\xi), h_j(\xi)) = (p, h_j(\xi)) \in U_j \times F$. (Note that the restriction $h_{j,p} := h_{j|F(p)}$ is a homeomorphism from the fiber F_p into F.)

III. Let $p \in U_j \cap U_k$. The mapping $h_{j,p} \circ [h_{k,p}]^{-1}$ is a homeomorphism of F into itself. Thus, $h_{j,p} \circ [h_{k,p}]^{-1}$ constitutes an element of the **structure group** G.

Remark: In the case of a trivial or product bundle, the group G contains only the *identity* element.

Example (A1.3). A circular cylinder can be constructed by glueing (or identifying) the opposite edges of a rectangular sheet of paper. (See example A1.2.) Mathematically speaking, the circular cylinder is the product manifold $S^1 \times \mathcal{I}$, where S^1 is a closed circle and $\mathcal{I}$ is an open linear segment. The product bundle $(S^1 \times \mathcal{I}, S^1, \Pi)$ is obtained by defining the projection mapping $\Pi(\xi) \equiv \Pi(p, y) := p$ for $p \in S^1$ and $y \in \mathcal{I}$. Let us choose the circle S^1 of unit radius and the interval $\mathcal{I} := (-1, 1) \subset \mathbb{R}$. A possible coordinate chart for the unit circle S^1 is furnished by the usual angle $\theta \equiv x = \chi(p)$, $x \in [-\pi, \pi]$. (*Caution*: Distinguish between the number π and the projection Π from the context!) An atlas for S^1 is provided by the union $U \cup \widehat{U}$, where $D = \chi(U) := \{x : -\pi < x < \pi/2\}$ and $\widehat{D} = \widehat{\chi}(\widehat{U}) := \{\widehat{x} : 0 < \widehat{x} < 3\pi/2\}$. In the intersection $U \cap \widehat{U}$, the corresponding coordinate transformation is furnished by $\widehat{x} = x$. (See fig. A1.4.)

The intersection $U \cap \widehat{U}$ is the intersection of subsets corresponding to $D \cap \widehat{D} = V := \{x : 0 < x < \pi/2\}$. Moreover,

$$W := \{x : -\pi < x < -\pi/2\} \subset D \quad \text{and} \quad \widehat{W} := \{\widehat{x} : \pi < \widehat{x} < 3\pi/2\} \subset \widehat{D}\,.$$

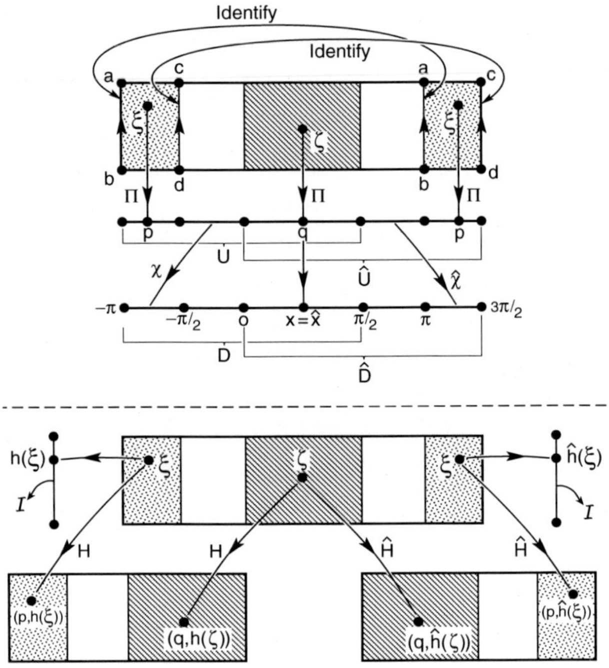

Figure A1.4: Fiber bundle $(S^1 \times \mathcal{I}, S^1, \Pi, G)$ and various mappings.

(Note that subsets W and $\widehat{W}$ are topologically identified.) The areas corresponding to $\chi^{-1}(V) \times \mathcal{I}$ are shaded. Moreover, the areas corresponding to $\chi^{-1}(W) \times \mathcal{I}$ and $\widehat{\chi}^{-1}(\widehat{W}) \times \mathcal{I}$, respectively, are dotted. (See fig. A1.4.) By the definition of various mappings, we obtain that

$$
\begin{aligned}
H(\xi) &= (\Pi(\xi), h(\xi)) = (p, h(\xi)), \\
\widehat{H}(\xi) &= (\Pi(\xi), \widehat{h}(\xi)) = (p, \widehat{h}(\xi)),
\end{aligned}
$$

for all $p \in U$ and all $p \in \widehat{U}$. Similarly, we get

$$
\begin{aligned}
H(\zeta) &= (\Pi(\zeta), h(\zeta)) = (q, h(\zeta)), \\
\widehat{H}(\zeta) &= (\Pi(\zeta), \widehat{h}(\zeta)) = (q, \widehat{h}(\zeta)),
\end{aligned}
$$

for all $q \in U \cap \widehat{U}$.

Since two coordinate charts coincide in $U \cap \widehat{U}$, the "vertical coordinates" satisfy $h(\zeta) = \widehat{h}(\zeta)$. For the other part, $\chi^{-1}(W) \times \mathcal{I}$ and $\chi^{-1}(\widehat{W}) \times \mathcal{I}$ inside $U \times \mathcal{I}$ and $\widehat{U} \times \mathcal{I}$ are identified. Therefore, the "vertical coordinates" again

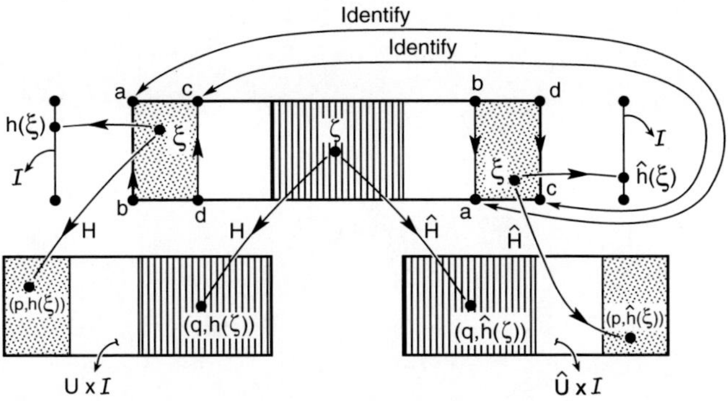

Figure A1.5: The Moebius strip as a fiber bundle.

satisfy $h(\xi) = \widehat{h}(\xi)$. Thus, in the union $U \cup \widehat{U}$, we always have

$$h \circ (\widehat{h})^{-1} = \widehat{h} \circ h^{-1} = e =: \text{identity}\,.$$

Therefore, the structural group is given by the singleton set $G = \{e\}$. □

Example (A1.4). Now we shall discuss the **Moebius strip** as a fiber bundle. It is obtained by twisting and glueing edges of a strip of paper. It cannot be described *globally* as a product manifold. However, *locally* it is a product of $U \times \mathcal{I}$, where $U \subset S^1$ and $\mathcal{I}$ is a line segment. We need some mechanism for twisting to occur. Let us choose the line interval $\mathcal{I} := (-1, 1) \subset \mathbb{R}$ and S^1 as the unit circle. We choose exactly the *same* chart as in the preceding example. (See figs A1.3, A1.4, and A1.5.)

The intersection $U \cap \widehat{U}$ is the intersection of subsets corresponding to $D \cap \widehat{D} =:$ $V = \{x : 0 < x < \pi/2\}$. Moreover, $W := \{x : -\pi < x < -\pi/2\} \subset D$ and $\widehat{W} := \{\widehat{x} : \pi < \widehat{x} < 3\pi/2\} \subset \widehat{D}$ (as in the preceding example).

From the definition of various mappings, we conclude that

$$H(\zeta) = (q, h(\zeta))$$

for all $\zeta \in \Pi^{-1} \circ \chi^{-1}(V) = \Pi^{-1} \circ \widehat{\chi}^{-1}(V)$. Since two charts match exactly in V, we must have identical "vertical" coordinates. Therefore,

$$h(\zeta) = \widehat{h}(\zeta) \quad \text{and} \quad h \circ \widehat{h}^{-1} = \widehat{h} \circ h^{-1} = e = \text{ identity}$$

for all $\zeta \in \Pi^{-1} \circ \chi^{-1}(V)$. However, in the other parts, $\chi^{-1}(W) \times \mathcal{I}$ in $U \times \mathcal{I}$ and $\widehat{\chi}^{-1}(\widehat{W}) \times \mathcal{I}$ in $\widehat{U} \times \mathcal{I}$ have opposite orientations incorporating the twist. Therefore, the "vertical" coordinates satisfy

$$h(\xi) = -\widehat{h}(\xi)$$

for all $\xi \in \Pi^{-1} \circ \chi^{-1}(W)$. Thus, in such cases

$$h \circ \widehat{h}^{-1} = \widehat{h} \circ h^{-1} = g\,,$$
$$g^2 = e\,.$$

Therefore, the structural group is furnished by $G = \{e, g\}$. (This group is isomorphic to the permutation or symmetric group S_2.) □

A bundle (E, M, Π, G) is said to be a C^k**-differentiable fiber bundle** provided E *and* M are C^k-differentiable manifolds, G is a **Lie group**, and a covering of M by a C^k-differentiable atlas exists.

Let us define the **tangent bundle** ${}^{1}_{0}\mathcal{T}(M)$ over the base manifold M. It is the space of ordered pairs $(p, \vec{\mathbf{v}}_p)$ for all p in M and all tangent vectors $\vec{\mathbf{v}}_p \in T_p(M)$. It constitutes the fiber bundle $({}^{1}_{0}\mathcal{T}(M), M, \Pi, G)$ in the following manner. The fiber at p is $T_p(M)$ and is isomorphic to the typical fiber $F = \mathcal{V}$, the N-dimensional (real) vector space. Projection Π is defined by $\Pi(p, \vec{\mathbf{v}}_p) = p$. Covering of M is by a typical atlas. The Lie group $G = G\ell(N, \mathbb{R})$ is the group of invertible mappings of N-dimensional real vector space $\mathcal{V}$ into itself.

A **cross section** of a general bundle (E, B, Π) is a mapping $\sigma : B \to E$ such that $\Pi \circ \sigma =$ identity. A **tangent** (or **contravariant**) **vector field** $\vec{\mathbf{V}}$ on M is a **cross section** of the tangent bundle ${}^{1}_{0}\mathcal{T}(M)$. A tangent vector field $\vec{\mathbf{V}}$ maps each point p in M into a tangent vector $\vec{\mathbf{V}}(p) \in T_p(M)$ by the rule $\vec{\mathbf{V}} : p \longmapsto (p, \vec{\mathbf{V}}(p)) = (\chi^{-1}(x)), \vec{\mathbf{V}}(\chi^{-1}(x))$. (See fig. A1.6.)

In a similar manner, we can construct a **tensor bundle** $({}^{r}_{s}\mathcal{T}(M), M, \Pi)$ and a tensor field ${}^{r}_{s}\mathbf{T} : p \longmapsto (p, {}^{r}_{s}\mathbf{T}(p)) = (\chi^{-1}(x)), {}^{r}_{s}\mathbf{T}(\chi^{-1}(x))$.

Remark: A tensor bundle always admits the zero cross section ${}^{r}_{s}\mathbf{0} =: p \longmapsto (p, {}^{r}_{s}\mathbf{0}(p))$.

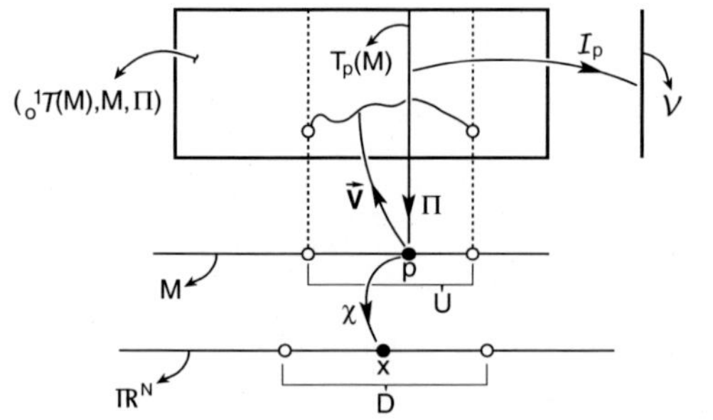

Figure A1.6: A tangent vector field $\vec{\mathbf{V}}$.

Appendix 2

Lie Derivatives

Let us consider a continuous vector field $\vec{\mathbf{V}}(x)$ in $D \subset \mathbb{R}^N$ and a system of ordinary differential equations:

$$\frac{d\mathcal{X}(t)}{dt} = \vec{\mathbf{V}}[\mathcal{X}(t)]\,, \tag{A2.1a}$$

$$\frac{d\mathcal{X}^j(t)}{dt} = V^j[\mathcal{X}^1(t), \ldots, \mathcal{X}^N(t)]\,; \tag{A2.1b}$$

$$t \in [-a, a] \subset \mathbb{R}\,.$$

In the case where the **Lipschitz condition** $\|\vec{\mathbf{V}}(x) - \vec{\mathbf{V}}(x_0)\| \leq K\|\vec{x} - \vec{x}_0\|$ is satisfied, the system of equations (A2.1a, b) admits *unique* solution of the initial value problem

$$\mathcal{X}(0) = x_0\,.$$

(See the book by Coddington and Levinson [6].) The solution is called the integral curve of (A2.1a, b) passing through x_0. We consider the family of integral curves passing through *various* initial points by putting

$$\begin{aligned} x &= \xi(t, x_0)\,, \\ x_0 &\equiv \xi(0, x_0)\,. \end{aligned} \tag{A2.2}$$

Changing the notation of (A2.2), we write

$$\begin{aligned} \widehat{x} &= \xi(t, x)\,, \\ x &\equiv \xi(0, x)\,, \\ \widehat{x}^j &= \xi^j(t, x)\,, \\ x^j &\equiv \xi^j(0, x)\,. \end{aligned} \tag{A2.3}$$

The original differential equations go over into

$$\begin{aligned} \frac{\partial \xi(t,x)}{\partial t} &= \vec{\mathbf{V}}[\xi(t,x)]\,, \\ \frac{\partial \xi^j(t,x)}{\partial t} &= V^j[\xi^1(t,x),\ldots,\xi^N(t,x)]\,. \end{aligned} \tag{A2.4}$$

The functions $\xi(t,\xi(s,x))$ and $\xi(t+s,x)$ satisfy the same differential equation as in (A2.4). The initial values of both of these functions $\xi(0,\xi(s,x)) \equiv \xi(s,x)$. By the *uniqueness* of the solution of the inital-value problem, we can infer that

$$\xi(t,\xi(s,x)) \equiv \xi(t+s,x)\,. \tag{A2.5}$$

Now, we introduce the mapping

$$\begin{aligned} \xi_t &\equiv \xi(t,\,\cdot\,) : x \longmapsto \widehat{x}\,, \\ \xi_0 &\equiv \xi(0,\,\cdot\,) = \text{identity}\,. \end{aligned} \tag{A2.6}$$

This mapping takes the point x into $\widehat{x}$ along the integral curve of (A2.4). Moreover, the mapping ξ_t takes the neighborhood $N_\delta(x)$ into the neighborhood $N_{\widehat{\delta}}(\widehat{x})$ along integral curves. By (A2.5) and (A2.6), we can write

$$\begin{aligned} \xi_t \circ \xi_s(x) &= \xi_t(\xi(s,x)) = \xi(t,\xi(s,x)) = \xi(t+s,x) = \xi_{t+s}(x)\,, \\ \xi_t \circ \xi_s &= \xi_{t+s}\,. \end{aligned} \tag{A2.7}$$

Therefore, $\xi_t \circ \xi_{-t} = \xi_0 =$ identity and $\xi_{-t} = (\xi_t)^{-1}$. These facts show us that the set of all mappings (diffeomorphisms) ξ_t with $t \in [-a,a]$ forms a **local pseudo-group**. The mapping ξ_t induces a *derivative mapping* $\vec{\boldsymbol{\xi}}'_t$ from the tangent space $T_x(\mathbb{R}^N)$ into the tangent space $T_{\widehat{x}}(\mathbb{R}^N) \equiv T_{\xi_t(x)}(\mathbb{R}^N)$. (See (3.41) and (3.43).) We define the **Lie derivative** of a vector field $\vec{\mathbf{U}}(x)$, restricted to an integral curve of the equation (A2.4), by the following rule:

$$L_{\vec{\mathbf{V}}}[\vec{\mathbf{U}}(x)] := \lim_{t\to 0}\left\{\frac{(\vec{\boldsymbol{\xi}}'_t)^{-1}[\vec{\mathbf{U}}(\widehat{x})] - \vec{\mathbf{U}}(x)}{t}\right\}. \tag{A2.8}$$

(See fig. A2.1.)

Example (A2.1). Let us work out the Lie derivative of the coordinate basis field $\frac{\partial}{\partial x^i}$. By (3.43), (A2.3), and (A2.4), we obtain

$$[\vec{\boldsymbol{\xi}}'_t]^{-1}\left(\frac{\partial}{\partial \widehat{x}^i}\right) = \frac{\partial \xi^j(-t,\widehat{x})}{\partial \widehat{x}^i}\,\frac{\partial}{\partial x^j}\,.$$

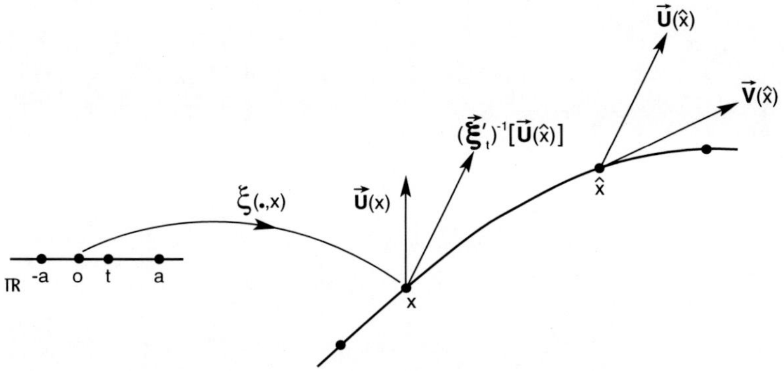

Figure A2.1: A vector field $\vec{\mathbf{U}}(x)$ along an integral curve $\xi(\,\cdot\,,x)$.

Therefore, from (A2.8),

$$\begin{aligned}
L_{\vec{\mathbf{V}}}\left[\frac{\partial}{\partial x^i}\right] &= \lim_{t\to 0}\left\{\frac{[\vec{\boldsymbol{\xi}}'_t]^{-1}\left(\frac{\partial}{\partial \widehat{x}^i}\right) - \frac{\partial}{\partial x^i}}{t}\right\} \\
&= \lim_{t\to 0}\left\{t^{-1}\left[\frac{\partial \xi^j(-t,\widehat{x})}{\partial \widehat{x}^i} - \delta^j_i\right]\right\}\frac{\partial}{\partial x^j} \\
&= \lim_{t\to 0}\left\{t^{-1}\left[\frac{\partial \widehat{x}^j}{\partial \widehat{x}^i} - t\left(\frac{\partial}{\partial(-t)}\,\frac{\partial \xi^j(-t,\widehat{x})}{\partial \widehat{x}^i}\right)_{|t=0} + 0(t^2) - \delta^j_i\right]\right\}\frac{\partial}{\partial x^j} \\
&= -\left\{\frac{\partial}{\partial \widehat{x}^i}\left[\frac{\partial \xi^j(-t,\widehat{x})}{\partial(-t)}\right]\right\}_{|t=0}\frac{\partial}{\partial x^j} = -\left\{\frac{\partial V^j[\xi^1(-t,\widehat{x}),\ldots]}{\partial \widehat{x}^i}\right\}_{|t=0}\frac{\partial}{\partial x^j} \\
&= -\frac{\partial V^j(x)}{\partial x^i}\,\frac{\partial}{\partial x^j}\,. \qquad \square
\end{aligned}$$

Using the example above and the definition

$$L_{\vec{\mathbf{V}}}[f(x)] := \vec{\mathbf{V}}(x)[f]\,, \tag{A2.9}$$

we can prove that

$$L_{\vec{\mathbf{V}}}[\vec{\mathbf{U}}(x)] = \left[\frac{\partial U^i(x)}{\partial x^j}V^j(x) - \frac{\partial V^i(x)}{\partial x^j}U^j(x)\right]\frac{\partial}{\partial x^i}\,. \tag{A2.10}$$

A consequence of the equation above is that

$$L_{\vec{\mathbf{V}}}[\vec{\mathbf{U}}(x)] = \vec{\mathbf{V}}(x)\vec{\mathbf{U}}(x) - \vec{\mathbf{U}}(x)\vec{\mathbf{V}}(x) = [\vec{\mathbf{V}},\vec{\mathbf{U}}]\,. \tag{A2.11}$$

(See (4.50) and (4.51).)

We generalize the definition of Lie derivatives in (A2.8) for a differentiable 1-form $\tilde{\mathbf{W}}(x)$ by (3.44) as

$$L_{\vec{\mathbf{V}}}[\tilde{\mathbf{W}}(x)] := \lim_{t\to 0}\left\{\frac{[\tilde{\boldsymbol{\xi}}'_t]^{-1}[\tilde{\mathbf{W}}(\widehat{x})] - \tilde{\mathbf{W}}(x)}{t}\right\}. \tag{A2.12}$$

Example (A2.2). Let us consider the Lie derivative of the basis 1-form dx^i. By (A2.12), (3.46), and (A2.4), we have

$$\begin{aligned} L_{\vec{\mathbf{V}}}[dx^i] &= \lim_{t\to 0}\left\{\frac{[\tilde{\boldsymbol{\xi}}'_t]^{-1}(d\widehat{x}^i) - dx^i}{t}\right\} \\ &= \lim_{t\to 0}\left\{t^{-1}\left[\frac{\partial \xi^i(t,x)}{\partial x^j} - \delta^i_j\right]\right\} dx^j \\ &= \left[\frac{\partial}{\partial x^j}\frac{\partial \xi^i(t,x)}{\partial t}\right]_{|t=0} dx^j = \frac{\partial V^i(x)}{\partial x^j}\, dx^j\,. \end{aligned}$$

□

By the example above and the definition (A2.9), we obtain for a differentiable 1-form $\tilde{\mathbf{W}}(x) = W_i(x)dx^i$,

$$L_{\vec{\mathbf{V}}}[\tilde{\mathbf{W}}(x)] = \left[\frac{\partial W_i(x)}{\partial x^j}V^j(x) + W_j(x)\frac{\partial V^j(x)}{\partial x^i}\right] dx^i\,. \tag{A2.13}$$

We are now in a position to define the Lie derivative of an arbitrary differentiable tensor field ${}^r_s\mathbf{T}(x)$. With the help of equations (A2.8), (A2.12), and (3.47), we define

$$\begin{aligned} &L_{\vec{\mathbf{V}}}[{}^r_s\mathbf{T}(x)] := \lim_{t\to 0}\left\{\frac{[{}^r_s\boldsymbol{\xi}'_t]^{-1}[{}^r_s\mathbf{T}(\widehat{x})] - {}^r_s\mathbf{T}(x)}{t}\right\}, \\ &\{[{}^r_s\boldsymbol{\xi}'_t]\,[{}^r_s\mathbf{T}(x)]\}\left(\tilde{\boldsymbol{\xi}}'_t(\tilde{\mathbf{W}}_1(x)),\ldots,\tilde{\boldsymbol{\xi}}'_t(\tilde{\mathbf{W}}_r(x)),\, \vec{\boldsymbol{\xi}}'_t(\vec{\mathbf{U}}_1(x)),\ldots,\vec{\boldsymbol{\xi}}'_t(\vec{\mathbf{U}}_s(x))\right) \\ &:= [{}^r_s\mathbf{T}(x)]\left(\tilde{\mathbf{W}}_1(x),\ldots,\tilde{\mathbf{W}}_r(x),\vec{\mathbf{U}}_1(x),\ldots,\vec{\mathbf{U}}_s(x)\right) \end{aligned} \tag{A2.14}$$

for all $\vec{\mathbf{U}}_1,\ldots,\vec{\mathbf{U}}_r$ in $T_x(\mathbb{R}^N)$ and all $\tilde{\mathbf{W}}_1,\ldots,\tilde{\mathbf{W}}_r$ in $\tilde{T}_x(\mathbb{R}^N)$.

We can work out the coordinate components of $L_{\vec{\mathbf{V}}}[{}^r_s\mathbf{T}(x)]$ by (A2.14).

These are furnished by

$$\{L_{\vec{\mathbf{V}}}[{}^r_s\mathbf{T}(x)]\}^{i_1\cdots i_r}_{j_1\cdots j_s} := \{L_{\vec{\mathbf{V}}}[{}^r_s\mathbf{T}(x)]\}\left(dx^{i_1},\ldots,dx^{i_r},\frac{\partial}{\partial x^{j_1}},\ldots,\frac{\partial}{\partial x^{j_s}}\right)$$

$$= (\partial_k T^{i_1\cdots i_r}_{j_1\cdots j_s})V^k(x) - \sum_{\alpha=1}^{r}\left[T^{i_1\cdots i_{\alpha-1}ki_{\alpha+1}\cdots i_r}_{j_1\cdots j_s}(x)\right]\partial_k V^{i_\alpha} \tag{A2.15}$$

$$+\sum_{\beta=1}^{s}\left[T^{i_1\cdots i_r}_{j_1\cdots j_{\beta-1}kj_{\beta+1}\cdots j_s}(x)\right]\partial_{j_\beta}V^k .$$

(The proof of the equation above is left to the reader.)

Example (A2.3). Consider a (0+2) th-order differentiable tensor field $\mathbf{T}_{..}(x)$. The components $\{L_{\vec{\mathbf{V}}}[\mathbf{T}_{..}(x)]\}_{ij}$ are provided from (A2.15) as

$$\{L_{\vec{\mathbf{V}}}[\mathbf{T}_{..}(x)]\}_{ij} = (\partial_k T_{ij})V^k(x) + T_{kj}(x)\partial_i V^k + T_{ik}(x)\partial_j V^k . \tag{A2.16}$$

For a differentiable mixed tensor field,

$$\{L_{\vec{\mathbf{V}}}[\mathbf{T}^{\cdot}_{\cdot}(x)]\}^i_j = (\partial_k T^i_j)V^k(x) - T^k_j(x)\partial_k V^i + T^i_k\partial_j V^k . \tag{A2.17}$$

From the equation above, we have the contraction

$$\{L_{\vec{\mathbf{V}}}[\mathbf{T}^{\cdot}_{\cdot}(x)]\}^i_i = (\partial_k T^i_i)V^k(x) = L_{\vec{\mathbf{V}}}[T^i_i(x)] .$$

□

The main theorem regarding the Lie derivatives of tensor fields is stated below.

Theorem (A2.4). *The differentiable tensor fields and the differentiable vector field $\vec{\mathbf{V}}(x)$ in D obey the following rules in regard to the Lie derivatives:*

(*i*) $L_{\vec{\mathbf{V}}}[{}^r_s\mathbf{T}(x)] \in {}^r_s\mathcal{T}_x(\mathbb{R}^N)$.

(*ii*) $L_{\vec{\mathbf{V}}}[\lambda({}^r_s\mathbf{T}(x)) + \mu({}^r_s\vec{\mathbf{U}}(x))] = \lambda\{L_{\vec{\mathbf{V}}}[{}^r_s\mathbf{T}(x)]\} + \mu\{L_{\vec{\mathbf{V}}}[{}^r_s\vec{\mathbf{U}}(x)]\}$.

(*iii*) $L_{\vec{\mathbf{V}}}[{}^r_s\mathbf{T}(x)\otimes{}^p_q\mathbf{U}(x)] = \{L_{\vec{\mathbf{V}}}[{}^r_s\mathbf{T}(x)]\}\otimes{}^p_q\mathbf{U}(x) + {}^r_s\mathbf{T}(x)\otimes\{L_{\vec{\mathbf{V}}}[{}^p_q\mathbf{U}(x)]\}$.

(*iv*) $L_{\vec{\mathbf{V}}}[\mathcal{C}^i_j({}^r_s\mathbf{T}(x))] = \mathcal{C}^i_j\{L_{\vec{\mathbf{V}}}[{}^r_s\mathbf{T}(x)]\}$.

(*v*) $L_{\vec{\mathbf{V}}}[d({}_p\mathbf{W}(x))] = d\{L_{\vec{\mathbf{V}}}[{}_p\mathbf{W}(x)]\}$.

The proof is left to the reader.

Note that the Lie derivative of a tensor field is defined on a general differentiable manifold *without* recourse to the existence of a connection. Moreover, Lie derivatives produce new tensor fields out of old ones.

In a Riemannian or pseudo-Riemannian manifold M_N, the symmetric metric tensor $\mathbf{g}_{..}(x)$ must exist. The symmetries or isometries of the manifold are investigated by obtaining the existence of **Killing vectors** $\vec{\mathbf{K}}(x)$ satisfying

$$L_{\vec{\mathbf{K}}}[\mathbf{g}_{..}(x)] = \mathbf{0}_{..}(x)\,. \tag{A2.18}$$

The equation above is called the **Killing equation**. Equation (A2.18), by the equation (A2.16), yields the component version

$$\{L_{\vec{\mathbf{K}}}[\mathbf{g}_{..}(x)]\}_{ij} = (\partial_\ell g_{ij})K^\ell(x) + g_{\ell j}(x)\partial_i K^\ell + g_{i\ell}(x)\partial_j K^\ell = 0\,. \tag{A2.19}$$

Introducing covariant derivatives involving Christoffel symbols, (A2.19) can be expressed neatly as

$$\nabla_i K_j + \nabla_j K_i = 0\,. \tag{A2.20}$$

Example (A2.5). Let us investigate the Killing equations (A2.19) in the simple case of the Euclidean plane $\mathbb{E}_2$. Using a global Cartesian chart, we obtain that

$$g_{ij}(x) = \delta_{ij}\,, \qquad \partial_\ell g_{ij} \equiv 0\,.$$

Equations (A2.19) yield three independent equations

$$\partial_1 K_1 = \partial_2 K_2 = 0, \quad \partial_1 K_2 + \partial_2 K_1 = 0\,. \tag{A2.21}$$

These are three linear, coupled, first-order partial differential equations for two unknown functions K_1 and K_2 in two independent variables $(x^1, x^2) \in \mathbb{R}^2$. Solving the first two equations in (A2.21), we get

$$\begin{aligned} K_1(x) &= F(x^2)\,, \\ K_2(x) &= H(x^1)\,. \end{aligned} \tag{A2.22}$$

Here, F and H are two *arbitrary* (differentiable) functions of integration. Substituting (A2.22) into the third equation in (A2.21), we derive that

$$F'(x^2) + H'(x^1) \equiv 0\,,$$
$$F'(x^2) = -H'(x^1) = \theta_0 = \text{const.}.$$

Solving the above ordinary differential equations, we finally get

$$\begin{aligned} K_1(x) &= \theta_0 x^2 + t_1\,, \\ K_2(x) &= -\theta_0 x^1 + t_2\,, \end{aligned} \tag{A2.23}$$

$$\vec{\mathbf{K}}(x) = \left[t_1 \frac{\partial}{\partial x^1} + t_2 \frac{\partial}{\partial x^2}\right] + \theta_0 \left[x^2 \frac{\partial}{\partial x^1} - x^1 \frac{\partial}{\partial x^2}\right].$$

Here, t_1, t_2 and θ_0 are three arbitrary constants of integration. The parameters t_1 and t_2 generate translation of coordinate axes, whereas the parameter θ_0 generates rotation of the axes. Therefore, the Killing vector field $\vec{\mathbf{K}}(x)$ generates the **group of isometry** in the Euclidean plane $\mathbb{E}_2$. □

Example (A2.6). Consider the surface S^2 of the unit sphere. (See examples 4.3.12 and 5.1.8.) The metric tensor is furnished by

$$\mathbf{g}_{..}(x) = dx^1 \otimes dx^1 + (\sin x^1)^2 dx^2 \otimes dx^2 .$$

The corresponding Killing equations (A2.20) yield

$$\begin{aligned} &\partial_1 K^1 = 0 , \\ &\partial_2 K^2 + (\cot x^1) K^1(x) = 0 , \\ &\partial_2 K^1 + (\sin x^1)^2 \partial_1 K^2 = 0 . \end{aligned} \tag{A2.24}$$

Solving the first equation, we obtain that

$$K^1(x) \equiv K^1(x^1, x^2) = f'(x^2) .$$

Here, f is some twice-differentiable function. Now, the integration the second Killing equation implies that

$$K^2(x) + (\cot x^1) f(x^2) = g(x^1) .$$

Here, g is some differentiable function. With these results, we deduce from the third equation that

$$\begin{aligned} &f''(x^2) + f(x^2) + (\sin x^1)^2 g'(x^1) = 0 , \\ &f''(x^2) + f(x^2) = C = -(\sin x^1)^2 g'(x^1) . \end{aligned} \tag{A2.25}$$

Here, C is a constant of separation. Integrating the equations above, we derive that

$$\begin{aligned} f(x^2) &= -C + A\cos x^2 + B \sin x^2 , \\ g(x^1) &= C \cot x^1 + D . \end{aligned}$$

Here, A, B, D are arbitrary constants of integration. Therefore, we obtain that

$$\begin{aligned} K^1(x) &= -A \sin x^2 + B \cos x^2 , \\ K^2(x) &= D - \cot x^1 (A \cos x^2 + B \sin x^2) , \\ \vec{\mathbf{K}}(x) &= -A\left(\sin x^2 \frac{\partial}{\partial x^1} + \cot x^1 \cos x^2 \frac{\partial}{\partial x^2}\right) \\ &\quad +B\left(\cos x^2 \frac{\partial}{\partial x^1} - \cot x^1 \sin x^2 \frac{\partial}{\partial x^2}\right) + D\frac{\partial}{\partial x^2} . \end{aligned}$$

There exist three linearly independent Killing vectors

$$\begin{aligned}\vec{\mathbf{K}}_{(1)}(x) &:= \sin x^2 \frac{\partial}{\partial x^1} + \cot x^1 \cos x^2 \frac{\partial}{\partial x^2}\,, \\ \vec{\mathbf{K}}_{(2)}(x) &:= -\cos x^2 \frac{\partial}{\partial x^1} + \cot x^1 \sin x^2 \frac{\partial}{\partial x^2}\,, \\ \vec{\mathbf{K}}_{(3)}(x) &:= -\frac{\partial}{\partial x^2}\,.\end{aligned} \tag{A2.26}$$

(Multiplying the above by the imaginary number i, we arrive at orbital angular momentum operators of quantum mechanics!) □

Answers and Hints to Selected Exercises

Exercises 1.1

2(i). Hint: For a non-zero element, the inverse is

$$(m + n\sqrt{2})^{-1} = \left(\frac{m}{m^2 - 2n^2}\right) - \left(\frac{n}{m^2 - 2n^2}\right)\sqrt{2}.$$

Exercises 1.2

1. Hint: Investigate the linear homogeneous equations $\alpha^1\lambda + \beta^1\mu = 0$, $\alpha^2\lambda + \beta^2\mu = 0$ for the unknown numbers λ and μ.

Exercises 1.3

3. Hint: Isomorphism is defined by

$$\mathcal{I}(\alpha^i\vec{\mathbf{e}}_i) := (\alpha^1, \ldots, \alpha^N).$$

Exercises 1.4

2.

$$\begin{aligned} w_1 &= \widehat{w}_1, \quad w_2 = \widehat{w}_2, \\ w_3 &= (\cosh\alpha)\widehat{w}_3 + (\sinh\alpha)\widehat{w}_4, \\ w_4 &= (\sinh\alpha)\widehat{w}_3 + (\cosh\alpha)\widehat{w}_4. \end{aligned}$$

Exercises 2.1

2. Hint: $\mathbf{T}_{..} \equiv \frac{1}{2}\left(\mathbf{T}_{..} + \mathbf{T}^T_{..}\right) + \frac{1}{2}\left(\mathbf{T}_{..} - \mathbf{T}^T_{..}\right).$

Exercises 2.2

1. Components of the vectors $\vec{\mathbf{a}}$, $\vec{\mathbf{b}}$, $\vec{\mathbf{c}}$ are given by $(\alpha^1, \alpha^2) = (-1, 1), (\beta^1, \beta^2) = (1, 2)$ and $(\gamma^1, \gamma^2) = (2, 1)$, respectively. The components of $\vec{\mathbf{a}} \otimes \vec{\mathbf{b}} \otimes \vec{\mathbf{c}}$ are provided by $\alpha^i \beta^j \gamma^k$; $i, j, k \in \{1, 2\}$.

Exercises 2.3

3(ii). Hint: Use (2.93).

Exercises 2.4

1. The metric is not positive-definite.

2(i). In the case where $\mathbf{g}..(\vec{\mathbf{a}}, \vec{\mathbf{a}}) > 0$, consider $\mathbf{g}..(\lambda\vec{\mathbf{a}} + \vec{\mathbf{b}}, \lambda\vec{\mathbf{a}} + \vec{\mathbf{b}}) \geq 0$ with $\lambda := -\mathbf{g}..(\vec{\mathbf{a}}, \vec{\mathbf{b}})/\mathbf{g}..(\vec{\mathbf{a}}, \vec{\mathbf{a}})$.

Exercises 3.1

2.

$$\frac{\partial^2}{\partial x^1 \partial x^2} f(x^1, x^2)_{|(0,0)} = 1,$$
$$\frac{\partial^2}{\partial x^2 \partial x^1} f(x^1, x^2)_{|(0,0)} = -1.$$

Exercises 3.2

4.

$$\mathcal{X}^\alpha(t) = t, \ \alpha \in \{1, 2, 3\},$$
$$\mathcal{X}^4(t) = \sqrt{3}t.$$

Exercises 3.3

4(i). Left-hand side

$$= \left(\delta^c{}_j \delta^i{}_c\right) dx^j \otimes \frac{\partial}{\partial x^i} = \delta_j{}^i dx^j \otimes \frac{\partial}{\partial x^i}$$

= right-hand side.

4(ii). Left-hand side

$$= \left[\delta^c{}_c \delta^i{}_j \frac{\partial}{\partial x^i} \otimes dx^j\right] \left(w_k dx^k, V^l \frac{\partial}{\partial x^l}\right)$$
$$= N\delta^i{}_j \delta^k_i \delta^j{}_l w_k V^l$$

= right-hand side.

Exercises 3.4

1. ${}_2\mathbf{W}(x) \wedge {}_2\mathbf{W}(x) \wedge {}_2\mathbf{W}(x) = {}_6\mathbf{O}(x)$.

Exercises 4.1

1. Hint: $\nabla_i V^k = \delta^k_i + N\left(x^k/x^i\right)$.

Exercises 4.3

4. Hint:

$$\begin{aligned}\widehat{R}^i{}_{jkl} =& R^i{}_{jkl} + \delta^i{}_j\left[\nabla_k \psi_l - \nabla_l \psi_k\right] \\ &+ \delta^i{}_l\left[\nabla_k \psi_j - \psi_j \cdot \psi_k\right] - \delta^i{}_k\left[\nabla_l \psi_j - \psi_j \cdot \psi_l\right].\end{aligned}$$

Exercises 5.1

1. Hint: See problem 1 in exercises 2.4.

4. Hint:

$$\begin{aligned}&\det\left[\widehat{A}_{pq}(x) - \widehat{\lambda}(x)\widehat{B}_{pq}(x)\right] \\ &= \left\{\det\left[\lambda^r{}_p(x)\right]\right\}\left\{\det\left[A_{rs}(x) - \widehat{\lambda}(x)B_{rs}(x)\right]\right\}\left\{\det\left[\lambda^s{}_q(x)\right]\right\}.\end{aligned}$$

Exercises 5.2

1. Hint: Use $\left\{ {j \atop jk} \right\} = \partial_k \ln\sqrt{|g|}$ and $\left\{ {l \atop jk} \right\} S^j_l = [jk,\, l]\, S^{jl}$.

3. Hint: Consider the thrice-differentiable coordinate transformation such that

$$\widehat{g}^{11} - \widehat{g}^{22} = \left[\frac{\partial \widehat{X}^1(x)}{\partial x^i} \frac{\partial \widehat{X}^1(x)}{\partial x^j} - \frac{\partial \widehat{X}^2(x)}{\partial x^i} \frac{\partial \widehat{X}^2(x)}{\partial x^j} \right] g^{ij}(x) = 0$$

and

$$\widehat{g}^{12} = \frac{\partial \widehat{X}^1(x)}{\partial x^i} \frac{\partial \widehat{X}^2(x)}{\partial x^j} g^{ij}(x) = 0.$$

Assuming the local existence of two solution functions $\widehat{X}^i(x)$ for two equations, prove the identities in the hatted coordinates.

8. Hint: Use the Ricci identities:

$$(\nabla_u \nabla_v - \nabla_v \nabla_u) R_{pqrs} = -\sum_{\beta=1}^{N} R_{q_1 q_{\beta-1} w q_{\beta+1}} R^{w}{}_{q_\beta uv}.$$

Exercises 5.3

2. Hint: The function $\chi^3(\tau)$ is implicitly given by

$$\int_{x_0^3}^{x^3} f(y)\, dy = \pm\sqrt{(k^1)^2 + (k^2)^2}\, (\tau - \tau_0)\,.$$

Here, k^1 and k^2 are arbitrary constants (with $|k^1| + |k^2| > 0$).

8(ii). $\exp\left[T(s) - T(s_0)\right] = \sec^2(s - s_0)$,

$\chi^\alpha(s) - \chi^\alpha(s_0) = k^{-2} k^\alpha \tan(s - s_0)$.

Here, arbitrary constants satisfy $k^2 := \delta_{\mu\nu} k^\mu k^\nu > 0$.

Exercises 5.4

6. By (5.282),

$$\begin{aligned} 0 &= \det\left[G_{ij} - \lambda g_{ij}\right]_{|T=c} \\ &= \left\{\det\left[G^{\#}_{\mu\nu} + \left(2A\ddot{A} + \dot{A}^2 - \lambda\right) g^{\#}_{\mu\nu}\right]\right\} \left\{A^{-2}\left[\frac{1}{2}R^{\#} - 3\dot{A}^2\right] + \lambda\right\}. \end{aligned}$$

Since $G^{\#}_{\mu\nu} = G^{\#}_{\nu\mu}$ and $g^{\#}_{\mu\nu}$ is positive-definite, the roots $\lambda_{(1)}, \ldots, \lambda_{N-1}$ of the equations above are real. The last root, $\lambda_N = A^{-2}\left[3\dot{A}^2 - \frac{1}{2}R^{\#}\right]$, is obviously real.

Exercises 6.1

1(i). Hint: Use example 6.1.1.

2. Hint: Use (5.17).

Exercises 6.2

2(i). Hint: Use (6.31).

Exercises 6.3

1. Hint: *Assume* that

$$R^h{}_{ijk} + \delta^h{}_j R_{ik} - \delta^h{}_k R_{ij} + g_{ik} R^h{}_j - g_{ij} R^h{}_k + \frac{R}{2}\left(\delta^h{}_k g_{ij} - \delta^h{}_j g_{ik}\right) \equiv 0.$$

Exercises 6.4

1. Hint: Use *orthogonal* coordinates. Since the tensor equation $\mathbf{C}^{\cdot}_{\cdot\cdot\cdot}(x) \equiv \mathbf{O}^{\cdot}_{\cdot\cdot\cdot}(x)$ can be proved in the orthogonal coordinates, it must be valid for any other chart. (Compare this with the hint for problem 1 in exercises 6.3.)

Exercises 7.1

1(ii). The Gaussian curvature is furnished by $K(u) = [1 + \delta^{\mu\nu}\,\partial_\mu f\,\partial_\nu f]^{-2}$ $\det[\partial_\alpha \partial_\beta f]$. The condition $K(u) = 0$ implies that the determinant is

$$\begin{vmatrix} \partial_1^2 f\,, & \partial_1\partial_2 f \\ \partial_1\partial_2 f\,, & \partial_2^2 f \end{vmatrix} = 0.$$

Thus, the functions $\partial_1 f$ and $\partial_2 f$ are functionally related. Therefore, there exists a non-constant differentiable function F such that $F(\partial_1 f,\, \partial_2 f) = 0$.

Exercises 7.2

4. Hint:

$$\begin{aligned} g^{\alpha\beta} &= e^{w} g'^{\alpha\beta}, \\ g^{\alpha 4} &= -\,e^{w} g'^{\alpha\beta} a_\beta, \\ g^{44} &= -\,e^{-w} + e^{w} g'^{\alpha\beta} a_\alpha a_\beta, \\ \det[g_{ij}] &= -\,e^{-2w} \det\left[g'_{\alpha\beta}\right]. \end{aligned}$$

Denoting the symbols $a'^{\lambda} := g'^{\lambda\gamma} a_{\gamma}$, the following equations can be established:

$$
\begin{aligned}
\left\{ \begin{matrix} \lambda \\ \alpha\beta \end{matrix} \right\} &= \left\{ \begin{matrix} \lambda \\ \alpha\beta \end{matrix} \right\}' - \frac{1}{2}\left[\delta^{\lambda}_{\ \beta}\partial_{\alpha} w + \delta^{\lambda}_{\ \alpha}\partial_{\beta} w - \left(g'_{\alpha\beta} + e^{2w} a_{\alpha} a_{\beta}\right) g'^{\lambda\gamma}\partial_{\gamma} w\right] \\
&\quad - \frac{1}{2} e^{2w}\left(a_{\alpha} f'^{\lambda}_{\ \beta} + a_{\beta} f'^{\lambda}_{\ \alpha}\right), \\
\left\{ \begin{matrix} 4 \\ \alpha\beta \end{matrix} \right\} &= \frac{1}{2}\left[\nabla'_{\beta} a_{\alpha} + \nabla'_{\alpha} a_{\beta} + 2a_{\beta}\partial_{\alpha} w + 2a_{\alpha}\partial_{\beta} w\right] \\
&\quad - \frac{1}{2}\left[\left(g'_{\alpha\beta} + e^{2w} a_{\alpha} a_{\beta}\right)\partial_{\gamma} w + e^{2w}\left(a_{\alpha} f_{\beta\gamma} + a_{\beta} f_{\alpha\gamma}\right)\right] a^{\gamma}, \\
\left\{ \begin{matrix} \gamma \\ 4\alpha \end{matrix} \right\} &= \frac{1}{2} e^{2w} g'^{\gamma\beta}\left(f_{\alpha\beta} + a_{\alpha}\partial_{\beta} w\right), \\
\left\{ \begin{matrix} \alpha \\ 44 \end{matrix} \right\} &= \frac{1}{2} e^{2w} g'^{\gamma\beta}\partial_{\beta} w, \\
\left\{ \begin{matrix} 4 \\ \alpha 4 \end{matrix} \right\} &= \frac{1}{2}\left[\partial_{\alpha} w - e^{2w} a'^{\beta}\left(f_{\alpha\beta} + a_{\alpha}\partial_{\beta} w\right)\right], \\
\left\{ \begin{matrix} 4 \\ 44 \end{matrix} \right\} &= -\frac{1}{2} e^{2w} a'^{\alpha}\partial_{\alpha} w.
\end{aligned}
$$

Exercises 7.3

1.

$$
\begin{aligned}
\partial_{\alpha}\xi^{1} &= \delta^{1}_{\ \alpha},\ \partial_{\alpha}\xi^{2} = \delta^{2}_{\ \alpha},\ \partial_{\alpha}\xi^{3} = \partial_{\ \alpha} V,\ \partial_{\alpha}\xi^{4} = \partial_{\ \alpha} V; \\
\delta^{\alpha\beta}\partial_{\alpha}\partial_{\beta} V &= \delta^{\alpha\beta}\partial_{\alpha}\partial_{\beta} W \equiv 0; \\
\overline{g}_{\mu\nu}(u) &= \delta_{\mu\nu} + \partial_{\mu} V \partial_{\nu} V + \partial_{\mu} W \partial_{\nu} W, \\
\overline{g}^{\mu\nu}(u) &= \delta^{\mu\nu}\left[1 + (\partial_1 V)^2 + (\partial_1 W)^2\right]^{-1} = \delta^{\mu\nu}\left[1 + (\partial_2 V)^2 + (\partial_2 W)^2\right]^{-1}, \\
\overline{\left\{ \begin{matrix} \lambda \\ \mu\nu \end{matrix} \right\}} &= \overline{g}^{\lambda\gamma}\left[\partial_{\mu}\partial_{\nu} V \cdot \partial_{\gamma} V + \partial_{\mu}\partial_{\nu} W \cdot \partial_{\gamma} W\right], \\
\overline{g}^{\mu\nu}\overline{\left\{ \begin{matrix} \lambda \\ \mu\nu \end{matrix} \right\}} &\equiv 0,\ \left\{ \begin{matrix} i \\ jk \end{matrix} \right\}_{|\xi(u)} \equiv 0.
\end{aligned}
$$

Using (7.92c), we obtain that $\delta^{AB} K_{A\mu\nu}\overline{g}^{\mu\nu} n^{i}_{\ B} \equiv 0$.

References

1. Bianchi, L., *Lezioni di geometria differenziale*, Spoerri, Pisa, 1902.
2. Bishop, R.L. and Goldberg, S.I., *Tensor analysis on manifolds*, Dover Publications, New York, 1980.
3. Bochner, S. and Yano, K., *Curvature and Betti numbers*, Annals of Mathematical Studies - 32, Princeton University Press, Princeton, 1953.
4. Bondi, H., Pirani, F.A.E., and Robinson, I., *Gravitational waves in general relativity III: Exact plane waves*, Proc. R. Soc. London A **251**, 519 (1959).
5. Choquet-Bruhat, Y., DeWitt-Morette, C., and Dillard-Bleick, M., *Analysis, manifolds and physics*, North-Holland, Amsterdam, 1978.
6. Coddington, E.A. and Levinson, N., *Theory of ordinary differential equations*, McGraw-Hill, New York, Toronto, and London, 1955.
7. Das, A., *The special theory of relativity: A mathematical exposition*, Springer, New York, Berlin, and London, 1996.
8. Das, A., *Short-ranged scalar gravity in Einstein's theory*, J. Math. Phys. **12**, 232 (1971).
9. Das, A. and Aruliah, D., *The spherically symmetric collapse of an inhomogeneous stellar dust ball into a black hole*, Review and research report No. 1996-02, Department of Mathematics, Simon Fraser University, B.C., Burnaby, Canada.
10. Eisenhart, L.P., *Riemannian geometry*, Princeton University Press, Princeton, 1966.
11. Eisenhart, L.P., *Non-Riemannian geometry*, American Mathematical Society, New York, 1964.
12. Faber, L., *Differential geometry and relativity theory: An introduction*, Marcel Dekker, New York and Basel, 1983.
13. Flanders, H., *Differential forms with applications to the physical sciences*, Academic Press, New York and London, 1963.

14. Gegenberg, J.D., *On the stationary Einstein-Maxwell-Klein-Gordon equations*, Ph. D. thesis, Simon Fraser University, 1981.

15. Gelbaum, B.R. and Olmsted, J.M.H., *Counter-examples in analysis*, Holden-Day, San Francisco, 1964.

16. Green, A.E. and Zerna, W., *Theoretical elasticity*, Oxford at the Clarendon Press, Oxford, 1968.

17. Halmos, P.R., *Finite-dimensional vector spaces*, Van Nostrand, Princeton, 1958.

18. Hawking, S.W. and Ellis, G.F.R., *The large scale structure of space-time*, Cambridge University Press, Cambridge, 1973.

19. Hicks, N.J., *Notes on differential geometry*, Van Nostrand, London, 1971.

20. Hocking, J.G. and Young, G.S., *Topology*, Addison-Wesley, Reading, Mass., 1961.

21. Hoffman, K. and Kunze, R., *Linear algebra*, Prentice-Hall, Englewood Cliffs, 1971.

22. Israel, W., *Differential forms in general relativity*, Commun. Dublin Inst. Adv. Stud. A, No. 19 (1970).

23. Kasner, E., *Geometrical theorems on Einstein's cosmological equations*, Am. J. Math. **43**, 217 (1921).

24. Kloster, S., Som, M.M., and Das, A., *On the stationary gravitational fields*, J. Math. Phys. **15**, 1096 (1973).

25. Kramer, D., Stephani, H., MacCallum, M., and Herlt, E., *Exact solutions of Einstein's field equations*, VEB Deutscher Verlag der Wissenschaften, Berlin, 1980.

26. Lovelock, D. and Rund, H., *Tensors, differential forms, and variational principles*, John Wiley and Sons, New York, 1975.

27. Misner, C.W., Thorne, K.S., and Wheeler, J.A., *Gravitation*, Freeman, San Francisco, 1973.

28. Noll, W., *Notes on tensor analysis* (prepared by Wang, C.C.), Mathematics Department, Johns Hopkins University, 1963.

29. Novotny, J. and Horsky, J., *On the plane gravitational condensor with the positive gravitational constant*, Czech. J. Phys. B**24**, 718 (1974).

30. O'Neill, B., *Elementary differential geometry*, Academic Press, New York, San Francisco, and London, 1966.

31. Petrov, A.Z., *One type of Einstein space*, Trudy Kazan. Aviats. Inst. **7**, 8 (1946).

32. Petrov, A.Z., *Einstein spaces*, Pergamon Press, London, New York, and Toronto, 1969.

33. Schouten, A., *Ricci-Calculus*, Springer-Verlag, Berlin, Göttingen, and Heidelberg, 1954.

34. Sokolnikoff, I.S., *Tensor analysis: Theory and applications to geometry and mechanics of continua*, John Wiley and Sons, New York and London, 1964.

35. Spivak, M., *Calculus on manifolds*, The Benjamin/Cunnings Publishing Company, 1965.

36. Sternberg, S., *Lectures on differential geometry*, Prentice-Hall, Englewood Cliffs, 1964.

37. Synge, J.L. and Schild, A., *Tensor calculus*, University of Toronto Press, Toronto, 1966.

38. Synge, J.L., *Relativity: The general theory*, North-Holland, Amsterdam, 1960.

39. Willmore, T.J., *An introduction to differential geometry*, Oxford University Press, Oxford, 1959.

List of Symbols

$A \cup B$	union of two sets
$A \cap B$	intersection of two sets
$A \times B$	Cartesian product of two sets
$A \subset B$	A is a subset of B
$(\vec{\mathbf{a}} \cdot \vec{\mathbf{b}})_g \equiv \mathbf{g}_{..}(\vec{\mathbf{a}}, \vec{\mathbf{b}})$	inner product between two vectors
$\lVert\vec{\mathbf{a}}\rVert$	norm or length of a vector
$[\vec{\mathbf{a}}, \vec{\mathbf{b}}] := \vec{\mathbf{a}}\,\vec{\mathbf{b}} - \vec{\mathbf{b}}\,\vec{\mathbf{a}}$	Lie bracket or commutator
B	bundle
$C^r(D \subset \mathbb{R}^N; \mathbb{R})$	set of all r-differentiable functions from $D \subset \mathbb{R}^N$ into $\mathbb{R}$
$\mathbb{C}$	set of all complex numbers
$C^i_{jk\ell}(x)$	components of Weyl's conformal tensor
(χ, U)	coordinate chart for a differentiable manifold
$\chi(p) = x \equiv (x^1, x^2, \ldots, x^N)$	local coordinates of a point p in a manifold
D	domain in $\mathbb{R}^N$ (open and connected)
∂D	$(N-1)$-dimensional boundary of D
∇_i, ∇_a	covariant derivatives

$\dfrac{D}{\partial t}$	covariant derivative along a curve
Δ	Laplacian in a manifold with a metric, also a finite difference
∇^2	Laplacian in a Euclidean space
δ^i_j, δ^a_b	components of a Kronecker delta (or identity matrix)
$[d_{ij}] = [d_{ab}]$ $:= \text{diag}\,[\underbrace{1,1,\ldots,1}_{p},\underbrace{-1,\ldots,-1}_{n}]$	flat metric components
$\square$	d'Alembertian operator; (also) completion of an example
$\blacksquare$	Q.E.D.
$\dfrac{\partial(\widehat{x}^1,\ldots,\widehat{x}^N)}{\partial(x^1,\ldots,x^N)}$	Jacobian of a coordinate transformation
$\mathbb{E}_N$	N-dimensional Euclidean space
$\{\vec{\mathbf{e}}_a\}_1^N := \{\vec{\mathbf{e}}_1,\ldots,\vec{\mathbf{e}}_N\}$	basis set for a vector space
$\in$	belongs to
(E, B, Π)	fiber bundle
$\varepsilon_{i_1 i_2 \ldots i_N}$	totally antisymmetric permutation symbol (Levi-Civita)
$\eta_{i_1 i_2 \ldots i_N}$	totally antisymmetric pseudo (or oriented) tensor (Levi-Civita)
$F_{ij}(x)$	electromagnetic tensor field
$\mathcal{F}$	field (in abstract algebra)
G	group
$\mathbf{g}_{..}(x), g_{ij}(x)$	metric tensor and corresponding components
$G^i_j(x), G^a_b(x)$	Einstein tensor components

γ	parametrized curve into a manifold
$\mathcal{X} := \chi \circ \gamma$	parametrized curve into $\mathbb{R}^N$
$\Gamma^p_{qr}(x)$	affine connection coefficients
$\gamma_{abc}(x)$	Ricci rotation coefficients
$\tilde{\mathbf{w}}(x)$	connection 1-form
$[jk, i]$	Christoffel symbol of the 1st kind
$\left\{ {i \atop jk} \right\}$	Christoffel symbol of the 2nd kind
$K(u)$	Gaussian curvature
$\vec{\mathbf{K}}(x), K_i(x)$	Killing vector and corresponding components
$K_{\mu\nu}(x)$	extrinsic curvature components of a hypersurface
$K^A_{\mu\nu}(x)$	extrinsic curvature components of a submanifold
$\mathbf{L}$, $[L] := [\ell^i_j]$	linear mapping and corresponding matrix
$[L]^T$	transposed matrix
$L(x, u)$	Lagrangian function
$L_{\vec{\mathbf{V}}}$	Lie derivative
M	differentiable manifold
$O(p, n; \mathbb{R})$	generalized Lorentz group
$\mathcal{I}O(p, n; \mathbb{R})$	generalized Poincaré group
$\mathbb{R}$	the set of real numbers
$\mathbb{R}^N := \underbrace{\mathbb{R} \times \mathbb{R} \times \cdots \times \mathbb{R}}_{N}$	Cartesian product of N copies of $\mathbb{R}$
$\mathbf{R}^{\cdot}_{\cdot\cdot\cdot}(x)$, $R^p_{qrs}(x)$	curvature tensor and corresponding general components

$R^i_{jk\ell}(x)$, $R^a_{bcd}(x)$	components of a Riemann-Christoffel tensor
$R^i_{jk}(x)$	components of a Schouten-Cotton tensor
$R_{ij}(x)$, $R_{ab}(x)$	components of a Ricci tensor
$R(x)$	curvature scalar (or invariant)
$S^i_{jk\ell}(x)$	components of a symmetrized curvature tensor
$*$	Hodge star operation
S^2	two-dimensional spherical surface
$\sigma(\vec{\mathbf{a}})$	separation of a vector
${}^r_s\mathbf{T}(x)$, $T^{p_1\cdots p_r}_{q_1\cdots q_s}(x)$	tensor field of order $(r+s)$ and its general components
$T^{i_1\cdots i_r}_{j_1\cdots j_s}(x)$	coordinate components of the (same) tensor field
$T^{a_1\cdots a_r}_{b_1\cdots b_s}(x)$	orthonormal components of the (same) tensor field
$\mathbf{T}^{\cdot}_{\cdot\cdot}(x)$, $T^p_{qr}(x)$	torsion tensor and the corresponding general components
${}^r_s\mathbf{T}(x) \otimes {}^p_q\mathbf{S}(x)$	tensor (or outer product) of two tensor fields
${}_{(w)}\theta^{i_1\cdots i_r}_{j_1\cdots j_s}(x)$	components of an oriented, relative tensor field of weight w
U	open subset of a manifold
$\mathcal{V}$	vector space in field $\mathcal{F}$
$\tilde{\mathcal{V}}$	dual vector space
${}^r_s\mathcal{T}(\mathcal{V}) := \underbrace{\mathcal{V}\otimes\cdots\otimes\mathcal{V}}_{r}\otimes\underbrace{\tilde{\mathcal{V}}\otimes\cdots\otimes\tilde{\mathcal{V}}}_{s}$	tensor space of order $(r+s)$
${}_p\mathbf{W}(x)$, $W_{i_1\cdots i_p}(x)$	p-form and its antisymmetric components

${}_p\mathbf{W}(x) \wedge {}_q\mathbf{A}(x)$	wedge product between a p-form and a q-form
$W^i_{jk\ell}(x)$	components of Weyl's projective tensor
$x = \mathcal{X}(t)$, $x^i = \mathcal{X}^i(t)$	parametrized curve in $\mathbb{R}^N$
$x = \xi(u)$, $x^i = \xi^i(u^1, \ldots, u^D)$	parametrized submanifold
η^i	components of the geodesic deviation vector
$\eta_{i_1 \cdots i_N}(x)$	totally antisymmetric pseudo (or oriented) tensor field (Levi-Civita)
z, $\overline{z}$	complex variable and its conjugate
$\mathbb{Z}$	the set of integers
$\mathbb{Z}^+$	the set of positive integers

Index